THE
GEOLOGICAL EVIDENCE
OF THE
ANTIQUITY OF MAN

THE
GEOLOGICAL EVIDENCE
OF THE
ANTIQUITY OF MAN

Charles Lyell

DOVER PUBLICATIONS, INC.
Mineola, New York

Bibliographical Note

This Dover edition, first published in 2004, is an unabridged republication of the edition first published by J. M. Dent & Sons Ltd., London, and E. P. Dutton & Co., New York, in 1914. The work was originally published as *The Geological Evidences of the Antiquity of Man: With Remarks on Theories of the Origin of Species by Variation* by J. Murray, London, in 1863.

Library of Congress Cataloging-in-Publication Data

Lyell, Charles, Sir, 1797-1875.
 [Geological evidences of the antiquity of man]
 The geological evidence of the antiquity of man / Charles Lyell.
  p. cm.
 This edition originally published : London : J.M. Dent & Sons ; New York : E.P. Dutton & Co., 1914.
 Includes bibliographical references.
 ISBN 0-486-43576-8 (pbk.)
  1. Human beings–Origin. 2. Glacial epoch. 3. Human evolution. 4. Evolution (Biology) 5. Species, Origin of. I. Title.

GN735.L95 2004
569.9–dc22

2003065491

Manufactured in the United States of America
Dover Publications, Inc., 31 East 2nd Street, Mineola, N.Y. 11501

INTRODUCTION

THE *Antiquity of Man* was published in 1863, and ran into a third edition in the course of that year. The cause of this is not far to seek. Darwin's *Origin of Species* appeared in 1859, only four years earlier, and rapidly had its effect in drawing attention to the great problem of the origin of living beings. The theories of Darwin and Wallace brought to a head and presented in a concrete shape the somewhat vague speculations as to development and evolution which had long been floating in the minds of naturalists. In the actual working out of Darwin's great theory it is impossible to overestimate the influence of Lyell. This is made abundantly clear in Darwin's letters, and it must never be forgotten that Darwin himself was a geologist. His training in this science enabled him to grasp the import of the facts so ably marshalled by Lyell in the *Principles of Geology*, a work which, as Professor Judd has clearly shown,[1] contributed greatly to the advancement of evolutionary theory in general.

From a study of the evolution of plants and of the lower animals it was an easy and obvious transition to man, and this step was soon taken. Since in his physical structure man shows so close a resemblance to the higher animals it was a natural conclusion that the laws governing the development of the one should apply also to the other, in spite of preconceived opinions derived from authority. Unfortunately the times were then hardly ripe for a calm and logical treatment of this question: prejudice in many cases took the place of argument, and the result was too often an undignified squabble instead of a scientific discussion. However, the dogmatism was not by any means all on one side. The disciples as usual went farther than the master, and their teaching when pushed to extremities resulted in a peculiarly dreary kind of materialism, a mental attitude which still survives to a certain extent among scientific and pseudo-scientific men of the old school. In more recent times this dogmatic agnosticism of the middle Victorian period has been

---

[1] Judd, *The Coming of Evolution* (Cambridge Manuals of Science and Literature), Cambridge, 1910, chaps. vi. and vii.

gradually replaced by speculations of a more positive type, such as those of the Mendelian school in biology and the doctrines of Bergson on the philosophical side. With these later developments we are not here concerned.

In dealing with the evolution and history of man as with that of any other animal, the first step is undoubtedly to collect the facts, and this is precisely what Lyell set out to do in the *Antiquity of Man*. The first nineteen chapters of the book are purely an empirical statement of the evidence then available as to the existence of man in pre-historic times: the rest of the book is devoted to a consideration of the connection between the facts previously stated and Darwin's theory of the origin of species by variation and natural selection. The keynote of Lyell's work, throughout his life, was observation. Lyell was no cabinet geologist; he went to nature and studied phenomena at first hand. Possessed of abundant leisure and ample means he travelled far and wide, patiently collecting material and building up the modern science of physical geology, whose foundations had been laid by Hutton and Playfair. From the facts thus collected he drew his inferences, and if later researches showed these inferences to be wrong, unlike some of his contemporaries, he never hesitated to say so. Thus and thus only is true progress in science attained.

Lyell is universally recognised as the leader of the Uniformitarian school of geologists, and it will be well to consider briefly what is implied in this term. The principles of Uniformitarianism may be summed up thus: *the present is the key to the past*. That is to say, the processes which have gone on in the past were the same in general character as those now seen in operation, though probably differing in degree. This theory is in direct opposition to the ideas of the *Catastrophic* school, which were dominant at the beginning of the nineteenth century. The catastrophists attributed all past changes to sudden and violent convulsions of nature, by which all living beings were destroyed, to be replaced by a fresh creation. At least such were the tenets of the extremists. In opposition to these views the school of Hutton and Lyell introduced the principle of continuity and development. There is no discrepancy between Uniformitarianism and evolution. The idea of Uniformitarianism does not imply that things have always been the same; only that they were similar, and between these two terms there is a wide distinction.

Evolution of any kind whatever naturally implies continuity, and this is the fundamental idea of Lyellian geology.

In spite, however, of this clear and definite conception of natural and organic evolution, in all those parts of his works dealing with earth-history, with the stratified rocks and with the organisms entombed in them, Lyell adopted a plan which has now been universally abandoned. He began with the most recent formations and worked backwards from the known to the unknown. To modern readers this is perhaps the greatest drawback to his work, since it renders difficult the study of events in their actual sequence. However, it must be admitted that, taking into account the state of geological knowledge before his time, this course was almost inevitable. The succession of the later rocks was fairly well known, thanks to the labours of William Smith and others, but in the lower part of the sequence of stratified rocks there were many gaps, and more important still, there was no definite base. Although this want of a starting point has been largely supplied by the labours of Sedgwick, Murchison, De la Beche, Ramsay, and a host of followers, still considerable doubt prevails as to which constitutes the oldest truly stratified series, and the difficulty has only been partially circumvented by the adoption of an arbitrary base-line, from which the succession is worked out both upwards and downwards. So the problem is only removed a stage further back. In the study of human origins a similar difficulty is felt with special acuteness; the beginnings must of necessity be vague and uncertain, and the farther back we go the fainter will naturally be the traces of human handiwork and the more primitive and doubtful those traces when discovered.

The reprinting of the *Antiquity of Man* is particularly appropriate at the present time, owing to the increased attention drawn to the subject by recent discoveries. Ever since the publication of the *Origin of Species* and the discussions that resulted from that publication, the popular imagination has been much exercised by the possible existence of forms intermediate between the apes and man; the so-called "Missing Link." Much has been written on this subject, some of it well-founded and some very much the reverse. The discovery of the Neanderthal skull is fully described in this volume, and this skull is certainly of a low type, but it is more human than ape-like. The same remark applies still more strongly to the Engis skull, the man of Spy, the recently discovered Sussex skull, and other well-known examples of early human remains. The *Pithecanthropus* of Java alone shows perhaps more affinity to the apes. The whole subject has been most ably discussed by Professor Sollas in his recent book entitled *Ancient Hunters*.

The study of Palæolithic flint implements has been raised to a fine art. Both in England and France a regular succession of primitive types has been established and correlated with the gravel terraces of existing rivers, and even with the deposits of rivers no longer existing and with certain glacial deposits. But with all of these the actual bodily remains of man are comparatively scanty. From this it may be concluded that primitive methods of burial were such as to be unfavourable to the actual preservation of human remains. Attempts have also been made to prove the existence of man in pre-glacial times, but hitherto none of these have met with general acceptance, since in no case is the evidence beyond doubt.

One of the most important results of recent research in this subject has been the establishment of the existence of man in interglacial times. When Lyell wrote, it was not fully recognised that the glaciation of Europe was not one continuous process, but that it could be divided into several episodes, glaciations, or advances of the ice, separated by a warm interglacial period. The monumental researches of Penck and Brückner in the Alps have there established four glaciations with mild interglacial periods, but all of these cannot be clearly traced in Britain. One very important point also is the recognition of the affinities of certain types of Palæolithic man to the Eskimo, the Australians, and the Bushmen of South Africa. However, it is impossible to give here a review of the whole subject. Full details of recent researches will be found in the works mentioned in the notes at the end of the book.

Another point of great interest and importance, arising directly from the study of early man is the nature of the events constituting the glacial period in Britain and elsewhere. This has been for many years a fertile subject of controversy, and is likely to continue such. Lyell, in common with most of the geologists of his day, assumes that during the glacial period the British Isles were submerged under the sea to a depth of many hundreds of feet, at any rate as regards the region north of a line drawn from London to Bristol. Later authors, however, explained the observed phenomena on the hypothesis of a vast ice-sheet of the Greenland type, descending from the mountains of Scotland and Scandinavia, filling up the North Sea and spreading over eastern England. This explanation is now accepted by the majority, but it must be recognised that it involves enormous mechanical difficulties. It is impossible to pursue the subject here; for a full discussion reference may be made to Professor Bonney's presidential address to the British Association at Sheffield in 1910.

It will be seen, therefore, that the *Antiquity of Man* opens up a wide field of speculation into a variety of difficult and obscure though interesting subjects. In the light of modern research it would be an easy task to pile up a mountain of criticism on points of detail. But, though easy, it would be a thankless task. It is scarcely too much to say that the dominant impression of most readers after perusing this book will be one of astonishment and admiration at the insight and breadth of view displayed by the author. When it was written the subject was a particularly thorny one to handle, and it undoubtedly required much courage to tackle the origin and development of the human race from a purely critical and scientific standpoint. It must be admitted on all hands that the result was eminently successful, taking into account the paucity of the available material, and the *Antiquity of Man* must ever remain one of the classics of prehistoric archæology.

This edition of the *Antiquity of Man* has been undertaken in order to place before the public in an easily accessible form one of the best known works of the great geologist Sir Charles Lyell; the book had an immense influence in its own day, and it still remains one of the best general accounts of an increasingly important branch of knowledge.

In order to avoid a multiplicity of notes and thus to save space, the nomenclature has been to a certain extent modernised: a new general table of strata has been inserted in the first chapter, in place of the one originally there printed, which was cumbrous and included many minor subdivisions of unnecessary minuteness.

The notes have been kept as short as possible, and they frequently contain little more than references to recent literature elucidating the points under discussion in the text.

1914. R. H. RASTALL.

BIBLIOGRAPHY

The passage of the Beresina (in verse), 1815; Principles of Geology, being an attempt to explain the former changes of the earth's surface, by reference to causes now in operation, 1830-33 (third edition, 1834; fourth, 1835; fifth, 1837; sixth, 1840; seventh, 1847; ninth, entirely revised edition, 1853; tenth, entirely revised edition, 1867, 1868; eleventh, entirely revised edition, 1872; twelfth, edited by L. Lyell, 1875); Elements of Geology, 1838 (second edition, 1841); A Manual of Elementary Geology (third and entirely revised edition of the former work, 1851; fourth and entirely revised edition, 1852; fifth, enlarged edition, 1855; Supplement to the fifth edition, 1857; second edition of the Supplement, revised, 1857); Elements of Geology, sixth edition, greatly enlarged, 1865; Travels in North America, with geological observations on the United States, Canada, and Nova Scotia, 1845; A Second Visit to the United States of North America, 1849; The Students' Elements of Geology, 1871 (second edition, revised and corrected, 1874; third, revised, with a table of British fossils [by R. Etheridge], 1878; fourth, revised by P. M. Duncan, with a table of British fossils [by R. Etheridge], 1884); The Geological Evidences of the Antiquity of Man, with remarks on theories of the origin of species by variation, 1863 (second edition, revised, 1863; third edition, revised, 1863; fourth edition, revised, 1873). There has also been published The Student's Lyell: a Manual of Elementary Geology, edited by J. W. Judd, 1896 (second edition revised and enlarged, 1911).

LECTURES, ADDRESSES, AND ARTICLES: On a Recent Formation of Freshwater Limestone in Forfarshire (*Trans. Geol. Soc.*, 2nd ser., vol. ii., 1826, pt. i.); On a Dike of Serpentine in the County of Forfar (*Edin. Journal Science*, 1825); English Scientific Societies (*Quarterly Review*, vol. xxxiv.; three papers with Sir Roderick and Mrs. Murchison (*Edin. Phil. Journal*, 1829; abstracts in *Proceedings Geol. Soc.*, i.; *Annales des Sciences Naturelles*, 1829; abstract in *Proceedings Geol. Soc.*, i.); Address delivered at the Geological Society of London, 1836; Lectures on Geology—Eight Lectures on Geology, delivered at the Broadway Tabernacle, New York (*New York Tribune*, 1842); a Paper on Madeira (*Quarterly Journal of the Geological Society*, x., 1853); On the Structure of Lavas which have Consolidated on Steep Slopes (*Phil. Trans.*, 1858); Address (to the British Association), 1864.

TRANSLATIONS: Antiquity of Man, translated into French by M. Chaper, 1864; and into German by L. Büchner, 1874; Elements of Geology (sixth edition), translated into French by M. J. Gineston, 1867; Report, extracted from the *Aberdeen Free Press* and translated into French, of Sir C. Lyell's address before the British Association, 1859, under the title of Antiquités antédiluviennes: L'homme fossile.

LIFE: Life, Letters, and Journals of Sir Charles Lyell, edited by his sister-in-law, Mrs. Lyell, 1881. *See also* Life and Letters of Charles Darwin, 1887; Life and Letters of Sedgwick, by Clark and Hughes, 1890.

CONTENTS

CHAPTER I

INTRODUCTORY

CHAPTER II

RECENT PERIOD—DANISH PEAT AND SHELL MOUNDS—SWISS LAKE-DWELLINGS

CHAPTER III

FOSSIL HUMAN REMAINS AND WORKS OF ART OF THE RECENT PERIOD—*continued*

CHAPTER IV

PLEISTOCENE PERIOD—BONES OF MAN AND EXTINCT MAMMALIA IN BELGIAN CAVERNS

Earliest Discoveries in Caves of Languedoc of Human Remains with Bones of extinct Mammalia—Researches in 1833 of Dr. Schmerling in the Liège Caverns—Scattered Portions of Human Skeletons

CHAPTER IX

WORKS OF ART IN PLEISTOCENE ALLUVIUM OF FRANCE AND ENGLAND

CHAPTER X

CAVERN DEPOSITS, AND PLACES OF SEPULTURE OF THE PLEISTOCENE PERIOD

CHAPTER XI

AGE OF HUMAN FOSSILS OF LE PUY IN CENTRAL FRANCE AND OF NATCHEZ ON THE MISSISSIPPI DISCUSSED

Contents xvii

CHAPTER XIX

RECAPITULATION OF GEOLOGICAL PROOFS OF MAN'S ANTIQUITY

CHAPTER XX

THEORIES OF PROGRESSION AND TRANSMUTATION

CHAPTER XXI

ON THE ORIGIN OF SPECIES BY VARIATION AND NATURAL SELECTION

CHAPTER XXII

OBJECTIONS TO THE HYPOTHESIS OF TRANSMUTATION CONSIDERED

CHAPTER XXIII

ORIGIN AND DEVELOPMENT OF LANGUAGES AND SPECIES COMPARED

CHAPTER XXIV

GEOLOGICAL EVIDENCE OF
THE ANTIQUITY OF MAN

CHAPTER I

INTRODUCTORY

Preliminary Remarks on the Subjects treated of in this Work—Definition
of the Terms Recent and Pleistocene—Tabular View of the entire
Series of Fossiliferous Strata.

No subject has lately excited more curiosity and general interest
among geologists and the public than the question of the Antiquity of the Human Race—whether or no we have sufficient
evidence in caves, or in the superficial deposits commonly called
drift or " diluvium," to prove the former co-existence of man
with certain extinct mammalia. For the last half-century the
occasional occurrence in various parts of Europe of the bones
of Man or the works of his hands in cave-breccias and stalag-
mites, associated with the remains of the extinct hyæna, bear,
elephant, or rhinoceros, has given rise to a suspicion that the
date of Man must be carried farther back than we had heretofore
imagined.   On the other hand extreme reluctance was naturally
felt on the part of scientific reasoners to admit the validity of
such evidence, seeing that so many caves have been inhabited
by a succession of tenants and have been selected by Man as a
place not only of domicile, but of sepulture, while some caves
have also served as the channels through which the waters of
occasional land-floods or engulfed rivers have flowed, so that the
remains of living beings which have peopled the district at more
than one era may have subsequently been mingled in such
caverns and confounded together in one and the same deposit.
But the facts brought to light in 1858, during the systematic
investigation of the Brixham cave, near Torquay in Devonshire,
which will be described in the sequel, excited anew the curiosity
of the British public and prepared the way for a general admission that scepticism in regard to the bearing of cave evidence

in favour of the antiquity of Man had previously been pushed to an extreme.

Since that period many of the facts formerly adduced in favour of the co-existence in ancient times of Man with certain species of mammalia long since extinct have been re-examined in England and on the Continent, and new cases bearing on the same question, whether relating to caves or to alluvial strata in valleys, have been brought to light. To qualify myself for the appreciation and discussion of these cases, I have visited in the course of the last three years many parts of England, France, and Belgium, and have communicated personally or by letter with not a few of the geologists, English and foreign, who have taken part in these researches. Besides explaining in the present volume the results of this inquiry, I shall give a description of the glacial formations of Europe and North America, that I may allude to the theories entertained respecting their origin, and consider their probable relations in a chronological point of view to the human epoch, and why throughout a great part of the northern hemisphere they so often interpose an abrupt barrier to all attempts to trace farther back into the past the signs of the existence of Man upon the earth.

In the concluding chapters I shall offer a few remarks on the recent modifications of the Lamarckian theory of progressive development and transmutation, which are suggested by Mr. Darwin's work on the *Origin of Species by Variation and Natural Selection*, and the bearing of this hypothesis on the different races of mankind and their connection with other parts of the animal kingdom.

Nomenclature.—Some preliminary explanation of the nomenclature adopted in the following pages will be indispensable, that the meaning attached to the terms Recent, Pleistocene, and Post-Tertiary may be correctly understood. [Note 1.]

Previously to the year 1833, when I published the third volume of the *Principles of Geology*, the strata called Tertiary had been divided by geologists into Lower, Middle, and Upper; the Lower comprising the oldest formations of the environs of Paris and London, with others of like age; the Middle, those of Bordeaux and Touraine; and the Upper, all that lay above or were newer than the last-mentioned group.

When engaged in 1828 in preparing for the press the treatise on geology above alluded to, I conceived the idea of classing the whole of this series of strata according to the different degrees of affinity which their fossil testacea bore to the living fauna. Having obtained information on this subject during my travels on the Continent, I learnt that M. Deshayes of Paris, already celebrated as a conchologist, had been led independently by the study of a large collection of Recent and fossil shells to very similar views respecting the possibility of arranging the Tertiary formations in chronological order, according to the proportional number of species of shells identical with living ones, which characterised each of the successive groups above mentioned. After comparing 3000 fossil species with 5000 living ones, the result arrived at was, that in the lower Tertiary strata there were about $3\frac{1}{2}$ per cent. identical with recent; in the middle Tertiary (the faluns of the Loire and Gironde), about 17 per cent.; and in the upper Tertiary, from 35 to 50, and sometimes in the most modern beds as much as 90 to 95 per cent.

For the sake of clearness and brevity, I proposed to give short technical names to these sets of strata, or the periods to which they respectively belonged. I called the first or oldest of them Eocene, the second Miocene, and the third Pliocene. The first of the above terms, Eocene, is derived from ἠώς, dawn, and καινός, recent; because an extremely small proportion of the fossil shells of this period could be referred to living species, so that this era seemed to indicate the dawn of the present testaceous fauna, no living species of shells having been detected in the antecedent or Secondary rocks.

Some conchologists are now unwilling to allow that any Eocene species of shell has really survived to our times so un-altered as to allow of its specific identification with a living species. I cannot enter in this place into this wide controversy. It is enough at present to remark that the character of the Eocene fauna, as contrasted with that of the antecedent Secondary formations, wears a very modern aspect, and that some able living conchologists still maintain that there are Eocene shells not specifically distinguishable from those now extant; though they may be fewer in number than was supposed in 1833.

The term Miocene (from μείων, less; and καινός, recent) is intended to express a minor proportion of recent species (of testacea); the term Pliocene (from πλείων, more; and καινός, recent), a comparative plurality of the same.

It has sometimes been objected to this nomenclature that certain species of infusoria found in the chalk are still existing, and, on the other hand, the Miocene and Older Pliocene deposits often contain the remains of mammalia, reptiles, and fish, exclusively of extinct species.   But the reader must bear in mind that the terms Eocene, Miocene, and Pliocene were originally invented with reference purely to conchological data, and in that sense have always been and are still used by me.

Since the first introduction of the terms above defined, the number of new living species of shells obtained from different parts of the globe has been exceedingly great, supplying fresh data for comparison, and enabling the palæontologist to correct many erroneous identifications of fossil and Recent forms. New species also have been collected in abundance from Tertiary formations of every age, while newly discovered groups of strata have filled up gaps in the previously known series. Hence modifications and reforms have been called for in the classifications first proposed. The Eocene, Miocene, and Pliocene periods have been made to comprehend certain sets of strata of which the fossils do not always conform strictly in the proportion of recent to extinct species with the definitions first given by me, or which are implied in the etymology of those terms. These innovations have been treated of in my *Elements or Manual of Elementary Geology*, and in the Supplement to the fifth edition of the same, published in 1859, where some modifications of my classification, as first proposed, are introduced; but I need not dwell on these on the present occasion, as the only formations with which we shall be concerned in the present volume are those of the most modern date, or the Post-Tertiary. It will be convenient to divide these into two groups, the Recent and the Pleistocene.   In the Recent we may comprehend those deposits in which not only all the shells but all the fossil mammalia are of living species; in the Pleistocene those strata in which, the shells being Recent, a portion, and often a considerable one, of the accompanying fossil quadrupeds belongs to extinct species.

Cases will occur where it may be scarcely possible to draw the line of demarcation between the Newer Pliocene and Pleistocene, or between the latter and the recent deposits; and we must expect these difficulties to increase rather than diminish with every advance in our knowledge, and in proportion as gaps are filled up in the series of geological records.

The annexed tabular view of the whole series of fossiliferous strata will enable the reader to see at a glance the chronological relation of the Recent and Pleistocene to the antecedent periods. [Note 2.]

TABLE OF STRATIFIED ROCKS

Kainozoic or Tertiary	Pleistocene and Recent Pliocene Miocene Oligocene Eocene
Mesozoic or Secondary	Cretaceous Jurassic Triassic
Palæozoic or Primary	Permian Carboniferous Devonian or Old Red Sandstone Silurian Ordovician Cambrian
Precambrian or Archæan	

PLATE 1.

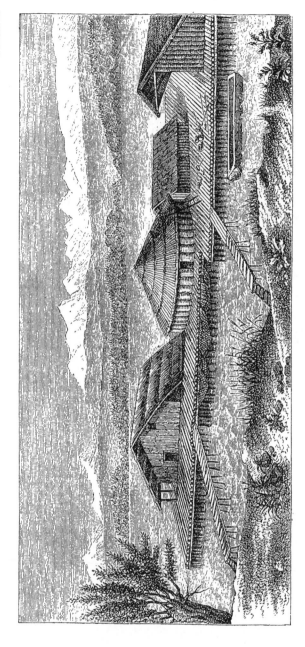

A VILLAGE BUILT ON PILES IN A SWISS LAKE

(Restored by Dr. F. Keller, partly from Dumont D'Urville's Sketch of similar habitations in New Guinea)

CHAPTER II

RECENT PERIOD—DANISH PEAT AND SHELL MOUNDS—SWISS
LAKE DWELLINGS

Works of Art in Danish Peat-Mosses—Remains of three Periods of Vegetation in the Peat—Ages of Stone, Bronze, and Iron—Shell-Mounds
or ancient Refuse-Heaps of the Danish Islands—Change in Geographical Distribution of Marine Mollusca since their Origin—Embedded Remains of Mammalia of recent Species—Human Skulls of
the same Period—Swiss Lake-Dwellings built on Piles—Stone and
Bronze Implements found in them—Fossil Cereals and other Plants
—Remains of Mammalia, wild and domesticated—No extinct Species
—Chronological Computations of the Date of the Bronze and Stone
Periods in Switzerland—Lake-Dwellings, or Artificial Islands called
" Crannoges," in Ireland.

Works of Art in Danish Peat

WHEN treating in the *Principles of Geology* of the changes of the
earth which have taken place in comparatively modern times, I
have spoken of the embedding of organic bodies and human
remains in peat, and explained under what conditions the growth
of that vegetable substance is going on in northern and humid
climates. Of late years, since I first alluded to the subject,
more extensive investigations have been made into the history
of the Danish peat-mosses. Of the results of these inquiries
I shall give a brief abstract in the present chapter, that we may
afterwards compare them with deposits of older date, which
throw light on the antiquity of the human race.

The deposits of peat in Denmark,[1] varying in depth from
10 to 30 feet, have been formed in hollows or depressions in the
northern drift or boulder formation hereafter to be described.
The lowest stratum, 2 to 3 feet thick, consists of swamp-peat
composed chiefly of moss or sphagnum, above which lies another
growth of peat, not made up exclusively of aquatic or swamp
plants. Around the borders of the bogs, and at various depths
in them, lie trunks of trees, especially of the Scotch fir (*Pinus
sylvestris*), often 3 feet in diameter, which must have grown on
the margin of the peat-mosses, and have frequently fallen into
them. This tree is not now, nor has ever been in historical

[1] An excellent account of these researches of Danish naturalists and
antiquaries has been drawn up by an able Swiss geologist, M. A. Morlot,
and will be found in the *Bulletin de la Société Vaudoise des Sci. Nat.*, tom. vi.,
Lausanne, 1860.

times, a native of the Danish Islands, and when introduced
there has not thriven; yet it was evidently indigenous in the
human period, for Steenstrup has taken out with his own hands
a flint instrument from below a buried trunk of one of these pines.
It appears clear that the same Scotch fir was afterwards supplanted by the sessile variety of the common oak, of which many
prostrate trunks occur in the peat at higher levels than the pines;
and still higher the pedunculated variety of the same oak (*Quer-
cus robur*, L.) occurs with the alder, birch(*Betula verrucosa*, Ehrh.),
and hazel. The oak has now in its turn been almost superseded
in Denmark by the common beech. Other trees, such as the
white birch (*Betula alba*), characterise the lower part of the bogs,
and disappear from the higher ; while others again, like the
aspen (*Populus tremula*), occur at all levels, and still flourish in
Denmark. All the land and fresh-water shells, and all the
mammalia as well as the plants, whose remains occur buried in
the Danish peat, are of Recent species. [Note 3.]

It has been stated, that a stone implement was found under
a buried Scotch fir at a great depth in the peat. By collecting
and studying a vast variety of such implements, and other
articles of human workmanship preserved in peat and in sand-
dunes on the coast, as also in certain shell-mounds of the aborigines
presently to be described, the Danish and Swedish antiquaries
and naturalists, MM. Nilsson, Steenstrup, Forchhammer,
Thomsen, Worsäae, and others, have succeeded in establishing
a chronological succession of periods, which they have called the
ages of stone, of bronze, and of iron, named from the materials which
have each in their turn served for the fabrication of implements.

The age of stone in Denmark coincided with the period of the
first vegetation, or that of the Scotch fir, and in part at least with
the second vegetation, or that of the oak. But a considerable
portion of the oak epoch coincided with " the age of bronze,"
for swords and shields of that metal, now in the Museum of
Copenhagen, have been taken out of peat in which oaks abound.
The age of iron corresponded more nearly with that of the beech
tree.[1] [Note 4.]

M. Morlot, to whom we are indebted for a masterly sketch of
the recent progress of this new line of research, followed up with so
much success in Scandinavia and Switzerland, observes that the introduction of the first tools made of bronze among a people previously ignorant of the use of metals, implies a great advance in
the arts, for bronze is an alloy of about nine parts of copper and

[1] Morlot, *Bulletin de la Société Vaudoise des Sci. Nat.*, tom. vi., p. 292.

one of tin; and although the former metal, copper, is by no means rare, and is occasionally found pure or in a native state, tin is not only scarce but never occurs native. To detect the existence of this metal in its ore, then to disengage it from the matrix, and finally, after blending it in due proportion with copper, to cast the fused mixture in a mould, allowing time for it to acquire hardness by slow cooling, all this bespeaks no small sagacity and skilful manipulation. Accordingly, the pottery found associated with weapons of bronze is of a more ornamental and tasteful style than any which belongs to the age of stone. Some of the moulds in which the bronze instruments were cast, and " tags," as they are called, of bronze, which are formed in the hole through which the fused metal was poured, have been found. The number and variety of objects belonging to the age of bronze indicates its long duration, as does the progress in the arts implied by the rudeness of the earlier tools, often mere repetitions of those of the stone age, as contrasted with the more skilfully worked weapons of a later stage of the same period.

It has been suggested that an age of copper must always have intervened between that of stone and bronze; but if so, the interval seems to have been short in Europe, owing apparently to the territory occupied by the aboriginal inhabitants having been invaded and conquered by a people coming from the East, to whom the use of swords, spears, and other weapons of bronze was familiar. Hatchets, however, of copper have been found in the Danish peat.

The next stage of improvement, or that manifested by the substitution of iron for bronze, indicates another stride in the progress of the arts. Iron never presents itself, except in meteorites, in a native state, so that to recognise its ores, and then to separate the metal from its matrix, demands no inconsiderable exercise of the powers of observation and invention. To fuse the ore requires an intense heat, not to be obtained without artificial appliances, such as pipes inflated by the human breath, or bellows, or some other suitable machinery.

Danish Shell-mounds, or Kjökkenmödding [1]

In addition to the peat-mosses, another class of memorials found in Denmark has thrown light on the pre-historical age.

[1] Mr. John Lubbock published, after these sheets were written, an able paper on the Danish " Shell-mounds " in the October number of the *Natural History Review*, 1861, p. 489, in which he has described the results of a recent visit to Denmark, made by him in company with Mr. Busk.

At certain points along the shores of nearly all the Danish islands, mounds may be seen, consisting chiefly of thousands of cast-away shells of the oyster, cockle, and other molluscs of the same species as those which are now eaten by Man. These shells are plentifully mixed up with the bones of various quadrupeds, birds, and fish, which served as the food of the rude hunters and fishers by whom the mounds were accumulated. I have seen similar large heaps of oysters, and other marine shells with interspersed stone implements, near the seashore, both in Massachusetts and in Georgia, U.S.A., left by the native North American Indians at points near to which they were in the habit of pitching their wigwams for centuries before the white man arrived.

Such accumulations are called by the Danes, Kjökkenmödding, or " kitchen-middens." Scattered all through them are flint knives, hatchets, and other instruments of stone, horn, wood, and bone, with fragments of coarse pottery, mixed with charcoal and cinders, but never any implements of bronze, still less of iron. The stone hatchets and knives had been sharpened by rubbing, and in this respect are one degree less rude than those of an older date, associated in France with the bones of extinct mammalia, of which more in the sequel. The mounds vary in height from 3 to 10 feet, and in area are some of them 1000 feet long, and from 150 to 200 wide. They are rarely placed more than 10 feet above the level of the sea, and are confined to its immediate neighbourhood, or if not (and there are cases where they are several miles from the shore), the distance is ascribable to the entrance of a small stream, which has deposited sediment, or to the growth of a peaty swamp, by which the land has been made to advance on the Baltic, as it is still doing in many places, aided, according to Puggaard, by a very slow upheaval of the whole country at the rate of 2 or 3 inches in a century.

There is also another geographical fact equally in favour of the antiquity of the mounds, viz. that they are wanting on those parts of the coast which border the Western Ocean, or exactly where the waves are now slowly eating away the land. There is every reason to presume that originally there were stations along the coast of the North Sea as well as that of the Baltic, but by the gradual undermining of the cliffs they have all been swept away.

Another striking proof, perhaps the most conclusive of all, that the " kitchen-middens " are very old, is derived from the character of their embedded shells. These consist entirely of living species; but, in the first place, the common eatable oyster is among them, attaining its full size, whereas the same *Ostrea edulis* cannot live at present in the brackish waters of the Baltic except near its entrance, where, whenever a north-westerly gale prevails, a current setting in from the ocean pours in a great body of salt water. Yet it seems that during the whole time of the accumulation of the " kitchen-middens " the oyster flourished in places from which it is now excluded. In like manner the eatable cockle, mussel, and periwinkle (*Cardium edule, Mytilus edulis,* and *Littorina littorea*), which are met with in great numbers in the " middens, " are of the ordinary dimensions which they acquire in the ocean, whereas the same species now living in the adjoining parts of the Baltic only attain a third of their natural size, being stunted and dwarfed in their growth by the quantity of fresh water poured by rivers into that inland sea.[1] Hence we may confidently infer that in the days of the aboriginal hunters and fishers, the ocean had freer access than now to the Baltic, communicating probably through the peninsula of Jutland, Jutland having been at no remote period an archipelago. Even in the course of the nineteenth century, the salt waters have made one irruption into the Baltic by the Lymfiord, although they have been now again excluded. It is also affirmed that other channels were open in historical times which are now silted up.[2]

If we next turn to the remains of vertebrata preserved in the mounds, we find that here also, as in the Danish peat-mosses, all the quadrupeds belong to species known to have inhabited Europe within the memory of Man. No remains of the mammoth, or rhinoceros, or of any extinct species appear, except those of the wild bull (*Bos urus,* Linn., or *Bos primigenius,* Bojanus), which are in such numbers as to prove that the species was a favourite food of the ancient people. But as this animal was seen by Julius Cæsar, and survived long after his time, its presence alone would not go far to prove the mounds to be of high antiquity. The Lithuanian aurochs or bison (*Bos bison,* L., *Bos priscus,* Boj., which has escaped extirpation only because protected by the Russian Czars, surviving in one forest in Lithuania) has not yet been met with, but will no doubt be detected hereafter, as it has been already found in the Danish peat. The beaver, long since destroyed in Denmark, occurs

---

[1] See *Principles of Geology,* ch. xxx.
[2] *See* Morlot, *Bulletin de la Société Vaudoise des Sci. Nat.,* tom. vi.

frequently, as does the seal (*Phoca Gryppus*, Fab.), now very rare on the Danish coast. With these are mingled bones of the red deer and roe, but the reindeer has not yet been found. There are also the bones of many carnivora, such as the lynx, fox, and wolf, but no signs of any domesticated animals except the dog. The long bones of the larger mammalia have been all broken as if by some instrument, in such a manner as to allow of the extraction of the marrow, and the gristly parts have been gnawed off, as if by dogs, to whose agency is also attributed the almost entire absence of the bones of young birds and of the smaller bones and softer parts of the skeletons of birds in general, even of those of large size. In reference to the latter, it has been proved experimentally by Professor Steenstrup, that if the same species of birds are now given to dogs, they will devour those parts of the skeleton which are missing, and leave just those which are preserved in the old " kitchen-middens."

The dogs of the mounds, the only domesticated animals, are of a smaller race than those of the bronze period, as shown by the peat-mosses, and the dogs of the bronze age are inferior in size and strength to those of the iron age. The domestic ox, horse, and sheep, which are wanting in the mounds, are confined to that part of the Danish peat which was formed in the ages of bronze and iron.

Among the bones of birds, scarcely any are more frequent in the mounds than those of the auk (*Alca impennis*), now extinct. The Capercailzie (*Tetrao urogallus*) is also met with, and may, it is suggested, have fed on the buds of the Scotch fir in times when that tree flourished around the peat-bogs. The different stages of growth of the roedeer's horns, and the presence of the wild swan, now only a winter visitor, have been appealed to as proving that the aborigines resided in the same settlements all the year round. That they also ventured out to sea in canoes such as are now found in the peat-mosses, hollowed out of the trunk of a single tree, to catch fish far from land, is testified by the bony relics of several deep-sea species, such as the herring, cod, and flounder. The ancient people were not cannibals, for no human bones are mingled with the spoils of the chase. Skulls, however, have been obtained not only from peat, but from tumuli of the stone period believed to be contemporaneous with the mounds. These skulls are small and round, and have a prominent ridge over the orbits of the eyes, showing that the ancient race was of small stature, with round heads and overhanging eyebrows—

in short, they bore a considerable resemblance to the modern Laplanders. The human skulls of the bronze age found in the Danish peat, and those of the iron period, are of an elongated form and larger size. There appear to be very few well-authenticated examples of crania referable to the bronze period—a circumstance no doubt attributable to the custom prevalent among the people of that era of burning their dead and collecting their bones in funeral urns.

No traces of grain of any sort have hitherto been discovered, nor any other indication that the ancient people had any knowledge of agriculture. The only vegetable remains in the mounds are burnt pieces of wood and some charred substance referred by Dr. Forchhammer to the *Zostera marina*, a sea plant which was perhaps used in the production of salt.

What may be the antiquity of the earliest human remains preserved in the Danish peat cannot be estimated in centuries with any approach to accuracy. In the first place, in going back to the bronze age, we already find ourselves beyond the reach of history or even of tradition. In the time of the Romans the Danish Isles were covered, as now, with magnificent beech forests. Nowhere in the world does this tree flourish more luxuriantly than in Denmark, and eighteen centuries seem to have done little or nothing towards modifying the character of the forest vegetation. Yet in the antecedent bronze period there were no beech trees, or at most but a few stragglers, the country being then covered with oak. In the age of stone again, the Scotch fir prevailed, and already there were human inhabitants in those old pine forests. How many generations of each species of tree flourished in succession before the pine was supplanted by the oak, and the oak by the beech, can be but vaguely conjectured, but the minimum of time required for the formation of so much peat must, according to the estimate of Steenstrup and other good authorities, have amounted to at least 4000 years; and there is nothing in the observed rate of the growth of peat opposed to the conclusion that the number of centuries may not have been four times as great, even though the signs of Man's existence have not yet been traced down to the lowest or amorphous stratum. As to the " kitchen-middens," they correspond in date to the older portion of the peaty record, or to the earliest part of the age of stone as known in Denmark.

Ancient Swiss Lake-dwellings, built on Piles

In the shallow parts of many Swiss lakes, where there is a depth of no more than from 5 to 15 feet of water, ancient wooden piles are observed at the bottom sometimes worn down to the surface of the mud, sometimes projecting slightly above it. These have evidently once supported villages, nearly all of them of unknown date, but the most ancient of which certainly belonged to the age of stone, for hundreds of implements resembling those of the Danish shell-mounds and peat-mosses have been dredged up from the mud into which the piles were driven.

The earliest historical account of such habitations is that given by Herodotus of a Thracian tribe, who dwelt, in the year 520 B.C., in Prasias, a small mountain-lake of Pæonia, now part of modern Roumelia.[1]

Their habitations were constructed on platforms raised above the lake, and resting on piles. They were connected with the shore by a narrow causeway of similar formation. Such platforms must have been of considerable extent, for the Pæonians lived there with their families and horses. Their food consisted largely of the fish which the lake produced in abundance.

In rude and unsettled times, such insular sites afforded safe retreats, all communication with the mainland being cut off, except by boats, or by such wooden bridges as could be easily removed.

The Swiss lake-dwellings seem first to have attracted attention during the dry winter of 1853-54, when the lakes and rivers sank lower than had ever been previously known, and when the inhabitants of Meilen, on the Lake of Zürich, resolved to raise the level of some ground and turn it into land, by throwing mud upon it obtained by dredging in the adjoining shallow water. During these dredging operations they discovered a number of wooden piles deeply driven into the bed of the lake, and among them a great many hammers, axes, celts, and other instruments. All these belonged to the stone period with two exceptions, namely, an armlet of thin brass wire, and a small bronze hatchet.

Fragments of rude pottery fashioned by the hand were abundant, also masses of charred wood, supposed to have formed parts of the platform on which the wooden cabins were built. Of this burnt timber, on this and other sites, subse-

---

[1] Herodotus, lib. v. cap. 16. Rediscovered by M. Deville, *Nat. Hist. Rev.*, vol. ii., 1862, p. 486.

quently explored, there was such an abundance as to lead to the
conclusion that many of the settlements must have perished by
fire. Herodotus has recorded that the Pæonians, above alluded
to, preserved their independence during the Persian invasion,
and defied the attacks of Darius by aid of the peculiar position
of their dwellings. "But their safety," observes Mr. Wylie,[1]
" was probably owing to their living in the middle of the lake,
ἐν μέσῃ τῇ λίμνῃ, whereas the ancient Swiss settlers were
compelled by the rapidly increasing depth of the water near the
margins of their lakes to construct their habitations at a short
distance from the shore, within easy bowshot of the land, and
therefore not out of reach of fiery projectiles, against which
thatched roofs and wooden walls could present but a poor
defence." To these circumstances and to accidental fires we are
probably indebted for the frequent preservation, in the mud
around the site of the old settlements, of the most precious tools
and works of art, such as would never have been thrown into the
Danish " kitchen-middens," which have been aptly compared
to a modern dusthole.

Dr. Ferdinand Keller of Zürich has drawn up a series of most
instructive memoirs, illustrated with well-executed plates, of
the treasures in stone, bronze, and bone brought to light in these
subaqueous repositories, and has given an ideal restoration of
part of one of the old villages (*see* p. 6),[2] such as he conceives
may have existed on the lakes of Zürich and Bienne. In this
view, however, he has not simply trusted to his imagination,
but has availed himself of a sketch published by M. Dumont
d'Urville, of similar habitations of the Papuans in New Guinea
in the Bay of Dorei. It is also stated by Dr. Keller, that on the
River Limmat, near Zürich, so late as the last century, there
were several fishing-huts constructed on this same plan.[3] It
will be remarked that one of the cabins is represented as circular. That such was the form of many in Switzerland is inferred
from the shape of pieces of clay which lined the interior, and
which owe their preservation apparently to their having been
hardened by fire when the village was burnt. In the sketch,
some fishing-nets are seen spread out to dry on the wooden

[1] W. M. Wylie, *Archæologia*, vol. xxxviii., 1859, a valuable paper on
the Swiss and Irish lake-habitations.
[2] Keller, *Pfahlbauten, Antiquarische Gesellschaft in Zürich*, Bd. xii.,
xiii., 1858-1861. In the fifth number of the *Natural History Review*,
January 9, 1862, Mr. Lubbock has published an excellent account of the
works of the Swiss writers on their lake-habitations.
[3] Keller, *ibid.*, Bd. ix., p. 81 note.

platform. The Swiss archæologist has found abundant evidence of fishing-gear, consisting of pieces of cord, hooks, and stones used as weights. A canoe also is introduced, such as are occasionally met with. One of these, made of the trunk of a single tree, fifty feet long and three and a half feet wide, was found capsized at the bottom of the Lake of Bienne. It appears to have been laden with stones, such as were used to raise the foundation of some of the artificial islands.

It is believed that as many as 300 wooden huts were sometimes comprised in one settlement, and that they may have contained about 1000 inhabitants. At Wangen, M. Lohle has calculated that 40,000 piles were used, probably not all planted at one time nor by one generation. Among the works of great merit devoted specially to a description of the Swiss lake-habitations is that of M. Troyon, published in 1860.[1] The number of sites which he and other authors have already enumerated in Switzerland is truly wonderful. They occur on the large lakes of Constance, Zürich, Geneva, and Neufchâtel, and on most of the smaller ones. Some are exclusively of the stone age, others of the bronze period. Of these last more than twenty are spoken of on the Lake of Geneva alone, more than forty on that of Neufchâtel, and twenty on the small Lake of Bienne.

One of the sites first studied by the Swiss antiquaries was the small lake of Mooseedorf, near Berne, where implements of stone, horn, and bone, but none of metal, were obtained. Although the flint here employed must have come from a distance (probably from the south of France), the chippings of the material are in such profusion as to imply that there was a manufactory of implements on the spot. Here also, as in several other settlements, hatchets and wedges of jade have been observed of a kind said not to occur in Switzerland or the adjoining parts of Europe, and which some mineralogists would fain derive from the East; amber also, which, it is supposed, was imported from the shores of the Baltic.

At Wangen near Stein, on the Lake of Constance, another of the most ancient of the lake-dwellings, hatchets of serpentine and greenstone, and arrow-heads of quartz have been met with. Here also remains of a kind of cloth, supposed to be of flax, not woven but plaited, have been detected. Professor Heer has recognised lumps of carbonised wheat, *Triticum vulgare,* and grains of another kind, *T. dicoccum,* and barley, *Hordeum dis-*

---

[1] *Sur les Habitations lacustres.*

tichum, and flat round cakes of bread; and at Robbenhausen and elsewhere *Hordeum hexastichum* in fine ears, the same kind of barley which is found associated with Egyptian mummies, showing clearly that in the stone period the lake-dwellers culti-vated all these cereals, besides having domesticated the dog, the ox, the sheep, and the goat.

Carbonised apples and pears of small size, such as still grow in the Swiss forests, stones of the wild plum, seeds of the rasp-berry and blackberry, and beech-nuts, also occur in the mud, and hazel-nuts in great plenty.

Near Morges, on the Lake of Geneva, a settlement of the bronze period, no less than forty hatchets of that metal have been dredged up, and in many other localities the number and variety of weapons and utensils discovered, in a fine state of preservation, is truly astonishing.

It is remarkable that as yet all the settlements of the bronze period are confined to Western and Central Switzerland. In the more eastern lakes those of the stone period alone have as yet been discovered.

The tools, ornaments, and pottery of the bronze period in Switzerland bear a close resemblance to those of corresponding age in Denmark, attesting the wide spread of a uniform civilisa-tion over Central Europe at that era. In some few of the Swiss aquatic stations a mixture of bronze and iron implements has been observed, but no coins. At Tiefenau, near Berne, in ground supposed to have been a battle-field, coins and medals of bronze and silver, struck at Marseilles, and of Greek manu-facture, and iron swords, have been found, all belonging to the first and pre-Roman division of the age of iron.

In the settlements of the bronze era the wooden piles are not so much decayed as those of the stone period; the latter having wasted down quite to the level of the mud, whereas the piles of the bronze age (as in the Lake of Bienne, for example) still project above it.

Professor Rütimeyer of Basle, well known to palæontologists as the author of several important memoirs on fossil vertebrata, has recently published a scientific description of great interest of the animal remains dredged up at various stations where they had been embedded for ages in the mud into which the piles were driven.[1]

These bones bear the same relation to the primitive inhabi-tants of Switzerland and some of their immediate successors

---

[1] *Die Fauna der Pfahlbauten in der Schweiz,* Basel, 1861.

as do the contents of the Danish " kitchen-middens " to the ancients fishing and hunting tribes who lived on the shores of the Baltic.

The list of wild mammalia enumerated in this excellent treatise contains no less than twenty-four species, exclusive of several domesticated ones: besides which there are eighteen species of birds, the wild swan, goose, and two species of ducks being among them; also three reptiles, including the eatable frog and fresh-water tortoise; and lastly, nine species of fresh-water fish. All these (amounting to fifty-four species) are with one exception still living in Europe. The exception is the wild bull (*Bos primigenius*), which, as before stated, survived in historical times. The following are the mammalia alluded to:—The bear (*Ursus arctos*), the badger, the common marten, the polecat, the ermine, the weasel, the otter, wolf, fox, wild cat, hedgehog, squirrel, field-mouse (*Mus sylvaticus*), hare, beaver, hog (comprising two races, namely, the wild boar and swamp-hog), the stag (*Cervus elaphus*), the roe-deer, the fallow-deer, the elk, the steinbock (*Capra ibex*), the chamois, the Lithuanian bison, and the wild bull. The domesticated species comprise the dog, horse, ass, pig, goat, sheep, and several bovine races.

The greater number, if not all, of these animals served for food, and all the bones which contained marrow have been split open in the same way as the corresponding ones found in the shell-mounds of Denmark before mentioned. The bones both of the wild bull and the bison are invariably split in this manner. As a rule, the lower jaws with teeth occur in greater abundance than any other parts of the skeleton—a circumstance which, geologists know, holds good in regard to fossil mammalia of all periods. As yet the reindeer is missing in the Swiss lake-settlements as in the Danish " kitchen-middens," although this animal in more ancient times ranged over France, together with the mammoth, as far south as the Pyrenees.

A careful comparison of the bones from different sites has shown that in settlements such as Wangen and Moosseedorf, belonging to the earliest age of stone, when the habits of the hunter state predominated over those of the pastoral, venison, or the flesh of the stag and roe, was more eaten than the flesh of the domestic cattle and sheep. This was afterwards reversed in the later stone period and in the age of bronze. At that later period also the tame pig, which is wanting in some of the oldest stations, had replaced the wild boar as a common article of food. In the beginning of the age of stone, in Switzerland, the goats outnumbered the sheep, but towards the close of the same period the sheep were more abundant than the goats.

The fox in the first era was very common, but it nearly disappears in the bronze age, during which period a large hunting-dog, supposed to have been imported into Switzerland from some foreign country, becomes the chief representative of the canine genus.

A single fragment of the bone of a hare (*Lepus timidus*) has been found at Moosseedorf. The almost universal absence of this quadruped is supposed to imply that the Swiss lake-dwellers were prevented from eating that animal by the same superstition which now prevails among the Laplanders, and which Julius Cæsar found in full force amongst the ancient Britons.[1]

That the lake-dwellers should have fed so largely on the fox, while they abstained from touching the hare, establishes, says Rütimeyer, a singular contrast between their tastes and ours.

Even in the earliest settlements, as already hinted, several domesticated animals occur, namely, the ox, sheep, goat, and dog. Of the three last, each was represented by one race only; but there were two races of cattle, the most common being of small size, and called by Rütimeyer *Bos brachyceros* (*Bos longifrons*, Owen), or the marsh cow, the other derived from the wild bull; though, as no skull has yet been discovered, this identification is not so certain as could be wished. It is, however, beyond question that at a later era, namely, towards the close of the stone and beginning of the bronze period, the lake-dwellers had succeeded in taming that formidable brute the *Bos primigenius*, the Urus of Cæsar, which he described as very fierce, swift, and strong, and scarcely inferior to the elephant in size. In a tame state its bones were somewhat less massive and heavy, and its horns were somewhat smaller than in wild individuals. Still in its domesticated form, it rivalled in dimensions the largest living cattle, those of Friesland, in North Holland, for example. When most abundant, as at Concise on the Lake of Neufchâtel, it had nearly superseded the smaller race, *Bos brachyceros,* and was accompanied there for a short time by a third bovine variety, called *Bos trochoceros*, an Italian race, supposed to have been imported from the southern side of the Alps.[2] This last-mentioned race, however, seems only to have lasted for a short time in Switzerland.

The wild bull (*Bos primigenius*) is supposed to have flourished for a while in a wild and tame state, just as now in Europe the

---

[1] *Commentaries*, lib. v. ch. 12.   [2] Cæsar's *Commentaries*, lib. v. ch. 12.

domestic pig co-exists with the wild boar; and Rütimeyer agrees with Cuvier and Bell,[1] in considering our larger domestic cattle of northern Europe as the descendants of this wild bull, an opinion which Owen disputes.[2]

In the later division of the stone period, there were two tame races of the pig, according to Rütimeyer; one large, and derived from the wild boar, the other smaller, called the " marsh-hog," or *Sus scrofa palustris*. It may be asked how the osteologist can distinguish the tame from the wild races of the same species by their skeletons alone. Among other characters, the diminished thickness of the bones and the comparative smallness of the ridges, which afford attachment to the muscles, are relied on; also the smaller dimensions of the tusks in the boar, and of the whole jaw and skull; and, in like manner, the diminished size of the horns of the bull and other modifications, which are the effects of a regular supply of food, and the absence of all necessity of exerting their activity and strength to obtain subsistence and defend themselves against their enemies.

A middle-sized race of dogs continued unaltered throughout the whole of the stone period; but the people of the bronze age possessed a larger hunting-dog, and with it a small horse, of which genus very few traces have been detected in the earlier settlements—a single tooth, for example, at Wangen, and only one or two bones at two or three other places.

In passing from the oldest to the most modern sites, the extirpation of the elk and beaver, and the gradual reduction in numbers of the bear, stag, roe, and fresh-water tortoise are distinctly perceptible. The aurochs, or Lithuanian bison, appears to have died out in Switzerland about the time when weapons of bronze came into use. It is only in a few of the most modern lake-dwellings, such as Noville and Chavannes in the Canton de Vaud (which the antiquaries refer to the sixth century), that some traces are observable of the domestic cat, as well as of a sheep with crooked horns, and with them bones of the domestic fowl.

After the sixth century, no extinction of any wild quadruped nor introduction of any tame one appears to have taken place, but the fauna was still modified by the wild species continuing to diminish in number and the tame ones to become more diversified by breeding and crossing, especially in the case of the dog, horse, and sheep. On the whole, however, the divergence of the domestic races from their aboriginal wild types, as ex-

[1] *British Quadrupeds*, p. 415.　　[2] *British Fossil Mammal.*, p. 500.

emplified at Wangen and Moosseedorf, is confined, according to Professor Rütimeyer, within narrow limits. As to the goat, it has remained nearly constant and true to its pristine form, and the small race of goat-horned sheep still lingers in some Alpine valleys in the Upper Rhine; and in the same region a race of pigs, corresponding to the domesticated variety of *Sus scrofa palustris*, may still be seen.

Amidst all this profusion of animal remains extremely few bones of Man have been discovered; and only one skull, dredged up from Meilen, on the Lake of Zürich, of the early stone period, seems as yet to have been carefully examined. Respecting this specimen, Professor His observes that it exhibits, instead of the small and rounded form proper to the Danish peat-mosses, a type much more like that now prevailing in Switzerland, which is intermediate between the long-headed and short-headed form.[1]

So far, therefore, as we can draw safe conclusions from a single specimen, there has been no marked change of race in the human population of Switzerland during the periods above considered.

It is still a question whether any of these subaqueous repositories of ancient relics in Switzerland go back so far in time as the kitchen-middens of Denmark, for in these last there are no domesticated animals except the dog, and no signs of the cultivation of wheat or barley; whereas we have seen that, in one of the oldest of the Swiss settlements, at Wangen, no less than three cereals make their appearance, with four kinds of domestic animals. Yet there is no small risk of error in speculating on the relative claims to antiquity of such ancient tribes, for some of them may have remained isolated for ages and stationary in their habits, while others advanced and improved.

We know that nations, both before and after the introduction of metals, may continue in very different stages of civilisation, even after commercial intercourse has been established between them, and where they are separated by a less distance than that which divides the Alps from the Baltic.

The attempts of the Swiss geologists and archæologists to estimate definitely in years the antiquity of the bronze and stone periods, although as yet confessedly imperfect, deserve notice, and appear to me to be full of promise. The most elaborate calculation is that made by M. Morlot, respecting the delta of the Tinière, a torrent which flows into the Lake of Geneva near Villeneuve. This small delta, to which the stream is annually

[1] Rütimeyer, *Die Fauna der Pfahlbauten in der Schweiz*, p. 181.

making additions, is composed of gravel and sand.   Its shape
is that of a flattened cone, and its internal structure has of late
been laid open to view in a railway cutting 1000 feet long and
32 feet deep.   The regularity of its structure throughout implies
that it has been formed very gradually, and by the uniform
action of the same causes.   Three layers of vegetable soil, each
of which must at one time have formed the surface of the cone,
have been cut through at different depths.   The first of these
was traced over a surface of 15,000 square feet, having an
average thickness of 5 inches, and being about 4 feet below the
present surface of the cone.   This upper layer belonged to the
Roman period, and contained Roman tiles and a coin.   The
second layer, followed over a surface of 25,000 square feet, was
6 inches thick, and lay at a depth of 10 feet.   In it were found
fragments of unvarnished pottery and a pair of tweezers in
bronze, indicating the bronze epoch.   The third layer, followed
for 35,000 square feet, was 6 or 7 inches thick and 19 feet deep.
In it were fragments of rude pottery, pieces of charcoal, broken
bones, and a human skeleton having a small, round, and very
thick skull.   M. Morlot, assuming the Roman period to represent an antiquity of from sixteen to eighteen centuries, assigns
to the bronze age a date of between 3000 and 4000 years, and to
the oldest layer, that of the stone period, an age of from 5000
to 7000 years.

Another calculation has been made by M. Troyon to obtain
the approximate date of the remains of an ancient settlement
built on piles and preserved in a peat-bog at Chamblon, near
Yverdun, on the Lake of Neufchâtel.   The site of the ancient
Roman town of Eburodunum (Yverdun), once on the borders
of the lake, and between which and the shore there now intervenes a zone of newly-gained dry land, 2500 feet in breadth,
shows the rate at which the bed of the lake has been filled up
with river sediment in fifteen centuries.   Assuming the lake to
have retreated at the same rate before the Roman period, the
pile-works of Chamblon, which are of the bronze period, must
be at the least 3300 years old.

For the third calculation, communicated to me by M. Morlot,
we are indebted to M. Victor Gilliéron, of Neuveville, on the
Lake of Bienne.   It relates to the age of a pile-dwelling, the
mammalian bones of which are considered by M. Rütimeyer to
indicate the earliest portion of the stone period of Switzerland,
and to correspond in age with the settlement of Moosseedorf.

The piles in question occur at the Pont de Thièle, between

the lakes of Bienne and Neufchâtel. The old convent of St. Jean, founded 750 years ago, and built originally on the margin of the Lake of Bienne, is now at a considerable distance from the shore, and affords a measure of the rate of the gain of land in seven centuries and a half. Assuming that a similar rate of the conversion of water into marshy land prevailed antecedently, we should require an addition of sixty centuries for the growth of the morass intervening between the convent and the aquatic dwelling of Pont de Thièle, in all 6750 years. M. Morlot, after examining the ground, thinks it highly probable that the shape of the bottom on which the morass rests is uniform; but this important point has not yet been tested by boring. The result, if confirmed, would agree exceedingly well with the chronological computation before mentioned of the age of the stone period of Tinière. As I have not myself visited Switzerland since these chronological speculations were first hazarded, I am unable to enter critically into a discussion of the objections which have been raised to the two first of them, or to decide on the merits of the explanations offered in reply.

Irish Lake-dwellings or Crannoges

The lake-dwellings of the British Isles, although not explored as yet with scientific zeal, as those of Switzerland have been in the last ten years, are yet known to be very numerous, and when carefully examined will not fail to throw great light on the history of the bronze and stone periods.

In the lakes of Ireland alone, no less than forty-six examples of artificial islands, called *crannoges*, have been discovered. They occur in Leitrim, Roscommon, Cavan, Down, Monaghan, Limerick, Meath, King's County, and Tyrone.[1] One class of these " stockaded islands," as they have been sometimes called, was formed, according to Mr. Digby Wyatt, by placing horizontal oak beams at the bottom of the lake, into which oak posts, from 6 to 8 feet high, were mortised, and held together by cross beams, till a circular enclosure was obtained.

A space of 520 feet diameter, thus enclosed at Lagore, was divided into sundry timbered compartments, which were found filled up with mud or earth, from which were taken " vast quantities of the bones of oxen, swine, deer, goats, sheep, dogs, foxes, horses, and asses." All these were discovered beneath 16 feet of bog, and were used for manure; but specimens of them are

---

[1] W. M. Wylie, *Archæologia*, vol. xxxviii., 1859, p. 8.

said to be preserved in the museum of the Royal Irish Academy. From the same spot were obtained a great collection of antiquities, which, according to Lord Talbot de Malahide and Mr. Wylie, were referable to the ages of stone, bronze, and iron.[1]

In Ardekillin Lake, in Roscommon, an islet of an oval form was observed, made of a layer of stones resting on logs of timber. Round this artificial islet or *crannoge* thus formed was a stone wall raised on oak piles. A careful description has been put on record by Captain Mudge, R.N., of a curious log-cabin discovered by him in 1833 in Drumkellin bog, in Donegal, at a depth of 14 feet from the surface. It was 12 feet square and 9 feet high, being divided into two stories each 4 feet high. The planking was of oak split with wedges of stone, one of which was found in the building. The roof was flat. A staked enclosure had been raised round the cabin, and remains of other similar huts adjoining were seen but not explored. A stone celt, found in the interior of the hut, and a piece of leather sandal, also an arrowhead of flint, and in the bog close at hand a wooden sword, give evidence of the remote antiquity of this building, which may be taken as a type of the early dwellings on the Crannoge islands.

" The whole structure," says Captain Mudge, " was wrought with the rudest kind of implements, and the labour bestowed on it must have been immense. The wood of the mortises was more bruised than cut, as if by a blunt stone chisel." [2] Such a chisel lay on the floor of the hut, and by comparing it with the marks of the tool used in forming the mortises, they were found " to correspond exactly, even to the slight curved exterior of the chisel; but the logs had been hewn by a larger instrument, in the shape of an axe. On the floor of the dwelling lay a slab of freestone, 3 feet long and 14 inches thick, in the centre of which was a small pit three quarters of an inch deep, which had been chiselled out. This is presumed to have been used for holding nuts to be cracked by means of one of the round shingle stones, also found there, which had served as a hammer. Some entire hazel-nuts and a great quantity of broken shells were strewed about the floor."

The foundations of the house were made of fine sand, such as is found with shingle on the seashore about 2 miles distant. Below the layer of sand the bog or peat was ascertained, on probing it with an instrument, to be at least 15 feet thick. Although the interior of the building when discovered was full of

[1] W. M. Wylie, *Archæologia*, vol. xxxviii., 1859, p. 8, who cites *Archæological Journal*, vol. vi., p. 101.
[2] Mudge, *Archæologia*, vol. xxvi.

" bog " or peaty matter, it seems when inhabited to have been surrounded by growing trees, some of the trunks and roots of which are still preserved in their natural position. The depth of overlying peat affords no safe criterion for calculating the age of the cabin or village, for I have shown in the *Principles of Geology* that both in England and Ireland, within historical times, bogs have burst and sent forth great volumes of black mud, which has been known to creep over the country at a slow pace, flowing somewhat at the rate of ordinary lava-currents, and sometimes overwhelming woods and cottages, and leaving a deposit upon them of bog-earth 15 feet thick.

None of these Irish lake-dwellings were built, like those of Helvetia, on platforms supported by piles deeply driven into the mud. " The Crannoge system of Ireland seems," says Mr. Wylie, " well nigh without a parallel in Swiss waters."

CHAPTER III

Delta and Alluvial Plain of the Nile

SOME new facts of high interest illustrating the geology of the
alluvial land of Egypt were brought to light between the years
1851 and 1854, in consequence of investigations suggested to the
Royal Society by Mr. Leonard Horner, and which were partly
carried out at the expense of the Society. The practical part
of the undertaking was entrusted by Mr. Horner to an Armenian
officer of engineers, Hekekyan Bey, who had for many years
pursued his scientific studies in England, and was in every way
highly qualified for the task.

It was soon found that to obtain the required information
respecting the nature, depth, and contents of the Nile mud in
various parts of the valley, a larger outlay was called for than
had been originally contemplated. This expense the late
viceroy, Abbas Pasha, munificently undertook to defray out of
his treasury, and his successor, after his death, continued the
operations with the same princely liberality.

Several engineers and a body of sixty workmen were employed
under the superintendence of Hekekyan Bey, men inured to
the climate and able to carry on the sinking of shafts and
borings during the hot months, after the waters of the Nile
had subsided, and in a season which would have been fatal to
Europeans.

The results of chief importance arising out of this inquiry were
obtained from two sets of shafts and borings sunk at intervals
in lines crossing the great valley from east to west. One of these

consisted of no fewer than fifty-one pits and artesian borings, made where the valley is 16 miles wide from side to side between the Arabian and Libyan deserts, in the latitude of Heliopolis, about 8 miles above the apex of the delta. The other line of borings and pits, twenty-seven in number, was in the parallel of Memphis, where the valley is only 5 miles broad.

Everywhere in these sections the sediment passed through was similar in composition to the ordinary Nile mud of the present day, except near the margin of the valley, where thin layers of quartzose sand, such as is sometimes blown from the adjacent desert by violent winds, were observed to alternate with the loam.

A remarkable absence of lamination and stratification was observed almost universally in the sediment brought up from all points except where the sandy layers above alluded to occurred. Mr. Horner attributes this want of all indication of successive deposition to the extreme thinness of the film of matter which is thrown down annually on the great alluvial plain during the season of inundation. The tenuity of this layer must indeed be extreme, if the French engineers are tolerably correct in their estimate of the amount of sediment formed in a century, which they suppose not to exceed on the average 5 inches. When the waters subside, this thin layer of new soil, exposed to a hot sun, dries rapidly, and clouds of dust are raised by the winds. The superficial deposit, moreover, is disturbed almost everywhere by agricultural labours, and even were this not the case, the action of worms, insects, and the roots of plants would suffice to confound together the deposits of two successive years.

All the remains of organic bodies, such as land-shells, and the bones of quadrupeds, found during the excavations belonged to living species. Bones of the ox, hog, dog, dromedary and ass were not uncommon, but no vestiges of extinct mammalia. No marine shells were anywhere detected; but this was to be expected, as the borings, though they sometimes reached as low as the level of the Mediterranean, were never carried down below it—a circumstance much to be regretted, since where artesian borings have been made in deltas, as in those of the Po and Ganges, to the depth of several hundred feet below the sea level it has been found, contrary to expectation, that the deposits passed through were fluviatile throughout, implying, probably, that a general subsidence of those deltas and alluvial formations has taken place. Whether there has been in like manner a sinking of the land in Egypt, we have as yet no means of proving;

but Sir Gardner Wilkinson infers it from the position in the delta on the shore near Alexandria of the tombs commonly called Cleopatra's Baths, which cannot, he says, have been originally built so as to be exposed to the sea which now fills them, but must have stood on land above the level of the Mediterranean. The same author adduces, as additional signs of subsidence, some ruined towns, now half under water, in the Lake Menzaleh, and channels of ancient arms of the Nile submerged with their banks beneath the waters of that same lagoon.

In some instances, the excavations made under the super-intendence of Hekekyan Bey were on a large scale for the first 16 or 24 feet, in which cases jars, vases, pots, and a small human figure in burnt clay, a copper knife, and other entire articles were dug up; but when water soaking through from the Nile was reached the boring instrument used was too small to allow of more than fragments of works of art being brought up. Pieces of burnt brick and pottery were extracted almost everywhere, and from all depths, even where they sank 60 feet below the surface towards the central parts of the valley. In none of these cases did they get to the bottom of the alluvial soil. It has been objected, among other criticisms, that the Arabs can always find whatever their employers desire to obtain. Even those who are too well acquainted with the sagacity and energy of Hekekyan Bey to suspect him of having been deceived, have suggested that the artificial objects might have fallen into old wells which had been filled up. This notion is inadmissible for many reasons. Of the ninety-five shafts and borings, seventy or more were made far from the sites of towns or villages; and allowing that every field may once have had its well, there would be but small chance of the borings striking upon the site even of a small number of them in seventy experiments.

Others have suggested that the Nile may have wandered over the whole valley, undermining its banks on one side and filling up old channels on the other. It has also been asked whether the delta with the numerous shifting arms of the river may not once have been at every point where the auger pierced.[1] To all these objections there are two obvious answers:—First, in historical times the Nile has on the whole been very stationary, and has not shifted its position in the valley; secondly, if the mud pierced through had been thrown down by the river in ancient channels, it would have been stratified, and would not

[1] For a detailed account of these sections, see Mr. Horner's paper in the *Philosophical Transactions* for 1855-1858.

have corresponded so closely with inundation mud. We learn from Captain Newbold that he observed in some excavations in the great plain alternations of sand and clay, such as are seen in the modern banks of the Nile; but in the borings made by Hekekyan Bey, such stratification seems scarcely in any case to have been detected.

The great aim of the criticisms above enumerated has been to get rid of the supposed anomaly of finding burnt brick and pottery at depths and places which would give them claim to an antiquity far exceeding that of the Roman domination in Egypt. For until the time of the Romans, it is said, no clay was burnt into bricks in the valley of the Nile. But a distinguished antiquary, Mr. S. Birch, assures me that this notion is altogether erroneous, and that he has under his charge in the British Museum, first, a small rectangular baked brick, which came from a Theban tomb which bears the name of Thothmes, a superintendent of the granaries of the god Amen Ra, the style of art, inscription, and name, showing that it is as old as the 18th dynasty (about 1450 B.C.); secondly, a brick bearing an inscription, partly obliterated, but ending with the words " of the temple of Amen Ra." This brick, decidedly long anterior to the Roman dominion, is referred conjecturally, by Mr. Birch, to the 19th dynasty, or 1300 B.C. Sir Gardner Wilkinson has also in his possession pieces of mortar, which he took from each of the three great pyramids, in which bits of broken pottery and of burnt clay or brick are imbedded.

M. Girard, of the French expedition to Egypt, supposed the average rate of the increase of Nile mud on the plain between Assouan and Cairo to be five English inches in a century. This conclusion, according to Mr. Horner, is very vague, and founded on insufficient data; the amount of matter thrown down by the waters in different parts of the plain varying so much that to strike an average with any approach to accuracy must be most difficult. Were we to assume six inches in a century, the burnt brick met with at a depth of 60 feet would be 12,000 years old.

Another fragment of red brick was found by Linant Bey, in a boring 72 feet deep, being 2 or 3 feet below the level of the Mediterranean, in the parallel of the apex of the delta, 200 metres distant from the river, on the Libyan side of the Rosetta branch.[1] M. Rosière, in the great French work on Egypt, has estimated the mean rate of deposit of sediment in the delta at $2\frac{1}{4}$ inches in

[1] Horner, *Philosophical Transactions*, 1858.

a century;[1] were we to take $2\frac{1}{2}$ inches, a work of art 72 feet deep must have been buried more than 30,000 years ago. But if the boring of Linant Bey was made where an arm of the river had been silted up at a time when the apex of the delta was somewhat farther south, or more distant from the sea than now, the brick in question might be comparatively very modern.

The experiments instituted by Mr. Horner at the pedestal of the fallen statue of King Rameses at Memphis, in the hope of obtaining an accurate chronometric scale for testing the age of a given thickness of Nile sediment, are held by some experienced Egyptologists not to be satisfactory, on the ground of the uncertainty of the rate of deposit accumulated at that locality. The point sought to be determined was the exact amount of Nile mud which had accumulated there since the time when that statue is supposed by some antiquaries to have been erected. Could we have obtained possession of such a measure, the rate of deposition might be judged of, approximately at least, whenever similar mud was observed in other places, or below the foundations of those same monuments. But the ancient Egyptians are known to have been in the habit of enclosing with embankments the areas on which they erected temples, statues, and obelisks, so as to exclude the waters of the Nile; and the point of time to be ascertained, in every case where we find a monument buried to a certain depth in mud, as at Memphis and Heliopolis, is the era when the city fell into such decay that the ancient embankments were neglected, and the river allowed to inundate the site of the temple, obelisk, or statue.

Even if we knew the date of the abandonment of such embankments, the enclosed areas would not afford a favourable opportunity for ascertaining the average rate of deposit in the alluvial plain; for Herodotus tells us that in his time those spots from which the Nile waters had been shut out for centuries appeared sunk, and could be looked down into from the surrounding grounds, which had been raised by the gradual accumulation over them of sediment annually thrown down. If the waters at length should break into such depressions, they must at first carry with them into the enclosure much mud washed from the steep surrounding banks, so that a greater quantity would be deposited in a few years than perhaps in as many centuries on the great plain outside the depressed area, where no such disturbing causes intervened.

[1] Description de l'Egypte (*Histoire Naturelle*, tom. ii. p. 494).

Ancient Mounds of the Valley of the Ohio

As I have already given several European examples of monuments of prehistoric date belonging to the Recent period, I will now turn to the American continent. Before the scientific investigation by Messrs. Squier and Davis of the " Ancient Monuments of the Mississippi Valley," [1] no one suspected that the plains of that river had been occupied, for ages before the French and British colonists settled there, by a nation of older date and more advanced in the arts than the Red Indians whom the Europeans found there. There are hundreds of large mounds in the basin of the Mississippi, and especially in the valleys of the Ohio and its tributaries, which have served, some of them for temples, others for outlook or defence, and others for sepulture. The unknown people by whom they were constructed, judging by the form of several skulls dug out of the burial-places, were of the Mexican or Toltec race. Some of the earthworks are on so grand a scale as to embrace areas of 50 or 100 acres within a simple enclosure, and the solid contents of one mould are estimated at 20,000,000 of cubic feet, so that four of them would be more than equal in bulk to the Great Pyramid of Egypt, which comprises 75,000,000. From several of these repositories pottery and ornamental sculpture have been taken, and various articles in silver and copper, also stone weapons, some composed of hornstone unpolished, and much resembling in shape some ancient flint implements found near Amiens and other places in Europe, to be alluded to in the sequel.

It is clear that the Ohio mound-builders had commercial intercourse with the natives of distant regions, for among the buried articles some are made of native copper from Lake Superior, and there are also found mica from the Alleghanies, sea-shells from the Gulf of Mexico, and obsidian from the Mexican mountains.

The extraordinary number of the mounds implies a long period, during which a settled agricultural population had made considerable progress in civilization, so as to require large temples for their religious rites, and extensive fortifications to protect them from their enemies. The mounds were almost all confined to fertile valleys or alluvial plains, and some at least are so ancient that rivers have had time since their construction to encroach on the lower terraces which support them, and again

---

[1] *Smithsonian Contributions*, vol. i., 1847.

to recede for the distance of nearly a mile, after having undermined and destroyed a part of the works. When the first European settlers entered the valley of the Ohio, they found the whole region covered with an uninterrupted forest, and tenanted by the Red Indian hunter, who roamed over it without any fixed abode, or any traditionary connection with his more civilised predecessors. The only positive data as yet obtained for calculating the minimum of time which must have elapsed since the mounds were abandoned, have been derived from the age and nature of the trees found growing on some of these earthworks. When I visited Marietta in 1842, Dr. Hildreth took me to one of the mounds, and showed me where he had seen a tree growing on it, the trunk of which when cut down displayed eight hundred rings of annual growth.[1] But the late General Harrison, President in 1841 of the United States, who was well skilled in woodcraft, has remarked, in a memoir on this subject, that several generations of trees must have lived and died before the mounds could have been overspread with that variety of species which they supported when the white man first beheld them, for the number and kinds of trees were precisely the same as those which distinguished the surrounding forest. " We may be sure," observed Harrison, " that no trees were allowed to grow so long as the earthworks were in use; and when they were forsaken, the ground, like all newly cleared land in Ohio, would for a time be monopolised by one or two species of tree, such as the yellow locust and the black or white walnut. When the individuals which were the first to get possession of the ground had died out one after the other, they would in many cases, instead of being replaced by the same species, be succeeded (by virtue of the law which makes a rotation of crops profitable in agriculture) by other kinds, till at last, after a great number of centuries (several thousand years, perhaps), that remarkable diversity of species characteristic of North America, and far exceeding what is seen in European forests, would be established.

Mounds of Santos in Brazil

I will next say a few words respecting certain human bones embedded in a solid rock at Santos in Brazil, to which I called attention in my *Travels in North America* in 1842.[2] I then imagined the deposit containing them to be of submarine origin, —an opinion which I have long ceased to entertain. We learn

[1] Lyell's *Travels in North America*, vol. ii. p. 29.   [2] Vol. i. p. 200.

from a memoir of Dr. Meigs that the River Santos has under-mined a large mound, 14 feet in height, and about 3 acres in area, covered with trees, near the town of St. Paul, and has exposed to view many skeletons, all inclined at angles between 20° and 25°, and all placed in a similar east and west position.[1] Seeing, in the Museum of Philadelphia, fragments of the cal-careous stone or tufa from this spot, containing a human skull with teeth, and in the same matrix, oysters with serpulæ attached, I at first concluded that the whole deposit had been formed beneath the waters of the sea, or at least, that it had been sub-merged after its origin, and again upheaved; also, that there had been time since its emergence for the growth on it of a forest of large trees. But after reading again, with more care, the original memoir of Dr. Meigs, I cannot doubt that the shells, like those of eatable kinds, so often accumulated in the mounds of the North American Indians not far from the sea, may have been brought to the place and heaped up with other materials at the time when the bodies were buried. Subsequently, the whole artificial earthwork, with its shells and skeletons, may have been bound together into a solid stone by the infiltration of carbonate of lime, and the mound may therefore be of no higher antiquity than some of those above alluded to on the Ohio, which, as we have seen, have in like manner been exposed in the course of ages to the encroachments and undermining action of rivers.

Delta of the Mississippi

I have shown in my *Travels in North America* that the deposits forming the delta and alluvial plain of the Mississippi consist of sedimentary matter, extending over an area of 30,000 square miles, and known in some parts to be several hundred feet deep. Although we cannot estimate correctly how many years it may have required for the river to bring down from the upper country so large a quantity of earthy matter—the data for such a computation being as yet incomplete—we may still approxi-mate to a minimum of the time which such an operation must have taken, by ascertaining experimentally the annual discharge of water by the Mississippi, and the mean annual amount of solid matter contained in its waters. The lowest estimate of the time required would lead us to assign a high antiquity, amount-ing to many tens of thousands of years (probably more than 100,000) to the existing delta.

[1] Meigs, *Trans. Amer. Phil. Soc.*, 1828, p. 285.

Whether all or how much of this formation may belong to the recent period, as above defined, I cannot pretend to decide, but in one part of the modern delta near New Orleans, a large excavation has been made for gas-works, where a succession of beds, almost wholly made up of vegetable matter, has been passed through, such as we now see forming in the cypress swamps of the neighbourhood, where the deciduous cypress (*Taxodium distichum*), with its strong and spreading roots, plays a conspicuous part. In this excavation, at the depth of sixteen feet from the surface, beneath four buried forests superimposed one upon the other, the workmen are stated by Dr. B. Dowler to have found some charcoal and a human skeleton, the cranium of which is said to belong to the aboriginal type of the Red Indian race. As the discovery in question had not been made when I saw the excavation in progress at the gas-works in 1846, I cannot form an opinion as to the value of the chronological calculations which have led Dr. Dowler to ascribe to this skeleton an antiquity of 50,000 years. In several sections, both natural in the banks of the Mississippi and its numerous arms, and where artificial canals had been cut, I observed erect stumps of trees, with their roots attached, buried in strata at different heights, one over the other. I also remarked, that many cypresses which had been cut through, exhibited many hundreds of rings of annual growth, and it then struck me that nowhere in the world could the geologist enjoy a more favourable opportunity for estimating in years the duration of certain portions of the Recent epoch.[1]

Coral Reefs of Florida

Professor Agassiz has described a low portion of the peninsula of Florida as consisting of numerous reefs of coral, which have grown in succession so as to give rise to a continual annexation of land, gained gradually from the sea in a southerly direction. This growth is still in full activity, and assuming the rate of advance of the land to be one foot in a century, the reefs being built up from a depth of 75 feet, and that each reef has in its turn added ten miles to the coast, Professor Agassiz calculates that it has taken 135,000 years to form the southern half of this peninsula. Yet the whole is of Post-Tertiary origin, the fossil zoophytes and shells being all of the same species as those now

[1] Dowler, cited by Dr. W. Usher, in Nott and Gliddon's *Types of Mankind*, p. 352.

inhabiting the neighbouring sea.[1] In a calcareous conglomerate forming part of the above-mentioned series of reefs, and supposed by Agassiz, in accordance with his mode of estimating the rate of growth of those reefs, to be about 10,000 years old, some fossil human remains were found by Count Pourtalès. They consisted of jaws and teeth, with some bones of the foot.

Recent Deposits of Seas and Lakes

I have shown, in the *Principles of Geology*, where the recent changes of the earth illustrative of geology are described at length, that the deposits accumulated at the bottom of lakes and seas within the last 4000 or 5000 years can neither be insignificant in volume or extent. They lie hidden, for the most part, from our sight; but we have opportunities of examining them at certain points where newly-gained land in the deltas of rivers has been cut through during floods, or where coral reefs are growing rapidly, or where the bed of a sea or lake has been heaved up by subterranean movements and laid dry.

As examples of such changes of level by which marine deposits of the Recent period have become accessible to human observation, I have adduced the strata near Naples in which the Temple of Serapis at Pozzuoli was entombed.[2] These upraised strata, the highest of which are about 25 feet above the level of the sea, form a terrace skirting the eastern shore of the Bay of Baiæ. They consist partly of clay, partly of volcanic matter, and contain fragments of sculpture, pottery, and the remains of buildings, together with great numbers of shells, retaining in part their colour, and of the same species as those now inhabiting the neighbouring sea. Their emergence can be proved to have taken place since the beginning of the sixteenth century. [Note 5.]

In the same work, as an example of a fresh-water deposit of the Recent period, I have described certain strata in Cashmere, a country where violent earthquakes, attended by alterations in the level of the ground, are frequent, in which fresh-water shells of species now inhabiting the lakes and rivers of that region are embedded, together with the remains of pottery, often at the depth of fifty feet, and in which a splendid Hindoo temple has lately been discovered, and laid open to view by the removal of the lacustrine silt which had enveloped it for four or five centuries.

[1] Agassiz, in Nott and Gliddon, *ibid.*, p. 352.
[2] *Principles of Geology*, Index, " Serapis."

In the same treatise it is stated that the west coast of South America, between the Andes and the Pacific, is a great theatre of earthquake movements, and that permanent upheavals of the land of several feet at a time have been experienced since the discovery of America. In various parts of the littoral region of Chile and Peru, strata have been observed enclosing shells in abundance, all agreeing specifically with those now swarming in the Pacific. In one bed of this kind, in the island of San Lorenzo, near Lima, Mr. Darwin found, at the altitude of 85 feet above the sea, pieces of cotton-thread, plaited rush, and the head of a stalk of Indian corn, the whole of which had evidently been embedded with the shells. At the same height, on the neighbouring mainland, he found other signs corroborating the opinion that the ancient bed of the sea had there also been uplifted 85 feet since the region was first peopled by the Peruvian race. But similar shelly masses are also met with at much higher elevations, at innumerable points between the Chilean and Peruvian Andes and the sea-coast, in which no human remains have as yet been observed. The preservation for an indefinite period of such perishable substances as thread is explained by the entire absence of rain in Peru. The same articles, had they been enclosed in the permeable sands of an European raised beach, or in any country where rain falls even for a small part of the year, would probably have disappeared entirely. [Note 6.]

In the literature of the eighteenth century, we find frequent allusion to the " era of existing continents," a period supposed to have coincided in date with the first appearance of Man upon the earth, since which event it was imagined that the relative level of the sea and land had remained stationary, no important geographical changes having occurred, except some slight additions to the deltas of rivers, or the loss of narrow strips of land where the sea had encroached upon its shores. But modern observations have tended continually to dispel this delusion, and the geologist is now convinced that at no given era of the past have the boundaries of land and sea, or the height of the one and depth of the other, or the geographical range of the species inhabiting them, whether of animals or plants, become fixed and unchangeable. Of the extent to which fluctuations have been going on since the globe had already become the dwelling-place of Man, some idea may be formed from the examples which I shall give in this and the next nine chapters.

Upheaval since the Human Period of the Central District of Scotland [Note 7]

It has long been a fact familiar to geologists, that, both on the east and west coasts of the central part of Scotland, there are lines of raised beaches, containing marine shells of the same species as those now inhabiting the neighbouring sea.[1] The two most marked of these littoral deposits occur at heights of about 50 and 25 feet above high-water mark, that of 50 feet being considered as the more ancient, and owing its superior elevation to a longer continuance of the upheaving movement. They are seen in some places to rest on the boulder clay of the glacial period, which will be described in future chapters.

In those districts where large rivers, such as the Clyde, Forth, and Tay, enter the sea, the lower of the two deposits, or that of 25 feet, expands into a terrace fringing the estuaries, and varying in breadth from a few yards to several miles. Of this nature are the flat lands which occur along the margin of the Clyde at Glasgow, which consist of finely laminated sand, silt, and clay. Mr. John Buchanan, a zealous antiquary, writing in 1855, informs us that in the course of the eighty years preceding that date, no less than seventeen canoes had been dug out of this estuarine silt, and that he had personally inspected a large number of them before they were exhumed. Five of them lay buried in silt under the streets of Glasgow, one in a vertical position with the prow uppermost as if it had sunk in a storm. In the inside of it were a number of marine shells. Twelve other canoes were found about 100 yards back from the river, at the average depth of about 19 feet from the surface of the soil, or 7 feet above high-water mark; but a few of them were only 4 or 5 feet deep, and consequently more than 20 feet above the sea-level. One was sticking in the sand at an angle of 45°, another had been capsized and lay bottom uppermost; all the rest were in a horizontal position, as if they had sunk in smooth water.[2]

Almost every one of these ancient boats was formed out of a single oak-stem, hollowed out by blunt tools, probably stone axes, aided by the action of fire; a few were cut beautifully smooth, evidently with metallic tools. Hence a gradation could be traced from a pattern of extreme rudeness to one showing

[1] R. Chambers, *Sea Margins*, 1848, and papers by Mr. Smith of Jordan Hill, *Mem. Wern. Soc.* vol. viii., and by Mr. C. Maclaren.
[2] J. Buchanan, *Rep. Brit. Ass.*, 1855, p. 80 ; also *Glasgow, Past and Present*, 1856.

great mechanical ingenuity.   Two of them were built of planks, one of the two, dug up on the property of Bankton in 1853, being 18 feet in length, and very elaborately constructed.   Its prow was not unlike the beak of an antique galley; its stern, formed of a triangular-shaped piece of oak, fitted in exactly like those of our day.   The planks were fastened to the ribs, partly by singularly shaped oaken pins, and partly by what must have been square nails of some kind of metal; these had entirely disappeared, but some of the oaken pins remained.   This boat had been upset, and was lying keel uppermost, with the prow pointing straight up the river.   In one of the canoes, a beautifully polished celt or axe of greenstone was found, in the bottom of another a plug of cork, which, as Mr. Geikie remarks, " could only have come from the latitudes of Spain, Southern France, or Italy." [1]

There can be no doubt that some of these buried vessels are of far more ancient date than others.   Those most roughly hewn, may be relics of the stone period; those more smoothly cut, of the bronze age; and the regularly built boat of Bankton may perhaps come within the age of iron.   The occurrence of all of them in one and the same upraised marine formation by no means implies that they belong to the same era, for in the beds of all great rivers and estuaries, there are changes continually in progress brought about by the deposition, removal, and redeposition of gravel, sand, and fine sediment, and by the shifting of the channel of the main currents from year to year, and from century to century.   All these it behoves the geologist and antiquary to bear in mind, so as to be always on their guard, when they are endeavouring to settle the relative date, whether of objects of art or of organic remains embedded in any set of alluvial strata.   Some judicious observations on this head occur in Mr. Geikie's memoir above cited, which are so much in point that I shall give them in full, and in his own words.

" The relative position in the silt, from which the canoes were exhumed, could help us little in any attempt to ascertain their relative ages, unless they had been found vertically above each other.   The varying depths of an estuary, its banks of silt and sand, the set of its currents, and the influence of its tides in scouring out alluvium from some parts of its bottom and redepositing it in others, are circumstances which require to be taken into account in all such calculations.   Mere coincidence of depth from the present surface of the ground, which is tolerably uni-

[1] Geikie, *Quart. Jour. Geol. Soc.*, vol. xviii., 1862, p. 224.

form in level, by no means necessarily proves contemporaneous deposition. Nor would such an inference follow even from the occurrence of the remains in distant parts of the very same stratum. A canoe might be capsized and sent to the bottom just beneath low-water mark; another might experience a similar fate on the following day, but in the middle of the channel. Both would become silted up on the floor of the estuary; but as that floor would be perhaps 20 feet deeper in the centre than towards the margin of the river, the one canoe might actually be twenty feet deeper in the alluvium than the other; and on the upheaval of the alluvial deposits, if we were to argue merely from the depth at which the remains were embedded, we should pronounce the canoe found at the one locality to be immensely older than the other, seeing that the fine mud of the estuary is deposited very slowly and that it must therefore have taken a long period to form so great a thickness as 20 feet. Again, the tides and currents of the estuary, by changing their direction, might sweep away a considerable mass of alluvium from the bottom, laying bare a canoe that may have foundered many centuries before. After the lapse of so long an interval, another vessel might go to the bottom in the same locality and be there covered up with the older one on the same general plane. These two vessels, found in such a position, would naturally be classed together as of the same age, and yet it is demonstrable that a very long period may have elapsed between the date of the one and that of the other. Such an association of these canoes, therefore, cannot be regarded as proving synchronous deposition; nor, on the other hand, can we affirm any difference of age from mere relative position, unless we see one canoe actually buried beneath another." [1]

At the time when the ancient vessels, above described, were navigating the waters where the city of Glasgow now stands, the whole of the low lands which bordered the present estuary of the Clyde formed the bed of a shallow sea. The emergence appears to have taken place gradually and by intermittent movements, for Mr. Buchanan describes several narrow terraces one above the other on the site of the city itself, with steep intervening slopes composed of the laminated estuary formation. Each terrace and steep slope probably mark pauses in the process of upheaval, during which low cliffs were formed, with beaches at their base. Five of the canoes were found within the precincts of the city at different heights on or near such terraces.

[1] Geikie, *Quart. Jour. Geol. Soc.*, vol. xviii., 1862, p. 222.

As to the date of the upheaval, the greater part of it cannot be assigned to the stone period, but must have taken place after tools of metal had come into use.

Until lately, when attempts were made to estimate the probable antiquity of such changes of level, it was confidently assumed, as a safe starting-point, that no alteration had occurred in the relative level of land and sea, in the central district of Scotland, since the construction of the Roman or Pictish wall (the " Wall of Antonine "), which reached from the Firth of Forth to that of the Clyde. The two extremities, it was said, of this ancient structure, bear such a relation to the present level of the two estuaries, that neither subsidence nor elevation of the land could have occurred for seventeen centuries at least.

But Mr. Geikie has lately shown that a depression of 25 feet on the Forth would not lay the eastern extremity of the Roman wall at Carriden under water, and he was therefore desirous of knowing whether the western end of the same would be submerged by a similar amount of subsidence. It has always been acknowledged that the wall terminated upon an eminence called the Chapel Hill, near the village of West Kilpatrick, on the Clyde. The foot of this hill, Mr. Geikie estimates to be about 25 or 27 feet above high-water mark, so that a subsidence of 25 feet could not lay it under water. Antiquaries have sometimes wondered that the Romans did not carry the wall farther west than this Chapel Hill; but Mr. Geikie now suggests, in explanation, that all the low land at present intervening between that point and the mouth of the Clyde, was sixteen or seventeen centuries ago, washed by the tides at high water.

The wall of Antonine, therefore, yields no evidence in favour of the land having remained stationary since the time of the Romans, but on the contrary, appears to indicate that since its erection the land has actually risen. Recent explorations by Mr. Geikie and Dr. Young, of the sites of the old Roman harbours along the southern margin of the Firth of Forth, lead to similar inferences. In the first place, it has long been known that there is a raised beach containing marine shells of living littoral species, at a height of about 25 feet, at Leith, as well as at other places along the coast above and below Edinburgh. Inveresk, a few miles below that city, is the site of an ancient Roman port, and if we suppose the sea at high water to have washed the foot of the heights on which the town stood, the tide would have ascended far up the valley of the Esk, and would have made the mouth of that river a safe and commodious harbour; whereas, had it been a shoaling estuary, as at present, it is difficult to see how the Romans should have made choice of it as a port.

At Cramond, at the mouth of the river Almond, above Edinburgh, was Alaterva, the chief Roman harbour on the southern coast of the Forth, where numerous coins, urns, sculptured stones and the remnant of a harbour have been detected. The old Roman quays built along what must then have been the sea margin, have been found on what is now dry land, and although some silt carried down in suspension by the waters of the Forth may account for a part of the gain of low land, we yet require an upward movement of about 20 feet to explain the growth of the dreary expanse of mud now stretching along the shore and extending outwards, where it attains its greatest breadth, well-nigh two miles, across which vessels, even of light burden, can now only venture at full tide. Had these shoals existed eighteen centuries ago, they would have prevented the Romans from selecting this as their chief port; whereas, if the land were now to sink 20 feet, Cramond would unquestionably be the best natural harbour along the whole of the south side of the Forth.[1]

Corresponding in level with the raised beach at Leith, above mentioned (or about 25 feet above high-water mark), is the Carse of Stirling, a low tract of land consisting of loamy and peaty beds, in which several skeletons of whales of large size have been found. One of these was dug up at Airthrie,[2] near Stirling, about a mile from the river, and 7 miles from the sea. Mr. Bald mentions that near it were found two pieces of stag's horn, artificially cut, through one of which a hole, about an inch in diameter, had been perforated. Another whale, 85 feet long, was found at Dunmore, a few miles below Stirling,[3] which, like that of Airthrie, lay about 20 feet above high-water mark. Three other skeletons of whales were found at Blair Drummond, between the years 1819 and 1824, 7 miles up the estuary above Stirling,[4] also at an elevation of between 20 and 30 feet above the sea. Near two of these whales, pointed instruments of deer's horn were found, one of which retained part of a wooden handle, probably preserved by having been enclosed in peat. This weapon is now in the museum at Edinburgh.

The position of these fossil whales and bone implements, and still more of an iron anchor found in the Carse of Falkirk, below

[1] Geikie, *Edinb. New Phil. Journ.* for July 1861.
[2] Bald, *Edinburgh Philosophical Journal*, i. p. 393, and *Memoirs, Wernerian Society*, iii. p. 327.
[3] *Edinburgh Philosophical Journal*, xi. pp. 220, 415.
[4] *Memoirs, Wernerian Society*, vol. v. p. 440.

Stirling, shows that the upheaval by which the raised beach of Leith was laid dry extended far westward probably as far as the Clyde, where, as we have seen, marine strata containing buried canoes rise to a similar height above the sea.

The same upward movement which reached simultaneously east and west from sea to sea was also felt as far north as the estuary of the Tay. This may be inferred from the Celtic name of *Inch* being attached to many hillocks, which rise above the general level of the alluvial plains, implying that these eminences were once surrounded by water or marshy ground. At various localities also in the silt of the Carse of Gowrie iron implements have been found.

The raised beach, also containing a great number of marine shells of recent species, traced up to a height of 14 feet above the sea by Mr. W. J. Hamilton at Elie, on the southern coast of Fife, is doubtless another effect of the same extensive upheaval.[1] A similar movement would also account for some changes which antiquaries have recorded much farther south, on the borders of the Solway Firth; though in this case, as in that of the estuary of the Forth, the conversion of sea into land has always been referred to the silting up of estuaries, and not to upheaval. Thus Horsley insists on the difficulty of explaining the position of certain Roman stations, on the Solway, the Forth, and the Clyde, without assuming that the sea has been excluded from certain areas which it formerly occupied.[2]

On a review of the whole evidence, geological and archæological, afforded by the Scottish coast-line, we may conclude that the last upheaval of 25 feet took place not only since the first human population settled in the island; but long after metallic implements had come into use, and there seems even a strong presumption in favour of the opinion that the date of the elevation may have been subsequent to the Roman occupation.

But the 25 feet rise is only the last stage of a long antecedent process of elevation, for examples of Recent marine shells have been observed 40 feet and upwards above the sea in Ayrshire. At one of these localities, Mr. Smith of Jordanhill informs me that a rude ornament made of cannel coal has been found on the coast in the parish of Dundonald, lying 50 feet above the sea-level, on the surface of the boulder-clay or till, and covered with gravel containing marine shells. If we suppose the upward movement to have been uniform in central Scotland before and

---

[1] *Proceedings of Geological Society*, vol. ii., 1833, p. 280.
[2] *Britannia*, p. 157, 1860.

after the Roman era, and assume that as 25 feet indicate seventeen centuries, so 50 feet imply a lapse of twice that number, or 3400 years, we should then carry back the date of the ornament in question to fifteen centuries before our era, or to the days of Pharaoh, and the period usually assigned to the exodus of the Israelites from Egypt. [Note 8.]

But all such estimates must be considered, in the present state of science, as tentative and conjectural, since the rate of movement of the land may not have been uniform, and its direction not always upwards, and there may have been long stationary periods, one of which of more than usual duration seems indicated by the 50-foot raised beach, which has been traced for vast distances along the western coast of Scotland.

Coast of Cornwall

Sir H. De la Beche has adduced several proofs of changes of level, in the course of the human period, in his *Report on the Geology of Cornwall and Devon*, 1839. He mentions (p. 406) that several human skulls and works of art, buried in an estuary deposit, were found in mining gravel for tin at Pentuan, near St. Austell, the skulls lying at the depth of 40 feet from the surface, and others at Carnon at the depth of 53 feet. The overlying strata were marine, containing sea-shells of living species, and bones of whales, besides the remains of several living species of mammalia.

Other examples of works of art, such as stone hatchets, canoes, and ships, buried in ancient river-beds in England, and in peat and shell-marl, I have mentioned in my work before cited.

Sweden and Norway

In the same work I have shown that near Stockholm, in Sweden, there occur, at slight elevations above the sea-level, horizontal beds of sand, loam, and marl, containing the same peculiar assemblage of testacea which now live in the brackish waters of the Baltic. Mingled with these, at different depths, have been detected various works of art implying a rude state of civilization, and some vessels built before the introduction of iron, and even the remains of an ancient hut, the marine strata containing it, which had been formed during a previous depression, having been upraised, so that the upper beds are now

60 feet higher than the surface of the Baltic. In the neighbourhood of these recent strata, both to the north-west and south of
Stockholm, other deposits similar in mineral composition occur,
which ascend to greater heights, in which precisely the same
assemblage of fossil shells is met with, but without any intermixture, so far as is yet known, of human bones or fabricated
articles.

On the opposite or western coast of Sweden, at Uddevalla,
Post-Tertiary strata, containing recent shells, not of that brackish
water character peculiar to the Baltic, but such as now live in
the Northern Ocean, ascend to the height of 200 feet; and beds
of clay and sand of the same age attain elevations of 300 and even
600 feet in Norway, where they have been usually described as
" raised beaches." They are, however, thick deposits of submarine origin, spreading far and wide, and filling valleys in the
granite and gneiss, just as the Tertiary formations, in different
parts of Europe, cover or fill depressions in the older rocks.

Although the fossil fauna characterising these upraised sands
and clays consists exclusively of existing northern species of
testacea, it is more than probable that they may not all belong
to that division of the Pleistocene strata which we are now
considering. If the contemporary mammalia were known, they
would, in all likelihood, be found to be referable, at least in part,
to extinct species; for, according to Lovén (an able living
naturalist of Norway), the species do not constitute such an
assemblage as now inhabits corresponding latitudes in the North
Sea. On the contrary, they decidedly represent a more arctic
fauna. In order to find the same species flourishing in equal
abundance, or in many cases to find them at all, we must go
northwards to higher latitudes than Uddevalla in Sweden, or
even nearer the pole than Central Norway.

Judging by the uniformity of climate now prevailing from
century to century, and the insensible rate of variation in the
geographical distribution of organic beings in our own times, we·
may presume that an extremely lengthened period was required
even for so slight a modification in the range of the molluscous
fauna, as that of which the evidence is here brought to light.
There are also other independent reasons for suspecting that the
antiquity of these deposits may be indefinitely great as compared
to the historical period. I allude to their present elevation
above the sea, some of them rising, in Norway, to the height of
600 feet or more. The upward movement now in progress in
parts of Norway and Sweden extends, as I have elsewhere

shown,[1] throughout an area about 1000 miles north and south, and for an unknown distance east and west, the amount of elevation always increasing as we proceed towards the North Cape, where it is said to equal 5 feet in a century. If we could assume that there had been an average of $2\frac{1}{2}$ feet in each hundred years for the last fifty centuries, this would give an elevation of 125 feet in that period. In other words, it would follow that the shores, and a considerable area of the former bed of the North Sea, had been uplifted vertically to that amount, and converted into land in the course of the last 5000 years. A mean rate of continuous vertical elevation of $2\frac{1}{2}$ feet in a century would, I conceive, be a high average; yet, even if this be assumed, it would require 24,000 years for parts of the sea-coast of Norway, where the Pleistocene marine strata occur, to attain the height of 600 feet. [Note 9.]

[1] *Principles*, 9th ed., ch. xxx.

CHAPTER IV

PLEISTOCENE PERIOD—BONES OF MAN AND EXTINCT MAMMALIA
IN BELGIAN CAVERNS

Earliest Discoveries in Caves of Languedoc of Human Remains with
Bones of extinct Mammalia—Researches in 1833 of Dr. Schmerling
in the Liège Caverns—Scattered Portions of Human Skeletons
associated with Bones of Elephant and Rhinoceros—Distribution
and probable Mode of Introduction of the Bones—Implements of
Flint and Bone—Schmerling's Conclusions as to the Antiquity of
Man ignored—Present State of the Belgian Caves—Human Bones
recently found in Cave of Engihoul—Engulfed Rivers—Stalagmitic
Crust—Antiquity of the Human Remains in Belgium how proved.

HAVING hitherto considered those formations in which both the
fossil shells and the mammalia are of living species, we may now
turn our attention to those of older date, in which the shells being
all recent, some of the accompanying mammalia are extinct, or
belong to species not known to have lived within the times of
history or tradition.

Discoveries of MM. Tournal and Christol in 1828 in the South of France

In the *Principles of Geology*, when treating of the fossil remains
found in alluvium and the mud of caverns, I gave an account
in 1832 of the investigations made by MM. Tournal and Christol
in the South of France.[1]

M. Tournal stated in his memoir that in the cavern of Bize,
in the department of the Aude, he had found human bones and
teeth, together with fragments of rude pottery, in the same mud
and breccia cemented by stalagmite in which land-shells of
living species were embedded, and the bones of mammalia, some
of extinct, others of recent species. The human bones were
declared by his fellow-labourer, M. Marcel de Serres, to be in
the same chemical condition as those of the accompanying
quadrupeds.[2]

Speaking of these fossils of the Bize cavern five years later,
M. Tournal observed that they could not be referred, as some

---

[1] 1st ed. vol. ii. ch. xiv., 1832, and 9th ed. p. 738, 1853.
[2] *Annales des Sciences Naturelles*, tom. xv., 1828, p. 348.

suggested, to a " diluvial catastrophe," for they evidently had not been washed in suddenly by a transient flood, but must have been introduced gradually, together with the enveloping mud and pebbles, at successive periods.[1]

M. Christol, who was engaged at the same time in similar researches in another part of Languedoc, published an account of them a year later, in which he described some human bones, as occurring in the cavern of Pondres, near Nîmes, in the same mud with the bones of an extinct hyæna and rhinoceros.[2] The cavern was in this instance filled up to the roof with mud and gravel, in which fragments of two kinds of pottery were detected, the lowest and rudest near the bottom of the cave, below the level of the extinct mammalia.

It has never been questioned that the hyæna and rhinoceros found by M. Christol were of extinct species; but whether the animals enumerated by M. Tournal might not all of them be referred to quadrupeds which are known to have been living in Europe in the historical period seems doubtful. They were said to consist of a stag, an antelope, and a goat, all named by M. Marcel de Serres as new; but the majority of palæontologists do not agree with this opinion. Still it is true, as M. Lartet remarks, that the fauna of the cavern of Bize must be of very high antiquity, as shown by the presence, not only of the Lithuanian aurochs (*Bison europæus*), but also of the reindeer, which has not been an inhabitant of the South of France in historical times, and which, in that country, is almost everywhere associated, whether in ancient alluvium or in the mud of caverns, with the mammoth.

In my work before cited,[3] I stated that M. Desnoyers, an observer equally well versed in geology and archæology, had disputed the conclusion arrived at by MM. Tournal and Christol, that the fossil rhinoceros, hyæna, bear, and other lost species had once been inhabitants of France contemporaneously with Man. " The flint hatchets and arrow-heads," he said, " and the pointed bones and coarse pottery of many French and English caves, agree precisely in character with those found in the tumuli, and under the dolmens (rude altars of unhewn stone) of the primitive inhabitants of Gaul, Britain, and Germany. The human bones, therefore, in the caves which are associated with such fabricated objects, must belong not to antediluvian

---

[1] *Annales de Chimie et de Physique*, 1833, p. 161.

[2] Christol, *Notice sur les Ossements humains des Cavernes du Gard*, Montpellier, 1829.

[3] *Principles*, 9th ed. p. 739.

periods, but to a people in the same stage of civilization as those who constructed the tumuli and altars."

" In the Gaulish monuments," he added, " we find, together with the objects of industry above mentioned, the bones of wild and domestic animals of species now inhabiting Europe, particularly of deer, sheep, wild boars, dogs, horses, and oxen. This fact has been ascertained in Quercy and other provinces; and it is supposed by antiquaries that the animals in question were placed beneath the Celtic altars in memory of sacrifices offered to the Gaulish divinity Hesus, and in the tombs to commemorate funeral repasts, and also from a superstition prevalent among savage nations, which induces them to lay up provisions for the manes of the dead in a future life. But in none of these ancient monuments have any bones been found of the elephant, rhinoceros, hyæna, tiger, and other quadrupeds, such as are found in caves, which might certainly have been expected had these species continued to flourish at the time that this part of Gaul was inhabited by Man." [1]

After giving no small weight to the arguments of M. Desnoyers, and the writings of Dr. Buckland on the same subject, and myself visiting several caves in Germany, I came to the opinion that the human bones mixed with those of extinct animals, in osseous breccias and cavern mud, in different parts of Europe, were probably not coeval. The caverns having been at one period the dens of wild beasts, and having served at other times as places of human habitation, worship, sepulture, concealment, or defence, one might easily conceive that the bones of Man and those of animals, which were strewed over the floors of subterranean cavities, or which had fallen into tortuous rents connecting them with the surface, might, when swept away by floods, be mingled in one promiscuous heap in the same ossiferous mud or breccia.[2]

That such intermixtures have really taken place in some caverns, and that geologists have occasionally been deceived, and have assigned to one and the same period fossils which had really been introduced at successive times, will readily be conceded. But of late years we have obtained convincing proofs, as we shall see in the sequel, that the mammoth, and many other extinct mammalian species very common in caves, occur also

---

[1] Desnoyers, *Bulletin de la Société Géologique de France*, tom. ii., p. 252; and article on Caverns, *Dictionnaire Universelle d'Histoire Naturelle*, Paris, 1845.

[2] *Principles*, 9th ed. p. 740.

in undisturbed alluvium, embedded in such a manner with works of art, as to leave no room for doubt that Man and the mammoth coexisted. Such discoveries have led me, and other geologists, to reconsider the evidence previously derived from caves brought forward in proof of the high antiquity of Man. With a view of re-examining this evidence, I have lately explored several caverns in Belgium and other countries, and re-read the principal memoirs and treatises treating of the fossil remains preserved in them, the results of which inquiries I shall now proceed to lay before the reader.

Researches, in 1833-1834, of Dr. Schmerling in the Caverns near Liège

The late Dr. Schmerling of Liège, a skilful anatomist and palæontologist, after devoting several years to the exploring of the numerous ossiferous caverns which border the valleys of the Meuse and its tributaries, published two volumes descriptive of the contents of more than forty caverns. One of these volumes consisted of an atlas of plates, illustrative of the fossil bones.[1]

Many of the caverns had never before been entered by scientific observers, and their floors were encrusted with unbroken stalagmite. At a very early stage of his investigations, Dr. Schmerling found the bones of Man so rolled and scattered as to preclude all idea of their having been intentionally buried on the spot. He also remarked that they were of the same colour, and in the same condition as to the amount of animal matter contained in them, as those of the accompanying animals, some of which, like the cave-bear, hyæna, elephant, and rhinoceros, were extinct; others, like the wild cat, beaver, wild boar, roe-deer, wolf, and hedgehog, still extant. The fossils were lighter than fresh bones, except such as had their pores filled with carbonate of lime, in which case they were often much heavier. The human remains of most frequent occurrence were teeth detached from the jaw, and the carpal, metacarpal, tarsal, metatarsal, and phalangeal bones separated from the rest of the skeleton. The corresponding bones of the cave-bear, the most abundant of the accompanying mammalia, were also found in the Liège caverns more commonly than any others, and in the same scattered condition. Occasionally, some of the long bones of

---

[1] *Recherches sur les Ossements fossiles découverts dans les Cavernes de la Province de Liège*, Liège, 1833-1834.

mammalia were observed to have been first broken across, and then reunited or cemented again by stalagmite, as they lay on the floor of the cave.

No gnawed bones nor any coprolites were found by Schmerling. He therefore inferred that the caverns of the province of Liège had not been the dens of wild beasts, but that their organic and inorganic contents had been swept into them by streams communicating with the surface of the country. The bones, he suggested, may often have been rolled in the beds of such streams before they reached their underground destination. To the same agency the introduction of many land-shells dispersed through the cave-mud was ascribed, such as *Helix nemoralis, H. lapicida, H. pomatia,* and others of living species. Mingled with such shells, in some rare instances, the bones of fresh-water fish, and of a snake (*Coluber*), as well as of several birds, were detected.

The occurrence here and there of bones in a very perfect state, or of several bones belonging to the same skeleton in natural juxtaposition, and having all their most delicate apophyses uninjured, while many accompanying bones in the same breccia were rolled, broken, or decayed, was accounted for by supposing that portions of carcasses were sometimes floated in during floods while still clothed with their flesh. No example was discovered of an entire skeleton, not even of one of the smaller mammalia, the bones of which are usually the least injured.

The incompleteness of each skeleton was especially ascertained in regard to the human subjects, Dr. Schmerling being careful, whenever a fragment of such presented itself, to explore the cavern himself, and see whether any other bones of the same skeleton could be found. In the Engis cavern, distant about eight miles to the south-west of Liège, on the left bank of the Meuse, the remains of at least three human individuals were disinterred. The skull of one of these, that of a young person, was embedded by the side of a mammoth's tooth. It was entire but so fragile, that nearly all of it fell to pieces during its extraction. Another skull, that of an adult individual, and the only one preserved by Dr. Schmerling in a sufficient state of integrity to enable the anatomist to speculate on the race to which it belonged, was buried 5 feet deep in a breccia, in which the tooth of a rhinoceros, several bones of a horse, and some of the reindeer, together with some ruminants, occurred. This skull, now in the museum of the University of Liège, is figured in Chap. V. (Fig. 2, p. 63), where further observations will be offered on its anatomical character, after a fuller account of the contents of the Liège caverns has been laid before the reader.

On the right bank of the Meuse, on the opposite side of the river to Engis, is the cavern of Engihoul. Bones of extinct animals mingled with those of Man were observed to abound in both caverns; but with this difference, that whereas in the Engis cave there were several human crania and very few other bones, in Engihoul there occurred numerous bones of the extremities belonging to at least three human individuals, and only two small fragments of a cranium. The like capricious distribution held good in other caverns, especially with reference to the cave-bear, the most frequent of the extinct mammalia. Thus, for example in the cave of Chokier, skulls of the bear were few, and other parts of the skeleton abundant, whereas in several other caverns these proportions were exactly reversed, while at Goffontaine skulls of the bear and other parts of the skeleton were found in their natural numerical proportions. Speaking generally, it may be said that human bones, where any were met with, occurred at all depths in the cave-mud and gravel, sometimes above and sometimes below those of the bear, elephant, rhinoceros, hyæna, etc.

Some rude flint implements of the kind commonly called flint knives or flakes, of a triangular form in the cross section (as in Fig. 14, p. 92), were found by Schmerling dispersed generally through the cave-mud, but he was too much engrossed with his osteological inquires to collect them diligently. He preserved some few of them, however, which I have seen in the museum at Liège. He also discovered in the cave of Chokier, $2\frac{1}{2}$ miles south-west from Liège, a polished and jointed needle-shaped bone, with a hole pierced obliquely through it at the base; such a cavity, he observed, as had never given passage to an artery. This instrument was embedded in the same matrix with the remains of a rhinoceros.[1]

Another cut bone and several artificially-shaped flints were found in the Engis cave, near the human skulls before alluded to. Schmerling observed, and we shall have to refer to the fact in the sequel (Chap. VIII.), that although in some forty fossiliferous caves explored by him human bones were the exception, yet these flint implements were universal, and he added that "none of them could have been subsequently introduced, being precisely in the same position as the remains of the accompanying animals." "I therefore," he continues, "attach great

[1] Schmerling, part ii., p. 177.

importance to their presence; for even if I had not found the human bones under conditions entirely favourable to their being considered as belonging to the antediluvian epoch, proofs of Man's existence would still have been supplied by the cut bones and worked flints." [1]

Dr. Schmerling, therefore, had no hesitation in concluding from the various facts ascertained by him, that Man once lived in the Liège district contemporaneously with the cave-bear and several other extinct species of quadrupeds. But he was much at a loss when he attempted to invent a theory to explain the former state of the fauna of the region now drained by the Meuse; for he shared the notion, then very prevalent among naturalists, that the mammoth and the hyæna [2] were beasts of a warmer climate than that now proper to Western Europe. In order to account for the presence of such " tropical species," he was half-inclined to imagine that they had been transported by a flood from some distant region; then again he raised the question whether they might not have been washed out of an older alluvium, which may have pre-existed in the neighbourhood. This last hypothesis was directly at variance with his own statements, that the remains of the mammoth and hyæna were identical in appearance, colour, and chemical condition with those of the bear and other associated fossil animals, none of which exhibited signs of having been previously enveloped in any dissimilar matrix. Another enigma which led Schmerling astray in some of his geological speculations was the supposed presence of the agouti, a South American rodent, " proper to the torrid zone." My friend M. Lartet, guided by Schmerling's figures of the teeth of this species, suggests, and I have little doubt with good reason, that they appertain to the porcupine, a genus found fossil in Pleistocene deposits of certain caverns in the south of France.

In the year 1833, I passed through Liège, on my way to the Rhine, and conversed with Dr. Schmerling, who showed me his splendid collection, and when I expressed some incredulity respecting the alleged antiquity of the fossil human bones, he pointedly remarked that if I doubted their having been con-temporaneous with the bear or rhinoceros, on the ground of Man being a species of more modern date, I ought equally to doubt the co-existence of all the other living species, such as the red deer, roe, wild cat, wild boar, wolf, fox, weasel, beaver, hare, rabbit, hedgehog, mole, dormouse, field-mouse, water-rat, shrew,

---

[1] Schmerling, part ii., p. 179.   [2] *Ibid.* part ii., pp. 70, 96.

and others, the bones of which he had found scattered everywhere indiscriminately through the same mud with the extinct quadrupeds. The year after this conversation I cited Schmerling's opinions, and the facts bearing on the antiquity of Man, in the 3rd edition of my *Principles of Geology* (p. 161, 1834), and in succeeding editions, without pretending to call in question their trustworthiness, but at the same time without giving them the weight which I now consider they were entitled to. He had accumulated ample evidence to prove that Man had been introduced into the earth at an earlier period than geologists were then willing to believe.

One positive fact, it will be said, attested by so competent a witness, ought to have outweighed any amount of negative testimony, previously accumulated, respecting the non-occurrence elsewhere of human remains in formations of the like antiquity. In reply, I can only plead that a discovery which seems to contradict the general tenor of previous investigations is naturally received with much hesitation. To have undertaken in 1832, with a view of testing its truth, to follow the Belgian philosopher through every stage of his observations and proofs, would have been no easy task even for one well-skilled in geology and osteology. To be let down, as Schmerling was, day after day, by a rope tied to a tree, so as to slide to the foot of the first opening of the Engis cave,[1] where the best-preserved human skulls were found; and, after thus gaining access to the first subterranean gallery, to creep on all fours through a contracted passage leading to larger chambers, there to superintend by torchlight, week after week and year after year, the workmen who were breaking through the stalagmitic crust as hard as marble, in order to remove piece by piece the underlying bone-breccia nearly as hard; to stand for hours with one's feet in the mud, and with water dripping from the roof on one's head, in order to mark the position and guard against the loss of each single bone of a skeleton; and at length, after finding leisure, strength, and courage for all these operations, to look forward, as the fruits of one's labour, to the publication of unwelcome intelligence, opposed to the prepossessions of the scientific as well as of the unscientific public;—when these circumstances are taken into account, we need scarcely wonder, not only that a passing traveller failed to stop and scrutinise the evidence, but that a quarter of a century should have elapsed before even the neighbouring professors of the University of Liège came

[1] Schmerling, part i., p. 30.

forth to vindicate the truthfulness of their indefatigable and clear-sighted countryman.

In 1860, when I revisited Liège, twenty-six years after my interview with Schmerling, I found that several of the caverns described by him had in the interval been annihilated. Not a vestige, for example, of the caves of Engis, Chokier, and Goffontaine remained. The calcareous stone, in the heart of which the cavities once existed, had been quarried away, and removed bodily for building and lime-making. Fortunately, a great part of the Engihoul cavern, situated on the right bank of the Meuse, was still in the same state as when Schmerling delved into it in 1831, and drew from it the bones of three human skeletons. I determined, therefore, to examine it, and was so fortunate as to obtain the assistance of a zealous naturalist of Liège, Professor Malaise, who accompanied me to the cavern, where we engaged some workmen to break through the crust of stalagmite, so that we could search for bones in the undisturbed earth beneath. Bones and teeth of the cave-bear were soon found, and several other extinct quadrupeds which Schmerling has enumerated. My companion, continuing the work perseveringly for weeks after my departure, succeeded at length in extracting from the same deposit, at the depth of 2 feet below the crust of stalagmite, three fragments of a human skull, and two perfect lower jaws with teeth, all associated in such a manner with the bones of bears, large pachyderms, and ruminants, and so precisely resembling these in colour and state of preservation, as to leave no doubt in his mind that Man was contemporary with the extinct animals. Professor Malaise has given figures of the human remains in the *Bulletin* of the Royal Academy of Belgium for 1860.[1]

The rock in which the Liège caverns occur belongs generally to the Carboniferous or Mountain Limestone, in some few cases only to the older Devonian formation. Whenever the work of destruction has not gone too far, magnificent sections, sometimes 200 and 300 feet in height, are exposed to view. They confirm Schmerling's doctrine, that most of the materials, organic and inorganic, now filling the caverns, have been washed into them through narrow vertical or oblique fissures, the upper extremities of which are choked up with soil and gravel, and would scarcely ever be discoverable at the surface, especially in so wooded a country. Among the sections obtained by quarrying, one of the finest which I saw was in the beautiful valley of

---

[1] Vol. x., p. 546.

Fond du Forêt, above Chaudefontaine, not far from the village of Magnée, where one of the rents communicating with the surface has been filled up to the brim with rounded and half-rounded stones, angular pieces of limestone and shale, besides sand and mud, together with bones, chiefly of the cave-bear. Connected with this main duct, which is from 1 to 2 feet in width, are several minor ones, each from 1 to 3 inches wide, also extending to the upper country or table-land, and choked up with similar materials. They are inclined at angles of 30° and 40°, their walls being generally coated with stalactite, pieces of which have here and there been broken off and mingled with the contents of the rents, thus helping to explain why we so often meet with detached pieces of that substance in the mud and breccia of the Belgian caves. It is not easy to conceive that a solid horizontal floor of hard stalagmite should, after its formation, be broken up by running water; but when the walls of steep and tortuous rents, serving as feeders to the principal fissures and to inferior vaults and galleries are encrusted with stalagmite, some of the incrustation may readily be torn up when heavy fragments of rock are hurried by a flood through passages inclined at angles of 30° or 40°.

The decay and decomposition of the fossil bones seem to have been arrested in most of the caves by a constant supply of water charged with carbonate of lime, which dripped from the roofs while the caves were becoming gradually filled up. By similar agency the mud, sand, and pebbles were usually consolidated.

The following explanation of this phenomenon has been suggested by the eminent chemist Liebig. On the surface of Franconia, where the limestone abounds in caverns, is a fertile soil in which vegetable matter is continually decaying. This mould or humus, being acted on by moisture and air, evolves carbonic acid, which is dissolved by rain. The rain water, thus impregnated, permeates the porous limestone, dissolves a portion of it, and afterwards, when the excess of carbonic acid evaporates in the caverns, parts with the calcareous matter and forms stalactite. So long as water flows, even occasionally, through a suite of caverns, no layer of pure stalagmite can be produced; hence the formation of such a layer is generally an event posterior in date to the cessation of the old system of drainage, an event which might be brought about by an earthquake causing new fissures, or by the river wearing its way down to a lower level, and thenceforth running in a new channel.

In all the subterranean cavities, more than forty in number, explored by Schmerling, he only observed one cave, namely that of Chokier, where there were two regular layers of stalagmite, divided by fossiliferous cave-mud. In this instance, we may suppose that the stream, after flowing for a long period at one level, cut its way down to an inferior suite of caverns, and, flowing through them for centuries, choked them up with debris; after which it rose once more to its original higher level: just as in the Mountain Limestone district of Yorkshire some rivers, habitually absorbed by a " swallow hole," are occasionally unable to discharge all their water through it; in which case they rise and rush through a higher subterranean passage, which was at some former period in the regular line of drainage, as is often attested by the fluviatile gravel still contained in it.

There are now in the basin of the Meuse, not far from Liège, several examples of engulfed brooks and rivers: some of them, like that of St. Hadelin, east of Chaudefontaine, which reappears after an underground course of a mile or two; others, like the Vesdre, which is lost near Goffontaine, and after a time re-emerges; some, again, like the torrent near Magnée, which, after entering a cave, never again comes to the day. In the season of floods such streams are turbid at their entrance, but clear as a mountain-spring where they issue again; so that they must be slowly filling up cavities in the interior with mud, sand, pebbles, snail-shells, and the bones of animals which may be carried away during floods.

The manner in which some of the large thigh and shank bones of the rhinoceros and other pachyderms are rounded, while some of the smaller bones of the same creatures, and of the hyæna, bear, and horse, are reduced to pebbles, shows that they were often transported for some distance in the channels of torrents, before they found a resting-place.

When we desire to reason or speculate on the probable antiquity of human bones found fossil in such situations as the caverns near Liège, there are two classes of evidence to which we may appeal for our guidance. First, considerations of the time required to allow of many species of carnivorous and herbivorous animals, which flourished in the cave period, becoming first scarce, and then so entirely extinct as we have seen that they had become before the era of the Danish peat and Swiss lake dwellings; secondly, the great number of centuries necessary for the conversion of the physical geography of the Liège district from its ancient to its present configuration; so many old underground channels, through which brooks and rivers flowed in the cave period, being now laid dry and choked up.

The great alterations which have taken place in the shape of the valley of the Meuse and some of its tributaries are often demonstrated by the abrupt manner in which the mouths of fossiliferous caverns open in the face of perpendicular precipices 200 feet or more in height above the present streams. There appears also, in many cases, to be such a correspondence in the openings of caverns on opposite sides of some of the valleys, both large and small, as to incline one to suspect that they originally belonged to a series of tunnels and galleries which were continuous before the present system of drainage came into play, or before the existing valleys were scooped out. Other signs of subsequent fluctuations are afforded by gravel containing elephant's bones at slight elevations above the Meuse and several of its tributaries. It may be objected that, according to the present rate of change, no lapse of ages would suffice to bring about such revolutions in physical geography as we are here contemplating. This may be true. It is more than probable that the rate of change was once far more active than it is now in the basin of the Meuse. Some of the nearest volcanoes, namely, those of the Lower Eifel about 60 miles to the eastward, seem to have been in eruption in Pleistocene times, and may perhaps have been connected and coeval with repeated risings or sinkings of the land in the Liège district. It might be said, with equal truth, that according to the present course of events, no series of ages would suffice to reproduce such an assemblage of cones and craters as those of the Eifel (near Andernach, for example); and yet some of them may be of sufficiently modern date to belong to the era when Man was contemporary with the mammoth and rhinoceros in the basin of the Meuse.

But, although we may be unable to estimate the minimum of time required for the changes in physical geography above alluded to, we cannot fail to perceive that the duration of the period must have been very protracted, and that other ages of comparative inaction may have followed, separating the Pleistocene from the historical periods, and constituting an interval no less indefinite in its duration.

CHAPTER V

PLEISTOCENE PERIOD—FOSSIL HUMAN SKULLS OF THE
NEANDERTHAL AND ENGIS CAVES

Human Skeleton found in Cave near Düsseldorf—Its Geological Position
and probable Age—Its abnormal and ape-like Characters—Fossil
Human Skull of the Engis Cave near Liège—Professor Huxley's Description of these Skulls—Comparison of each, with extreme Varieties
of the native Australian Race—Range of Capacity in the Human and
Simian Brains—Skull from Borreby in Denmark—Conclusions of
Professor Huxley—Bearing of the peculiar Characters of the Neanderthal Skull on the Hypothesis of Transmutation.

Fossil human Skeleton of the Neanderthal Cave near Düsseldorf

BEFORE I speak more particularly of the opinions which anatomists have expressed respecting the osteological characters
of the human skull from Engis, near Liège, mentioned in the last
chapter and described by Dr. Schmerling, it will be desirable to
say something of the geological position of another skull, or
rather skeleton, which, on account of its peculiar conformation,
has excited no small sensation in the last few years. I allude to
the skull found in 1857 in a cave situated in that part of the
valley of the Düssel, near Düsseldorf, which is called the Neanderthal. The spot is a deep and narrow ravine about 70 English
miles north-east of the region of the Liège caverns treated of in
the last chapter, and close to the village and railway station of
Hochdal between Düsseldorf and Elberfeld. The cave occurs
in the precipitous southern or left side of the winding ravine,
about sixty feet above the stream, and a hundred feet below the
top of the cliff. The accompanying section will give the reader
an idea of its position.

When Dr. Fuhlrott of Elberfeld first examined the cave, he
found it to be high enough to allow a man to enter. The width
was 7 or 8 feet, and the length or depth 15. I visited the spot
in 1860, in company with Dr. Fuhlrott, who had the kindness to
come expressly from Elberfeld to be my guide, and who brought
with him the original fossil skull, and a cast of the same, which he
presented to me. In the interval of three years, between 1857
and 1860, the ledge of rock, *f*, on which the cave opened, and
which was originally 20 feet wide, had been almost entirely

quarried away, and, at the rate at which the work of dilapidation was proceeding, its complete destruction seemed near at hand.

In the limestone are many fissures, one of which, still partially filled with mud and stones, is represented in the section at *a c* as continuous from the cave to the upper surface of the country.

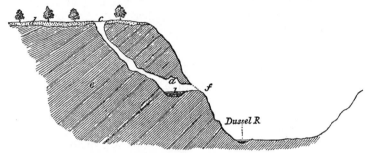

Fig. 1.

Section of the Neanderthal Cave near Düsseldorf.

a Cavern 60 feet above the Düssel, and 100 feet below the surface of the country at *c.*
b Loam covering the floor of the cave near the bottom of which the human skeleton was found.
b, c Rent connecting the cave with the upper surface of the country.
d Superficial sandy loam.
e Devonian limestone.
f Terrace, or ledge of rock.

Through this passage the loam, and possibly the human body to which the bones belonged, may have been washed into the cave below. The loam, which covered the uneven bottom of the cave, was sparingly mixed with rounded fragments of chert, and was very similar in composition to that covering the general surface of that region.

There was no crust of stalagmite overlying the mud in which the human skeleton was found, and no bones of other animals in the mud with the skeleton; but just before our visit in 1860 the tusk of a bear had been met with in some mud in a lateral embranchment of the cave, in a situation precisely similar to *b*, Fig. 1, and on a level corresponding with that of the human skeleton. This tusk, shown us by the proprietor of the cave, was $2\frac{1}{2}$ inches long and quite perfect; but whether it was referable to a recent or extinct species of bear, I could not determine.

From a printed letter of Dr. Fuhlrott we learn that on removing the loam, which was five feet thick, from the cave, the human skull was first noticed near the entrance, and, further in, the other

bones lying in the same horizontal plane. It is supposed that
the skeleton was complete, but the workmen, ignorant of its
value, scattered and lost most of the bones, preserving only the
larger ones.[1]

The cranium, which Dr. Fuhlrott showed me, was covered
both on its outer and inner surface, and especially on the latter,
with a profusion of dendritical crystallisations, and some other
bones of the skeleton were ornamented in the same way. These
markings, as Dr. Hermann von Meyer observes, afford no sure
criterion of antiquity, for they have been observed on Roman
bones. Nevertheless, they are more common in bones that have
been long embedded in the earth. The skull and bones, moreover, of the Neanderthal skeleton had lost so much of their
animal matter as to adhere strongly to the tongue, agreeing in
this respect with the ordinary condition of fossil remains of the
Pleistocene period. On the whole, I think it probable that this
fossil may be of about the same age as those found by Schmerling in the Liège caverns; but, as no other animal remains were
found with it, there is no proof that it may not be newer. Its
position lends no countenance whatever to the supposition of its
being more ancient.

When the skull and other parts of the skeleton were first
exhibited at a German scientific meeting at Bonn, in 1857, some
doubts were expressed by several naturalists, whether it was
truly human. Professor Schaaffhausen, who, with the other
experienced zoologists, did not share these doubts, observed
that the cranium, which included the frontal bone, both parietals,
part of the squamous, and the upper third of the occipital, was
of unusual size and thickness, the forehead narrow and very low,
and the projection of the supra-orbital ridges enormously great.
He also stated that the absolute and relative length of the thigh
bone, humerus, radius, and ulna, agreed well with the dimensions
of a European individual of like stature at the present day; but
that the thickness of the bones was very extraordinary, and the
elevations and depressions for the attachment of muscles were
developed in an unusual degree. Some of the ribs, also, were of
a singularly rounded shape and abrupt curvature, which was
supposed to indicate great power in the thoracic muscles.[2]

In the same memoir, the Prussian anatomist remarks that the

---

[1] Fuhlrott, Letter to Professor Schaaffhausen, cited *Natural History
Review*, No. 2, p. 156. See also *Naturhistorischer Verein*, Bonn, 1859.
[2] Professor Schaaffhausen's *Memoir*, translated, *Natural History
Review*, April 1861.

depression of the forehead (*See* Fig. 3, p. 64), is not due to any
artificial flattening, such as is practised in various modes by
barbarous nations in the Old and New World, the skull being
quite symmetrical, and showing no indication of counter-
pressure at the occiput; whereas, according to Morton, in the
Flat-heads of the Columbia, the frontal and parietal bones are
always unsymmetrical.[1] On the whole, Professor Schaaffhausen
concluded that the individual to whom the Neanderthal skull
belonged must have been distinguished by small cerebral development, and uncommon strength of corporeal frame.

When on my return to England I showed the cast of the
cranium to Professor Huxley, he remarked at once that it was
the most ape-like skull he had ever beheld. Mr. Busk, after
giving a translation of Professor Schaaffhausen's memoir in the
Natural History Review, added some valuable comments of his
own on the characters in which this skull approached that of the
gorilla and chimpanzee.

Professor Huxley afterwards studied the cast with the object
of assisting me to give illustrations of it in this work, and in
doing so discovered what had not previously been observed,
that it was quite as abnormal in the shape of its occipital as in
that of its frontal or superciliary region. Before citing his words
on the subject, I will offer a few remarks on the Engis skull
which the same anatomist has compared with that of the
Neanderthal. [Note 10.]

Fossil Skull of the Engis Cave near Liège

Among six or seven human skeletons, portions of which were
collected by Dr. Schmerling from three or four caverns near
Liège, embedded in the same matrix with the remains of the
elephant, rhinoceros, bear, hyæna, and other extinct quadrupeds,
the most perfect skull, as I have before stated, was that of an
adult individual found in the cavern of Engis. This skull, Dr.
Schmerling figured in his work, observing that it was too imperfect to enable the anatomist to determine the facial angle, but
that one might infer, from the narrowness of the frontal portion,
that it belonged to an individual of small intellectual development. He speculated on its Ethiopian affinities, but not confidently, observing truly that it would require many more
specimens to enable an anatomist to arrive at sound conclusions

[1] *Natural History Review*, No. 2, p. 160.

on such a point. M. Geoffroy St. Hilaire and other osteologists, who examined the specimen, denied that it resembled a negro's skull. When I saw the original in the museum at Liège, I invited Dr. Spring, one of the professors of the university, to whom we are indebted for a valuable memoir on the human bones found in the cavern of Chauvaux, near Namur, to have a cast made of this Engis skull. He not only had the kindness to comply with my request, but rendered a service to the scientific world by adding to the original cranium several detached fragments which Dr. Schmerling had obtained from Engis, and which were found to fit in exactly, so that the cast represented at Fig. 2 is more complete than that given in the first plate of Schmerling's work. It exhibits on the right side the position of the auditory foramen (see Fig. 6), which was not included in Schmerling's figure. Mr. Busk, when he saw this cast, remarked to me that, although the forehead was, as Schmerling had truly stated, somewhat narrow, it might nevertheless be matched by the skulls of individuals of European race, an observation since fully borne out by measurements, as will be seen in the sequel.

OBSERVATIONS BY PROFESSOR HUXLEY ON THE HUMAN SKULLS OF
ENGIS AND THE NEANDERTHAL

" The Engis skull, as originally figured by Professor Schmerling, was in a very imperfect state; but other fragments have since been added to it by the care of Dr. Spring, and the cast upon which my observations are based (Fig. 2) exhibits the frontal, parietal, and occipital regions, as far as the middle of the occipital foramen, with the squamous and mastoid portions of the right temporal bone entire, or nearly so, while the left temporal bone is wanting. From the middle of the occipital foramen to the middle of the roof of each orbit, the base of the skull is destroyed, and the facial bones are entirely absent.

" The extreme length of the skull is 7.7 inches, and as its extreme breadth is not more than 5.25, its form is decidedly dolichocephalic. At the same time its height (4¾ inches from the plane of the glabello-occipital line (a d) to the vertex) is good, and the forehead is well arched; so that while the horizontal circumference of the skull is about 20½ inches, the longitudinal arc from the nasal spine of the frontal bone to the occipital protuberance (d) measures about 13¾ inches. The transverse arc from one auditory foramen to the other across the middle of the sagittal suture measures about 13 inches. The sagittal suture (b c) is 5½ inches in length. The superciliary prominences are well, but not excessively, developed, and are separated by a median depression in the region of the glabella. They indicate large frontal sinuses. If a line joining the glabella and the occipital protuberance (a d) be made horizontal, no part of the occiput projects more than ₁₀th of an inch behind the posterior extremity of that line; and the upper edge of the auditory foramen is almost in contact with the same line, or rather with one drawn parallel to it on the outer surface of the skull.

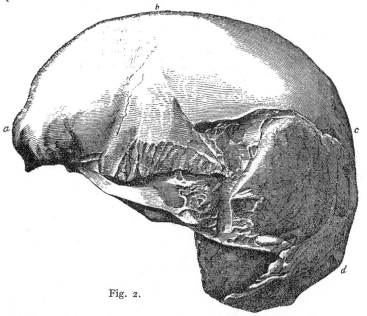

Fig. 2.

Side view of the cast of part of a human skull found by Dr. Schmerling embedded amongst the remains of extinct mammalia in the cave of Engis, near Liège.

a Superciliary ridge and glabella.   c The apex of the lambdoidal suture.
b Coronal suture.   d The occipital protuberance.

" The Neanderthal skull, with which also I am acquainted only by means of Professor Schaaffhausen's drawings of an excellent cast and of photographs, is so extremely different in appearance from the Engis cranium, that it might well be supposed to belong to a distinct race of mankind. It is 8 inches in extreme length and 5.75 inches in extreme breadth, but only measures 3.4 inches from the glabello-occipital line to the vertex. The longitudinal arc, measured as above, is 12 inches; the transverse arc cannot be exactly ascertained, in consequence of the absence of the temporal bones, but was probably about the same, and certainly exceeded $10\frac{1}{4}$ inches. The horizontal circumference is 23 inches. This great circumference arises largely from the vast development of the superciliary ridges, which are occupied by great frontal sinuses whose inferior apertures are displayed exceedingly well in one of Dr. Fuhlrott's photographs, and form a continuous transverse prominence, somewhat excavated in the middle line, across the lower part of the brows. In consequence of this structure, the forehead appears still lower and more retreating than it really is. To an anatomical eye the posterior part of the skull is even more striking than the anterior. The occipital protuberance occupies the extreme posterior end of the skull when the glabello-occipital line is made horizontal, and so far from any part of the occipital region extending beyond it, this region of the skull slopes obliquely upward and forward, so that the lamb-doidal suture is situated well upon the upper surface of the cranium. At the same time, notwithstanding the great length of the skull, the sagittal suture is remarkably short (4½ inches), and the squamosal suture is very straight.

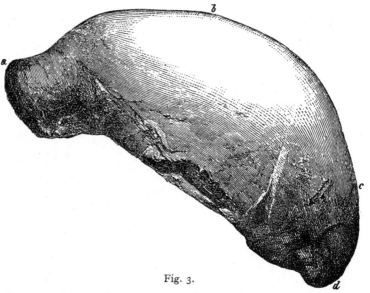

Fig. 3.

Side view of the cast of a part of a human skull from a cave in the Neanderthal, near Düsseldorf.

a The superciliary ridge and glabella.   c The apex of the lambdoidal suture.
b The coronal suture.   d The occipital protuberance.

" In human skulls, the superior curved ridge of the occipital bone and the occipital protuberance correspond, approximatively, with the level of the tentorium and with the lateral sinuses, and conse-quently with the inferior limit of the posterior lobes of the brain. At first, I found some difficulty in believing that a human brain could have its posterior lobes so flattened and diminished as must have been the case in the Neanderthal man, supposing the ordinary relation to obtain between the superior occipital ridges and the tentorium; but on my application, through Sir Charles Lyell, Dr. Fuhlrott, the possessor of the skull, was good enough not only to ascertain the existence of the lateral sinuses in their ordinary posi-tion, but to send convincing proofs of the fact, in excellent photographic views of the interior of the skull, exhibiting clear indications of these sinuses.

" There can be no doubt that, as Professor Schaaffhausen and Mr. Busk have stated, this skull is the most brutal of all known human skulls, resembling those of the apes not only in the prodigious development of the superciliary prominences and the forward extension of the orbits, but still more in the depressed form of the brain-case, in the straightness of the squamosal suture, and in the complete retreat of the occiput forwards and upward, from the superior occipital ridges.

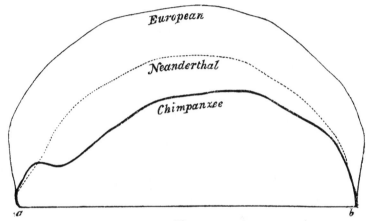

Fig. 4.

Outline of the skull of an adult Chimpanzee, of that from the Neanderthal, and of that of a European, drawn to the same absolute size, in order better to exhibit their relative differences. The superciliary region of the Neanderthal skull appears less prominent than in Fig. 3, as the contours are all taken along the middle line where the superciliary projection of the Neanderthal skull is least marked.

a The glabella.
b The occipital protuberance, or the point on the exterior of each skull which corresponds roughly with the attachment of the tentorium, or with the inferior boundary of the posterior cerebral lobes.

" But the cranium, in its present condition, is stated by Professor Schaaffhausen to contain 1033.24 cubic centimetres of water, or, in other words, about 63 English cubic inches. As the entire skull could hardly have held less than 12 cubic inches more, its minimum capacity may be estimated at 75 cubic inches. The most capacious healthy European skull yet measured had a capacity of 114 cubic inches, the smallest (as estimated by weight of brain) about 55 cubic inches, while, according to Professor Schaaffhausen, some Hindoo skulls have as small a capacity as about 46 cubic inches (27 oz. of water). The largest cranium of any Gorilla yet measured contained 34.5 cubic inches. The Neanderthal cranium stands, therefore, in capacity, very nearly on a level with the mean of the two human extremes, and very far above the pithecoid maximum.

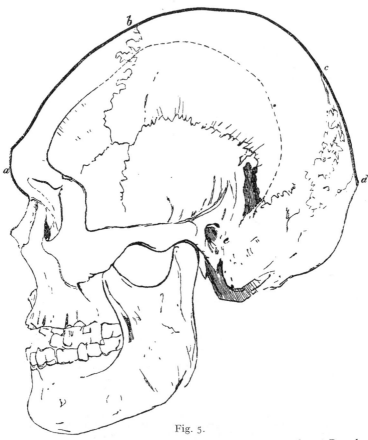

Fig. 5.

Skull associated with ground flint implements, from a tumulus at Borreby in Denmark, after a camera lucida drawing by Mr. G. Busk, F.R.S. The thick dark line indicates so much of the skull as corresponds with the fragment from the Neanderthal.

a Superciliary ridge.	c The apex of the lambdoidal suture.
b Coronal suture.	d The occipital protuberance.

e The auditory foramen.

" Hence, even in the absence of the bones of the arm and thigh, which, according to Professor Schaaffhausen, had the precise proportions found in Man, although they were stouter than ordinary human bones, there could be no reason for ascribing this cranium to anything but a man; while the strength and development of the muscular ridges of the limb-bones are characters in perfect accordance with those exhibited, in a minor degree, by the bones of such hardy savages, exposed to a rigorous climate, as the Patagonians.

" The Neanderthal cranium has certainly not undergone compression, and, in reply to the suggestion that the skull is that of an idiot, it may be urged that the *onus probandi* lies with those who adopt the hypothesis. Idiotcy is compatible with very various forms and capacities of the cranium, but I know of none which present the least resemblance to the Neanderthal skull; and, furthermore, I shall proceed to show that the latter manifests but an extreme degree of a stage of degradation exhibited, as a natural condition, by the crania of certain races of mankind.

" Mr. Busk drew my attention, some time ago, to the resemblance between some of the skulls taken from tumuli of the stone period at Borreby in Denmark, of which Mr. Busk possesses numerous accurate figures, and the Neanderthal cranium. One of the Borreby skulls in particular (Fig. 5) has remarkably projecting superciliary ridges, a retreating forehead, a low flattened vertex, and an occiput which shelves upward and forward. But the skull is relatively higher and broader, or more brachycephalic, the sagittal suture longer, and the superciliary ridges less projecting, than in the Neanderthal skull. Nevertheless, there is, without doubt, much resemblance in character between the two skulls—a circumstance which is the more interesting, since the other Borreby skulls have better foreheads and less prominent superciliary ridges, and exhibit altogether a higher conformation.

" The Borreby skulls belong to the stone period of Denmark, and the people to whom they appertained were probably either contemporaneous with, or later than, the makers of the ' refuse-heaps ' of that country. In other words, they were subsequent to the last great physical changes of Europe, and were contemporaries of the urus and bison, not of the *Elephas primigenius, Rhinoceros tichorhinus,* and *Hyæna spelæa.*

" Supposing for a moment, what is not proven, that the Neanderthal skull belonged to a race allied to the Borreby people and was as modern as they, it would be separated by as great a distance of time as of anatomical character from the Engis skull, and the possibility of its belonging to a distinct race from the latter might reasonably appear to be greatly heightened.

" To prevent the possibility of reasoning in a vicious circle, however, I thought it would be well to endeavour to ascertain what amount of cranial variation is to be found in a pure race at the present day; and as the natives of Southern and Western Australia are probably as pure and homogeneous in blood, customs, and language, as any race of savages in existence, I turned to them, the more readily as the Hunterian museum contains a very fine collection of such skulls.

" I soon found it possible to select from among these crania two (connected by all sorts of intermediate gradations), the one of which should very nearly resemble the Engis skull, while the other should somewhat less closely approximate the Neanderthal cranium in form, size, and proportions. And at the same time others of these skulls presented no less remarkable affinities with the low type of Borreby skull.

" That the resemblances to which I allude are by no means of a merely superficial character, is shown by the accompanying diagram (Fig. 6), which gives the contours of the two ancient and of one of the Australian skulls, and by the following table of measurements.

	A	B	C	D	E	F
Engis . . .	$20\frac{1}{2}$	$13\frac{3}{4}$	$12\frac{1}{2}$	$4\frac{3}{4}$	$7\frac{3}{4}$	$5\frac{1}{4}$
Australian, No. 1 .	$20\frac{1}{2}$	13	12	$4\frac{3}{4}$	$7\frac{1}{2}$	$5\frac{4}{10}$
Australian, No. 2 .	22	$12\frac{1}{2}$	$10\frac{3}{4}$	$3\frac{8}{10}$	7·9	$5\frac{3}{4}$
Neanderthal . .	23	12	10	$3\frac{3}{4}$	8	$5\frac{3}{4}$

A The horizontal circumference in the plane of a line joining the glabella with the occipital protuberance.

B The longitudinal arc from the nasal depression along the middle line of the skull to the occipital tuberosity.

C From the level of the glabello-occipital line on each side, across the middle of the sagittal suture to the same point on the opposite side.

D The vertical height from the glabello-occipital line.

E The extreme longitudinal measurement.

F The extreme transverse measurement.[1]

" The question whether the Engis skull has rather the character of one of the high races or of one of the lower has been much disputed, but the following measurements of an English skull, noted in the catalogue of the Hunterian museum as typically Caucasian (*see* Fig. 4) will serve to show that both sides may be right, and that cranial measurements alone afford no safe indication of race.

	A	B	C	D	E	F
English . . .	21	$13\frac{3}{4}$	$12\frac{1}{2}$	$4\frac{4}{10}$	$7\frac{7}{8}$	$5\frac{1}{3}$

" In making the preceding statement, it must be clearly understood that I neither desire to affirm that the Engis and Neanderthal skulls belong to the Australian race, nor to assert even that the ancient skulls belong to one and the same race, so far as race is measured by language, colour of skin, or character of hair. Against the conclusion that they are of the same race as the Australians various minor anatomical differences of the ancient skulls, such as the geat development of the frontal sinuses, might be urged; while against the supposition of either the identity, or the diversity, of race of the two arises the known independence of the variation of cranium on the one hand, and of hair, colour, and language on the other.

" But the amount of variation of the Borreby skulls, and the fact that the skulls of one of the purest and most homogeneous of existing races of men can be proved to differ from one another in the same

1 I have taken the glabello-occipital line as a base in these measurements, simply because it enables me to compare all the skulls, whether fragments or entire, together. The greatest circumference of the English skull lies in a plane considerably above that of the glabello-occipital line, and amounts to 22 inches.

characters, though perhaps not quite to the same extent, as the Engis and Neanderthal skulls, seem to me to prohibit any cautious reasoner from affirming the latter to have been necessarily of distinct races.

"The marked resemblances between the ancient skulls and their modern Australian analogues, however, have a profound interest, when it is recollected that the stone axe is as much the weapon and the implement of the modern as of the ancient savage; that the former turns the bones of the kangaroo and of the emu to the same account as the latter did the bones of the deer and the urus; that

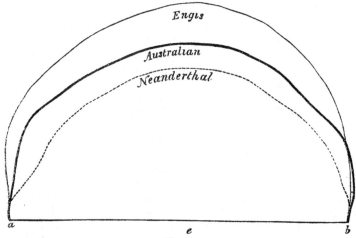

Fig. 6.

Outlines of the skull from the Neanderthal, of an Australian skull from Port Adelaide, and of the skull from the Cave of Engis, drawn to the same absolute length, in order the better to contrast their proportions.

a b As in Fig. 4, p. 65.
e The position of the auditory foramen of the Engis skull.

the Australian heaps up the shells of devoured shellfish in mounds which represent the "refuse-heaps" or "Kjökkenmödding," of Denmark; and, finally, that, on the other side of Torres Straits, a race akin to the Australians are among the few people who now build their houses on pile-works, like those of the ancient Swiss lakes.

"That this amount of resemblance in habit and in the conditions of existence is accompanied by as close a resemblance in cranial configuration, illustrates on a great scale that what Cuvier demonstrated of the animals of the Nile valley is no less true of men; circumstances remaining similar, the savage varies little more, it would seem, than the ibis or the crocodile, especially if we take into account the enormous extent of the time over which our knowledge of man now extends, as compared with that measured by the duration of the sepulchres of Egypt.

" Finally, the comparatively large cranial capacity of the Neanderthal skull, overlaid though it may be by pithecoid bony walls, and
the completely human proportions of the accompanying limb-bones,
together with the very fair development of the Engis skull, clearly
indicate that the first traces of the primordial stock whence Man has
proceeded need no longer be sought, by those who entertain any
form of the doctrine of progressive development, in the newest
Tertiaries; but that they may be looked for in an epoch more distant
from the age of the *Elephas primigenius* than that is from us."

The two skulls which form the subject of the preceding comments and illustrations have given rise to nearly an equal amount
of surprise for opposite reasons; that of Engis because being so
unequivocally ancient, it approached so near to the highest or
Caucasian type; that of the Neanderthal, because, having no
such decided claims to antiquity, it departs so widely from the
normal standard of humanity.   Professor Huxley's observation
regarding the wide range of variation, both as to shape and
capacity, in the skulls of so pure a race as the native Australian,
removes to no small extent this supposed anomaly, assuming
what though not proved is very probable, that both varieties
co-existed in the Pleistocene period in Western Europe.

As to the Engis skull, we must remember that although
associated with the elephant, rhinoceros, bear, tiger, and hyæna,
all of extinct species, it nevertheless is also accompanied by a
bear, stag, wolf, fox, beaver, and many other quadrupeds of
species still living.   Indeed many eminent palæontologists, and
among them Professor Pictet, think that, numerically considered,
the larger portion of the mammalian fauna agrees specifically
with that of our own period, so that we are scarcely entitled to
feel surprised if we find human races of the Pleistocene epoch undistinguishable from some living ones.   It would merely tend to
show that Man has been as constant in his osteological characters
as many other mammalia now his contemporaries.   The expectation of always meeting with a lower type of human skull,
the older the formation in which it occurs, is based on the theory
of progressive development, and it may prove to be sound; nevertheless we must remember that as yet we have no distinct
geological evidence that the appearance of what are called the
inferior races of mankind has always preceded in chronological
order that of the higher races.

It is now admitted that the differences between the brain of
the highest races of Man and that of the lowest,[1] though less in
degree, are of the same order as those which separate the human

[1] *Natural History Review*, 1861, p. 8.

from the simian brain; and the same rule holds good in regard to the shape of the skull. The average Negro skull differs from that of the European in having a more receding forehead, more prominent superciliary ridges, and more largely developed prominences and furrows for the attachment of muscles; the face also, and its lines, are larger proportionally. The brain is somewhat less voluminous on the average in the lower races of mankind, its convolutions rather less complicated, and those of the two hemispheres more symmetrical, in all which points an approach is made to the simian type. It will also be seen, by reference to the late Dr. Morton's works, and by the foregoing statements of Professor Huxley, that the range of size or capacity between the highest and lowest human brain is greater than that between the highest simian and lowest human brain; but the Neanderthal skull, although in several respects it is more ape-like than any human skull previously discovered, is, in regard to volume, by no means contemptible.

Eminent anatomists have shown that in the average proportions of some of the bones the Negro differs from the European, and that in most of these characters, he makes a slightly nearer approach to the anthropoid quadrumana; [1] but Professor Schaaffhausen has pointed out that in these proportions the Neanderthal skeleton does not differ from the ordinary standard, so that the skeleton by no means indicates a transition between *Homo* and *Pithecus*.

There is doubtless, as shown in the diagram Fig. 4, a nearer resemblance in the outline of the Neanderthal skull to that of a chimpanzee than had ever been observed before in any human cranium; and Professor Huxley's description of the occipital

[1] " The inferior races of mankind exhibit proportions which are in many respects intermediate between the higher, or European, orders, and the monkeys. In the Negro, for instance, the stature is less than in the European. The cranium, as is well known, bears a small proportion to the face. Of the extremities the upper are proportionately longer, and there is, in both upper and lower, a less marked preponderance of the proximal over the distal segments. For instance, in the Negro, the thigh and arm are rather shorter than in the European; the leg is actually of equal length in both races, and is therefore, relatively, a little longer in the Negro; the fore-arm in the latter is actually, as well as relatively, a little longer; the foot is an eighth, and the hand a twelfth longer than in the European. It is well known that the foot is less well formed in the Negro than in the European. The arch of the instep, the perfect conformation of which is essential to steadiness and ease of gait, is less elevated in the former than in the latter. The foot is thereby rendered flatter as well as longer, more nearly resembling the monkey's, between which and the European there is a marked difference in this particular."—From *A Treatise on the Human Skeleton*, by Dr. Humphry, Lecturer on Surgery and Anatomy in the Cambridge University Medical School, p. 91.

region shows that the resemblance is not confined to the mere excessive prominence of the superciliary ridges.

The direct bearing of the ape-like character of the Neanderthal skull on Lamarck's doctrine of progressive development and transmutation, or on that modification of it which has of late been so ably advocated by Mr. Darwin, consists in this, that the newly observed deviation from a normal standard of human structure is not in a casual or random direction, but just what might have been anticipated if the laws of variation were such as the transmutationists require. For if we conceive the cranium to be very ancient, it exemplifies a less advanced stage of progressive development and improvement. If it be a comparatively modern race, owing its peculiarities of conformation to degeneracy, it is an illustration of what botanists call " atavism," or the tendency of varieties to revert to an ancestral type, which type, in proportion to its antiquity, would be of lower grade. To this hypothesis, of a genealogical connection between Man and the lower animals, I shall again allude in the concluding chapters. [Note 11.]

CHAPTER VI

PLEISTOCENE ALLUVIUM AND CAVE DEPOSITS WITH FLINT IMPLEMENTS

General Position of Drift with extinct Mammalia in Valleys—Discoveries of M. Boucher de Perthes at Abbeville—Flint Implements found also at St. Acheul, near Amiens—Curiosity awakened by the systematic Exploration of the Brixham Cave—Flint Knives in same, with Bones of extinct Mammalia—Superposition of Deposits in the Cave—Visits of English and French Geologists to Abbeville and Amiens.

Pleistocene Alluvium containing Flint Implements in the Valley of the Somme

THROUGHOUT a large part of Europe we find at moderate elevations above the present river-channels, usually at a height of less than 40 feet, but sometimes much higher, beds of gravel, sand, and loam containing bones of the elephant, rhinoceros, horse, ox, and other quadrupeds, some of extinct, others of living, species, belonging for the most part to the fauna already alluded to in the fourth chapter as characteristic of the interior of caverns. The greater part of these deposits contain fluviatile shells, and have undoubtedly been accumulated in ancient river-beds. These old channels have long since been dry, the streams which once flowed in them having shifted their position, deepening the valleys, and often widening them on one side.

It has naturally been asked, if Man co-existed with the extinct species of the caves, why were his remains and the works of his hands never embedded outside the caves in ancient river-gravel containing the same fossil fauna? Why should it be necessary for the geologist to resort for evidence of the antiquity of our race to the dark recesses of underground vaults and tunnels which may have served as places of refuge or sepulture to a succession of human beings and wild animals, and where floods may have confounded together in one breccia the memorials of the fauna of more than one epoch? Why do we not meet with a similar assemblage of the relics of Man, and of living and extinct quadrupeds, in places where the strata can be thoroughly scrutinised in the light of day?

Recent researches have at length demonstrated that such memorials, so long sought for in vain, do in fact exist, and their recognition is the chief cause of the more favourable reception now given to the conclusions which MM. Tournal, Christol, Schmerling, and others, arrived at thirty years ago respecting the fossil contents of caverns. [Note 12.]

A very important step in this new direction was made thirteen years after the publication of Schmerling's researches, by M. Boucher de Perthes, who found in ancient alluvium at Abbeville, in Picardy, some flint implements, the relative antiquity of which was attested by their geological position. The antiquarian knowledge of their discoverer enabled him to recognise in their rude and peculiar type a character distinct from that of the polished stone weapons of a later period, usually called " celts." In the first volume of his *Antiquités Celtiques,* published in 1847, M. Boucher de Perthes styled these older tools " antediluvian," because they came from the lowest beds of a series of ancient alluvial strata bordering the valley of the Somme, which geologists had termed " diluvium." He had begun to collect these implements in 1841. From that time they had been annually dug out of the drift or deposits of gravel and sand, of which fine sections were laid open from 20 to 35 feet in depth, whenever excavations were made in repairing the fortifications of Abbeville; or as often as flints were wanted for the roads, or loam for making bricks. For years previously bones of quadrupeds of the genera elephant, rhinoceros, bear, hyæna, stag, ox, horse, and others, had been collected there, and sent from time to time to Paris to be examined and named by Cuvier, who had described them in his *Ossements Fossiles.* A correct account of the associated flint tools and of their position was given in 1847 by M. Boucher de Perthes in his work above cited, and they were stated to occur at various depths, often 20 or 30 feet from the surface, in sand and gravel, especially in those strata which were nearly in contact with the subjacent white Chalk. But the scientific world had no faith in the statement that works of art, however rude, had been met with in undisturbed beds of such antiquity. Few geologists visited Abbeville in winter, when the sand-pits were open, and when they might have opportunities of verifying the sections, and judging whether the instruments had really been embedded by natural causes in the same strata with the bones of the mammoth, rhinoceros, and other extinct mammalia. Some of the tools figured in the *Antiquités Celtiques* were so rudely shaped, that many imagined them to have owed their peculiar forms to accidental fracture in a river's bed; others suspected frauds on the part of the workmen, who might have fabricated them for sale, or that the gravel had been disturbed, and that the worked flints had got mingled with the bones of the mammoth long after that animal and its associates had disappeared from the earth.

No one was more sceptical than the late eminent physician of Amiens, Dr. Rigollot, who had long before (in the year 1819) written a memoir on the fossil mammalia of the valley of the Somme. He was at length induced to visit Abbeville, and, having inspected the collection of M. Boucher de Perthes, returned home resolved to look for himself for flint tools in the gravel-pits near Amiens. There, accordingly, at a distance of about 30 miles from Abbeville, he immediately found abundance of similar flint implements, precisely the same in the rudeness of their make, and the same in their geological position; some of them in gravel nearly on a level with the Somme, others in similar deposits resting on Chalk at a height of about 90 feet above the river.

Dr. Rigollot having in the course of four years obtained several hundred specimens of these tools, most of them from St. Acheul in the south-east suburbs of Amiens, lost no time in communicating an account of them to the scientific world, in a memoir illustrated by good figures of the worked flints and careful sections of the beds. These sections were executed by M. Buteux, an engineer well qualified for the task, who had written a good description of the geology of Picardy. Dr. Rigollot, in this memoir, pointed out most clearly that it was not in the vegetable soil, nor in the brick-earth with land and fresh-water shells next below, but in the lower beds of coarse flint-gravel, usually 12, 20, or 25 feet below the surface, that the implements were met with, just as they had been previously stated by M. Boucher de Perthes to occur at Abbeville. The conclusion, therefore, which was legitimately deduced from all the facts, was that the flint tools and their fabricators were coeval with the extinct mammalia embedded in the same strata.

Brixham Cave, near Torquay, Devonshire

Four years after the appearance of Dr. Rigollot's paper, a sudden change of opinion was brought about in England respecting the probable co-existence, at a former period, of Man and many extinct mammalia, in consequence of the results obtained from a careful exploration of a cave at Brixham, near Torquay, in Devonshire. As the new views very generally adopted by English geologists had no small influence on the subsequent progress of opinion in France, I shall interrupt my account of the researches made in the valley of the Somme, by a brief notice of those which were carried on in 1858 in Devonshire with more than usual care and scientific method. Dr. Buckland, in his celebrated work, entitled *Reliquiæ Diluvianæ*, published in 1823, in which he treated of the organic remains contained in caves, fissures, and " diluvial gravel " in England, had given a clear statement of the results of his own original observations, and had declared that none of the human bones or stone implements met with by him in any of the caverns could be considered to be as old as the mammoth and other extinct quadrupeds. Opinions in harmony with this conclusion continued until very lately to be generally in vogue in England; although about the time that Schmerling was exploring the Liège caves, the Rev. Mr. M'Enery, a Catholic priest, residing near Torquay, had found in a cave one mile east of that town, called " Kent's Hole," in red loam covered with stalagmite, not only bones of the mammoth, tichorhine rhinoceros, hippopotamus, cave-bear, and other mammalia, but several remarkable flint tools, some of which he supposed to be of great antiquity, while there were also remains of Man in the same cave of a later date.[1]

About ten years afterwards, in a *Memoir on the Geology of South Devon*, published in 1842 by the Geological Society of London,[2] an able geologist, Mr. Godwin-Austen, declared that he had obtained in the same cave (Kent's Hole) works of Man from undisturbed loam or clay, under stalagmite, mingled with the remains of extinct animals, and that all these must have been introduced " before the stalagmite flooring had been formed." He maintained that such facts could not be explained away by the hypothesis of sepulture, as in Dr. Buckland's well-known case of the human skeleton of Paviland, because in the Devon cave the flint implements were widely distributed through the loam, and lay beneath the stalagmite.

[1] The MS. and plates prepared for a joint memoir on Kent's Hole, by Mr. M'Enery and Dr. Buckland, have recently been published by Mr. Vivian of Torquay, from which, as well as from some of the unprinted MS., I infer that Mr. M'Enery only refrained out of deference to Dr. Buckland from declaring his belief in the contemporaneousness of certain flint implements of an antique type and the bones of extinct animals. Two of these implements from Kent's Hole, figured in Plate 12 of the posthumous work above alluded to, approach very closely in form and size to the common Abbeville implements.

[2] *Transactions of the Geological Society*, 2nd series, vol. vi., p. 444.

As the osseous and other contents of Kent's Hole had, by repeated diggings, been thrown into much confusion, it was thought desirable in 1858, when a new and intact bone-cave was discovered at Brixham, about four miles south of Torquay, to have a thorough and systematic examination made of it. The Royal Society, chiefly at the instance of Dr. Falconer, made two grants towards defraying the expenses, and Miss Burdett-Coutts contributed liberally towards the same object. A committee of geologists was charged with the investigations, among whom Dr. Falconer and Mr. Prestwich took a prominent part, visiting Torquay while the excavations were in progress. Mr. Pengelly, another member of the committee, well qualified for the task by nearly twenty years' previous experience in cave explorations, zealously directed and superintended the work. By him, in 1859, I was conducted through the subterranean galleries after they had been cleared out; and Dr. Falconer, who was also at Torquay, showed me the numerous fossils which had been discovered, and which he was then studying, all numbered and labelled, with reference to a journal in which the geological position of each specimen was recorded with scrupulous care.

The discovery of the existence of this suite of caverns near the sea at Brixham was made accidentally by the roof of one of them being broken through in quarrying. None of the four external openings now exposed to view in steep cliffs or in the sloping side of a valley were visible before the breccia and earthy matter which blocked them up were removed during the late exploration. According to a ground-plan drawn up by Professor Ramsay, it appears that some of the passages which run nearly north and south are fissures connected with the vertical dislocation of the rocks, while another set, running nearly east and west, are tunnels, which have the appearance of having been to a great extent hollowed out by the action of running water. The central or main entrance, leading to what is called the " reindeer gallery," because a perfect antler of that animal was found sticking in the stalagmitic floor, is 95 feet above the level of the sea, being also 78 above the bottom of the adjoining valley. The united length of the galleries which were cleared out amounted to several hundred feet. Their width never exceeded 8 feet. They were sometimes filled up to the roof with mud, but occasionally there was a considerable space between the roof and floor. The latter, in the case of the fissure-caves, was covered with stalagmite, but in the tunnels it was usually free from any such incrustation. The following was the general succession of the deposits forming the contents of the underground passages and channels:—

First. At the top, a layer of stalagmite varying in thickness from 1 to 15 inches, which sometimes contained bones, such as the reindeer's horn, already mentioned, and an entire humerus of the cave-bear.

Secondly. Next below, loam or bone-earth, of an ochreous red colour, with angular stones and some pebbles, from 2 to 13 feet in thickness.

Thirdly. At the bottom of all, gravel with many rounded pebbles in it. This was everywhere removed so long as the tunnels which narrowed downwards were wide enough to be worked. It proved to be almost entirely barren of fossils.

The mammalia obtained from the bone-earth consisted of *Elephas primigenius*, or mammoth; *Rhinoceros tichorhinus*; *Ursus spelæus*; *Hyæna spelæa*; *Felis spelæa*, or the cave-lion; *Cervus tarandus*, or the reindeer; a species of horse, ox, and several rodents, and others not yet determined.

No human bones were obtained anywhere during these excavations, but many flint knives, chiefly from the lowest part of the bone-earth; and one of the most perfect lay at the depth of 13 feet from the surface, and was covered with bone-earth of that thickness. Neglecting the less perfect specimens, some of which were met with even in the lowest gravel, about fifteen knives, recognised as artificially formed by the most experienced antiquaries, were taken from the bone-earth, and usually from near the bottom. Such knives, considered apart from the associated mammalia, afford in themselves no safe criterion of antiquity, as they might belong to any part of the age of stone, similar tools being sometimes met with in tumuli posterior in date to the era of the introduction of bronze. But the contemporaneity of those at Brixham with the extinct animals is demonstrated not only by the occurrence at one point in overlying stalagmite of the bone of a cave-bear, but also by the discovery at the same level in the bone-earth, and in close proximity to a very perfect flint tool, of the entire left hind-leg of a cave-bear. This specimen, which was shown me by Dr. Falconer and Mr. Pengelly, was exhumed from the earthy deposit in the reindeer gallery, near its junction with the flint-knife gallery, at the distance of about sixty-five feet from the main entrance. The mass of earth containing it was removed entire, and the matrix cleared away carefully by Dr. Falconer in the presence of Mr. Pengelly. Every bone was in its natural place, the femur, tibia, fibula, ankle-bone, or astragalus, all in juxtaposition. Even the patella or detached bone of the knee-pan was searched for, and not in vain. Here, therefore, we have evidence of an entire limb not having been washed in a fossil state out of an older alluvium, and then swept afterwards into a cave, so as to be mingled with flint implements, but having been introduced when clothed with its flesh, or at least when it had the separate bones bound together by their natural ligaments, and in that state buried in mud.

If they were not all of contemporary date, it is clear from this case, and from the humerus of the *Ursus spelæus*, before cited, as found in a floor of stalagmite, that the bear lived after the flint tools were manufactured, or in other words, that Man in this district preceded the cave-bear.

A glance at the position of Windmill Hill, in which the caverns are situated, and a brief survey of the valleys which bound it on three sides, are enough to satisfy a geologist that the drainage and geographical features of this region have undergone great changes since the gravel and bone-earth were carried by streams into the subterranean cavities above described. Some worn pebbles of hæmatite, in particular, can only have come from their nearest parent rock, at a period when the valleys immediately adjoining the caves were much shallower than they now are. The reddish loam in which the bones are embedded is such as may be seen on the surface of limestone in the neighbourhood, but the currents which were formerly charged with such mud must have run at a level 78 feet above that of the stream now flowing in the same valley. It was remarked by Mr. Pengelly that the stones and bones in the loam had their longest axes parallel to the direction of the tunnels and fissures, showing that they were deposited by the action of a stream.[1]

It appears that so long as the flowing water had force enough to propel stony fragments, no layer of fine mud could accumulate, and so long as there was a regular current capable of carrying in fine mud and bones, no superficial crust of stalagmite. In some passages, as before stated, stalagmite was wanting, while in one place seven or eight alternations of stalagmite and loam were observed, seeming to indicate a prevalence of more rainy seasons, succeeded by others, when the water was for a time too low to flood the area where the calcareous incrustation accumulated.

If the regular sequence of the three deposits of pebbles, mud,

[1] Pengelly, *Geologist*, vol. iv., 1861, p. 153.

and stalagmite was the result of the causes above explained, the order of superposition would be constant, yet we could not be sure that the gravel in one passage might not sometimes be coeval with the bone-earth or stalagmite in another.

If therefore the flint knives had not been very widely dispersed, and if one of them had not been at the bottom of the bone-earth, close to the leg of the bear above described, their antiquity relatively to the extinct mammalia might have been questioned. No coprolites were found in the Brixham excavations, and very few gnawed bones. These few may have been brought from some distance before they reached their place of rest. Upon the whole, the same conclusion which Dr. Schmerling came to, respecting the filling up of the caverns near Liège, seems applicable to the caves of Brixham.

Dr. Falconer, after aiding in the investigations above alluded to near Torquay, stopped at Abbeville on his way to Sicily, in the autumn of 1858, and saw there the collection of M. Boucher de Perthes. Being at once satisfied that the flints called hatchets had really been fashioned by the hand of Man, he urged Mr. Prestwich, by letter, thoroughly to explore the geology of the valley of the Somme. This he accordingly accomplished, in company with Mr. John Evans [Note 13], of the Society of Antiquaries, and, before his return that same year, succeeded in dissipating all doubts from the minds of his geological friends by extracting, with his own hands, from a bed of undisturbed gravel, at St. Acheul, a well-shaped flint hatchet. This implement was buried in the gravel at a depth of 17 feet from the surface, and was lying on its flat side. There were no signs of vertical rents in the enveloping matrix, nor in the overlying beds of sand and loam, in which were many land and fresh-water shells; so that it was impossible to imagine that the tool had gradually worked its way downwards, as some had suggested, through the incumbent soil, into an older formation.[1]

There was no one in England whose authority deserved to have so much weight in overcoming incredulity in regard to the antiquity of the implements in question. For Mr. Prestwich, besides having published a series of important memoirs on the Tertiary formations of Europe, had devoted many years specially to the study of the drift and its organic remains. His report, therefore, to the Royal Society, accompanied by a photograph showing the position of the flint tool *in situ* before it was removed

---

[1] Prestwich, *Proceedings of the Royal Society*, 1859, and *Philosophical Transactions*, 1860.

from its matrix, not only satisfied many inquirers, but induced others to visit Abbeville and Amiens; and one of these, Mr. Flower, who accompanied Mr. Prestwich on his second excursion to St. Acheul, in June 1859, succeeded, by digging into the bank of gravel, in disinterring, at the depth of 22 feet from the surface, a fine, symmetrically-shaped weapon of an oval form, lying in and beneath strata which were observed by many witnesses to be perfectly undisturbed.[1]

Shortly afterwards, in the year 1859, I visited the same pits, and obtained seventy flint tools, one of which was taken out while I was present, though I did not see it before it had fallen from the matrix. I expressed my opinion in favour of the antiquity of the flint tools to the meeting of the British Association at Aberdeen, in the same year.[2] On my way through Rouen, I stated my convictions on this subject to M. George Pouchet, who immediately betook himself to St. Acheul, commissioned by the municipality of Rouen, and did not quit the pits till he had seen one of the hatchets extracted from gravel in its natural position.[3]

M. Gaudry also gave the following account of his researches in the same year to the Royal Academy of Sciences at Paris. " The great point was not to leave the workmen for a single instant, and to satisfy oneself by actual inspection whether the hatchets were found *in situ*. I caused a deep excavation to be made, and found nine hatchets, most distinctly *in situ* in the diluvium, associated with teeth of *Equus fossilis* and a species of *Bos*, different from any now living, and similar to that of the diluvium and of caverns." [4] In 1859, M. Hébert, an original observer of the highest authority, declared to the Geological Society of France that he had, in 1854, or four years before Mr. Prestwich's visit to St. Acheul, seen the sections at Abbeville and Amiens, and had come to the opinion that the hatchets were imbedded in the " lower diluvium," and that their origin was as ancient as that of the mammoth and the rhinoceros. M. Desnoyers also made excavations after M. Gaudry, at St. Acheul, in 1859, with the same results.[5]

After a lively discussion on the subject in England and France, it was remembered, not only that there were numerous recorded cases leading to similar conclusions in regard to cavern deposits,

---

[1] *Quart. Jour. Geol. Soc.*, vol. xvi., 1860, p. 190.
[2] See *Report of British Association for* 1859.
[3] *Actes du Musée d'Histoire Naturelle de Rouen*, 1860, p. 33.
[4] *Comptes rendus*, September 26 and October 3, 1859.
[5] *Bulletin*, vol. xvii., p. 18.

but, also, that Mr. Frere had, so long ago as 1797, found flint weapons, of the same type as those of Amiens, in a fresh-water formation in Suffolk, in conjunction with elephant remains; and nearly a hundred years earlier (1715), another tool of the same kind had been exhumed from the gravel of London, together with bones of an elephant; to all which examples I shall allude more fully in the sequel.

I may conclude this chapter by quoting a saying of Professor Agassiz, " that whenever a new and startling fact is brought to light in science, people first say, ' it is not true,' then that ' it is contrary to religion,' and lastly, ' that everybody knew it before."

If I were considering merely the cultivators of geology, I should say that the doctrine of the former co-existence of Man with many extinct mammalia had already gone through these three phases in the progress of every scientific truth towards acceptance. But the grounds of this belief have not yet been fully laid before the general public, so as to enable them fairly to weigh and appreciate the evidence. I shall therefore do my best in the next three chapters to accomplish this task.

CHAPTER VII

PEAT AND ALLUVIUM OF THE VALLEY OF THE SOMME

Geological Structure of the Valley of the Somme and of the surrounding Country—Position of Alluvium of different Ages—Peat near Abbeville—Its Animal and Vegetable Contents—Works of Art in Peat—Probable Antiquity of the Peat, and Changes of Level since its Growth began—Flint Implements of Antique Type in older Alluvium—Their various Forms and great Numbers.

Geological Structure of the Somme Valley

THE valley of the Somme in Picardy, alluded to in the last chapter, is situated geologically in a region of white Chalk with flints, the strata of which are nearly horizontal. The Chalk hills which bound the valley are almost everywhere between 200 and 300 feet in height. On ascending to that elevation, we find ourselves on an extensive table-land, in which there are slight elevations and depressions. The white Chalk itself is scarcely ever exposed at the surface on this plateau, although seen on the slopes of the hills, as at b and c (Fig. 7). The general surface of the upland region is covered continuously for miles in every direction by loam or brick-earth (No. 4), about 5 feet thick, devoid of fossils. To the wide extent of this loam the soil of Picardy chiefly owes its great fertility. Here and there we also observe, on the Chalk, outlying patches of Tertiary sand and clay (No. 5, Fig. 7), with Eocene fossils, the remnants of a formation once more extensive, and which probably once spread in one continuous mass over the Chalk, before the present system of valleys had begun to be shaped out. It is necessary to allude to these relics of Tertiary strata, of which the larger part is missing, because their denudation has contributed largely to furnish the materials of gravels in which the flint implements and bones of extinct mammalia are entombed. From this source have been derived not only the regular-formed egg-shaped pebbles, so common in the old fluviatile alluvium at all levels, but those huge masses of hard sandstone, several feet in diameter, to which I shall allude in the sequel. The upland loam also (No. 4) has often, in no slight degree, been formed at the expense of the same Tertiary sands and clays, as is attested by its becoming more or less sandy or argillaceous, according to the nature of the nearest Eocene outlier in the neighbourhood.

The average width of the valley of the Somme between Amiens and Abbeville is one mile. The height, therefore, of the hills, in relation to the river-plain, could not be correctly represented in the annexed diagram (Fig. 7), as they would have to be reduced in altitude; or if not, it would be necessary to make the space between *c* and *b* four times as great. The dimensions also of the masses, of drift or alluvium, 2 and 3, have been exaggerated, in order to render them sufficiently conspicuous; for, all important as we shall find them to be as geological monuments of the Pleistocene period, they form a truly insignificant feature in the general structure of the country, so much so, that they might easily be overlooked in a cursory survey of the district, and are usually unnoticed in geological maps not specially devoted to the superficial formations.

Fig. 7.

Section across the Valley of the Somme in Picardy.

1 Peat, 20 to 30 feet thick, resting on gravel, *a*.
2 Lower level gravel with elephants' bones and flint tools, covered with fluviatile loam, 20 to 40 feet thick.
3 Higher level gravel with similar fossils, and with overlying loam, in all 30 feet thick.
4 Upland loam without shells (*Iimon des plateaux*), 5 or 6 feet thick.
5 Eocene strata, resting on the Chalk in patches.

It will be seen by the description given of the section (Fig. 7) that No. 2 indicates the lower level gravels, and No. 3 the higher ones, or those rising to elevations of 80 or 100 feet above the river. Newer than these is the peat No. 1, which is from 10 to 30 feet in thickness, and which is not only of later date than the alluvium, 2 and 3, but is also posterior to the denudation of those gravels, or to the time when the valley was excavated through them. Underneath the peat is a bed of gravel, *a*, from 3 to 14 feet thick, which rests on undisturbed Chalk. This gravel was probably formed, in part at least, when the valley was scooped out to its present depth, since which time no geological change has taken place, except the growth of the peat, and certain oscillations in the general level of the country, to which we shall allude by and by. A thin layer of impervious clay separates the gravel *a* from the peat No. 1, and seems to have been a necessary preliminary to the growth of the peat.

Peat of the Valley of the Somme

As hitherto, in our retrospective survey, we have been obliged, for the sake of proceeding from the known to the less known, to reverse the natural order of history, and to treat of the newer before the older formations, I shall begin my account of the geological monuments of the valley of the Somme by saying something of the most modern of all of them, the peat. This substance occupies the lower parts of the valley far above Amiens, and below Abbeville as far as the sea. It has already been stated to be in some places 30 feet thick, and is even occasionally more than 30 feet, corresponding in that respect to the Danish mosses before described (Chap. II.). Like them, it belongs to the Recent period; all the embedded mammalia, as well as the shells, being of the same species as those now inhabiting Europe. The bones of quadrupeds are very numerous, as I can bear witness, having seen them brought up from a considerable depth near Abbeville, almost as often as the dredging instrument was used. Besides remains of the beaver, I was shown, in the collection of M. Boucher de Perthes, two perfect lower jaws with teeth of the bear, *Ursus arctos ;* and in the Paris Museum there is another specimen, also from the Abbeville peat.

The list of mammalia already comprises a large proportion of those proper to the Swiss lake-dwellings, and to the shell-mounds and peat of Denmark; but unfortunately as yet no special study has been made of the French fauna, like that by which the Danish and Swiss zoologists and botanists have enabled us to compare the wild and tame animals and the vegetation of the age of stone with that of the age of iron.

Notwithstanding the abundance of mammalian bones in the peat, and the frequency of stone implements of the Celtic and Gallo-Roman periods, M. Boucher de Perthes has only met with three or four fragments of human skeletons.

At some depth in certain places in the valley near Abbeville, the trunks of alders have been found standing erect as they grew, with their roots fixed in an ancient soil, afterwards covered with peat. Stems of the hazel, and nuts of the same, abound; trunks, also, of the oak and walnut. The peat extends to the coast, and is there seen passing under the sand-dunes and below the sea-level. At the mouth of the river Canche, which joins the sea near the embouchure of the Somme, yew trees, firs, oaks, and hazels have been dug out of peat, which is there worked for fuel, and is about three feet thick.[1] During great storms, large masses of compact peat, enclosing trunks of flattened trees, have been thrown up on the coast at the mouth of the Somme; seeming to indicate that there has been a subsidence of the land and a consequent submergence of what was once a westward continuation of the valley of the Somme into what is now a part of the English Channel.

Whether the vegetation of the lowest layers of peat differed as to the geographical distribution of some of the trees from the middle, and this from the uppermost peat, as in Denmark, has not yet been ascertained; nor have careful observations been made with a view of calculating the minimum of time which the accumulation of so dense a mass of vegetable matter must have taken. A foot in thickness of highly compressed peat, such as is sometimes reached in the bottom of the bogs, is obviously the equivalent in time of a much greater thickness of peat of spongy and loose texture, found near the surface. The workmen who cut peat, or dredge it up from the bottom of swamps and ponds, declare that in the course of their lives none of the hollows which they have found, or caused by extracting peat, have ever been refilled, even to a small extent. They deny, therefore, that the peat grows. This, as M. Boucher de Perthes observes, is a mistake; but it implies that the increase in one generation is not very appreciable by the unscientific.

The antiquary finds near the surface Gallo-Roman remains, and still deeper Celtic weapons of the stone period. [Note 14.] But the depth at which Roman works of art occur varies in different places, and is no sure test of age; because in some parts of the swamps, especially near the river, the peat is often so fluid that heavy substances may sink through it, carried down by their own gravity. In one case, however, M. Boucher de Perthes observed several large flat dishes of Roman pottery, lying in a horizontal position in the peat, the shape of which must have prevented them from sinking or penetrating through the underlying peat. Allowing about fourteen centuries for the growth of the superincumbent vegetable matter, he calculated that the thickness gained in a hundred years would be no more than three centimetres.[2] This rate of increase would demand so many thousands of years for the formation of the entire thickness of 30 feet, that we must hesitate before adopting it as a chronometric scale. Yet, by multiplying observations of this kind,

---

[1] D'Archiac, *Hist. des Progrès*, vol. ii., p. 154.
[2] *Antiquités Celtiques*, vol. ii., p. 134.

and bringing one to bear upon and check another, we may eventually succeed in obtaining data for estimating the age of the peaty deposit. [Note 15.]

The rate of increase in Denmark may not be applicable to France; because differences in the humidity of the climate, or in the intensity and duration of summer's heat and winter's cold, as well as diversity in the species of plants which most abound, would cause the peat to grow more or less rapidly, not only when we compare two distinct countries in Europe, but the same country at two successive periods.

I have already alluded to some facts which favour the idea that there has been a change of level on the coast since the peat began to grow. This conclusion seems confirmed by the mere thickness of peat at Abbeville, and the occurrence of alder and hazel-wood near the bottom of it. If 30 feet of peat were now removed, the sea would flow up and fill the valley for miles above Abbeville. Yet this vegetable matter is all of supra-marine origin, for where shells occur in it they are all of terrestrial or fluviatile kinds, so that it must have grown above the sea-level when the land was more elevated than now. We have already seen what changes in the relative level of sea and land have occurred in Scotland subsequently to the time of the Romans, and are therefore prepared to meet with proofs of similar movements in Picardy. In that country they have probably not been confined simply to subsidence, but have comprised oscillations in the level of the land, by which marine shells of the Pleistocene period have been raised some 10 feet or more above the level of the sea.

Small as is the progress hitherto made in interpreting the pages of the peaty record, their importance in the valley of the Somme is enhanced by the reflection that, whatever be the number of centuries to which they relate, they belong to times posterior to the ancient implement-bearing beds, which we are next to consider, and are even separated from them, as we shall see, by an interval far greater than that which divides the earliest strata of the peat from the latest.

Flint Implements of the Pleistocene Period in the Valley of the Somme

The alluvium of the valley of the Somme exhibits nothing extraordinary or exceptional in its position or external appearance, nor in the arrangement or composition of its materials, nor

in its organic remains; in all these characters it might be matched
by the drift of a hundred other valleys in France or England.
Its claim to our peculiar attention is derived from the wonderful
number of flint tools, of a very antique type, which, as stated in
the last chapter, occur in undisturbed strata, associated with the
bones of extinct quadrupeds.

As much doubt has been cast on the question, whether the
so-called flint hatchets have really been shaped by the hands of
Man, it will be desirable to begin by satisfying the reader's mind
on that point, before inviting him to study the details of sections
of successive beds of mud, sand, and gravel, which vary considerably even in contiguous localities.

Since the spring of 1859, I have paid three visits to the valley
of the Somme, and examined all the principal localities of these
flint tools.   In my excursions around Abbeville, I was accompanied by M. Boucher de Perthes, and during one of my explorations in the Amiens district, by Mr. Prestwich.   The first
time I entered the pits at St. Acheul, I obtained seventy flint
instruments, all of them collected from the drift in the course
of the preceding five or six weeks.   The two prevailing forms
of these tools are represented in the annexed Figs. 8 and 9,
each of which are half the size of the originals; the first being
the spear-headed form, varying in length from six to eight inches;
the second, the oval form, which is not unlike some stone implements, used to this day as hatchets and tomahawks by natives of
Australia, but with this difference, that the edge in the Australian
weapons (as in the case of those called celts in Europe) has been
produced by friction, whereas the cutting edge in the old tools of
the valley of the Somme was always gained by the simple fracture
of the flint, and by the repetition of many dexterous blows.

The oval-shaped Australian weapons, however, differ in being
sharpened at one end only.   The other, though reduced by
fracture to the same general form, is left rough, in which state
it is fixed into a cleft stick, which serves as a handle.   To this it
is firmly bound by thin straps of opossum's hide.   One of these
tools, now in my possession, was given me by Mr. Farquhar-
son of Haughton, who saw a native using it in 1854 on the
Auburn river, in Burnet district, North Australia.

Out of more than a hundred flint implements which I obtained
at St. Acheul, not a few had their edges more or less fractured or
worn, either by use as instruments before they were buried in
gravel, or by being rolled in the river's bed.

Some of these tools were probably used as weapons, both of

war and of the chase, others to grub up roots, cut down trees, and scoop out canoes. Some of them may have served, as Mr. Prestwich has suggested, for cutting holes in the ice both for fishing and for obtaining water, as will be explained in the eighth chapter when we consider the arguments in favour of the higher level drift having belonged to a period when the rivers were frozen over for several months every winter.

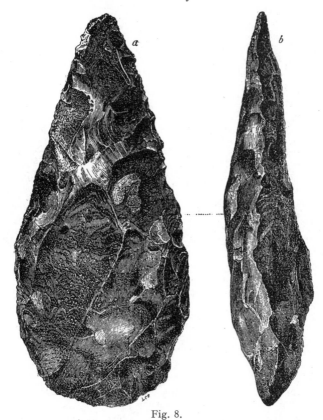

Fig. 8.

Flint implement from St. Acheul, near Amiens, of the spear-head shape (half the size of the original, which is $7\frac{1}{2}$ inches long).

a Side view.   *b* Same seen edgewise.

These spear-headed implements have been found in greater number, proportionally to the oval ones, in the upper level gravel at St. Acheul, than in any of the lower gravels in the valley of the Somme. In these last the oval form predominates, especially at Abbeville.

When the natural form of a Chalk-flint presented a suitable handle at one end, as in the specimen, Fig. 10, that part was left as found. The portion, for example, between *b* and *c* has probably not been altered; the protuberances which are fractured having been broken off by river action before the flint was chipped artificially. The other extremity, *a*, has been worked till it acquired a proper shape and cutting edge.

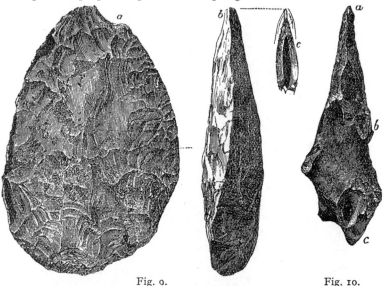

Fig. 9.   Fig. 10.

Flint implements from the Pleistocene Drift of Abbeville and Amiens.

Fig. 9. *a* Oval-shaped flint hatchet from Mautort, near Abbeville, half size of original, which is 5½ inches long, from a bed of gravel underlying the fluvio-marine stratum.

b Same seen edgewise.

c Shows a recent fracture of the edge of the same at the point *a*, or near the top. This portion of the tool, *c*, is drawn of the natural size, the black central part being the unaltered flint, the white outer coating, the layer which has been formed by discoloration or bleaching since the tool was first made.

The entire surface of No. 9 must have been black when first shaped, and the bleaching to such a depth must have been the work of time, whether produced by exposure to the sun and air before it was embedded, or afterwards when it lay deep in the soil.

Fig. 10. Flint tool from St. Acheul, seen edgewise; original, 6½ inches long, and 3 inches wide.

b, c Portion not artificially shaped.

a, b Part chipped into shape, and having a cutting edge at *a*.

Many of the hatchets are stained of an ochreous-yellow colour, when they have been buried in yellow gravel, others have acquired white or brown tints, according to the matrix in which they have been enclosed.

This accordance in the colouring of the flint tools with the character of the bed from which they have come, indicates, says Mr. Prestwich, not only a real derivation from such strata, but also a sojourn therein of equal duration to that of the naturally broken flints forming part of the same beds.[1]

The surface of many of the tools is encrusted with a film of carbonate of lime, while others are adorned by those ramifying crystallisations called dendrites (*see* Figs. 11-13), usually consisting of the mixed oxides of iron and manganese, forming extremely delicate blackish brown sprigs, resembling the smaller

Fig. 11.   Fig. 12.   Fig. 13.

Dendrites on surfaces of flint hatchets in the drift of St. Acheul, near Amiens.

Fig. 11. *a* Natural size.   Fig. 12. *b* Natural size. *c* Magnified.
Fig. 13. *d* Natural size. *e* Magnified.

kinds of sea weed. They are a useful test of antiquity when suspicions are entertained of the workmen having forged the hatchets which they offer for sale. The most general test, however, of the genuineness of the implements obtained by purchase is their superficial varnish-like or vitreous gloss, as contrasted with the dull aspect of freshly fractured flints. I also remarked, during each of my three visits to Amiens, that there were some extensive gravel-pits, such as those of Montiers and St. Roch, agreeing in their geological character with those of St. Acheul, and only a mile or two distant, where the workmen, although familiar with the forms, and knowing the marketable value of the articles above described, assured me that they had never been able to find a single implement.

Respecting the authenticity of the tools as works of art, Professor Ramsay, than whom no one could be a more com-

[1] *Philosophical Transactions*, 1861, p. 297.

petent judge, observes: "For more than twenty years, like others of my craft, I have daily handled stones, whether fashioned by nature or art; and the flint hatchets of Amiens and Abbeville seem to me as clearly works of art as any Sheffield whittle." [1]

Mr. Evans classifies the implements under three heads, two of which, the spear heads and the oval or almond-shaped kinds, have already been described. The third form (Fig. 14) consists of flakes, apparently intended for knives or some of the smaller ones for arrow heads.

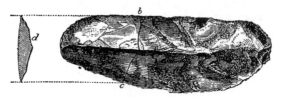

Fig. 14.

Flint knife or flake from below the sand containing *Cyrena fluminalis*. Menchecourt, Abbeville.

d Transverse section along the line of fracture, *b*, *c*. Size, two-thirds of the original.

In regard to their origin, Mr. Evans observes that there is a uniformity of shape, a correctness of outline, and a sharpness about the cutting edges and points, which cannot be due to anything but design.[2]

Of these knives and flakes, I obtained several specimens from a pit which I caused to be dug at Abbeville, in sand in contact with the Chalk, and below certain fluvio-marine beds, which will be alluded to in the next chapter.

Between the spear-head and oval shapes, there are various intermediate gradations, and there are also a vast variety of very rude implements, many of which may have been rejected as failures, and others struck off as chips in the course of manufacturing the more perfect ones. Some of these chips can only be recognised by an experienced eye as bearing marks of human workmanship.

It has often been asked, how, without the use of metallic hammers, so many of these oval and spear-headed tools could have been wrought into so uniform a shape. Mr. Evans, in order experimentally to illustrate the process, constructed a stone

[1] *Athenæum*, July 16, 1859.   [2] *Archæologia*, vol. xxxviii.

hammer, by mounting a pebble in a wooden handle, and with this tool struck off flakes from the edge on both sides of a Chalk flint, till it acquired precisely the same shape as the oval tool, Fig. 9, p. 90.

If I were invited to estimate the probable number of the more perfect tools found in the valley of the Somme since 1842, rejecting all the knives, and all that might be suspected of being spurious or forged, I should conjecture that they far exceeded a thousand.   Yet it would be a great mistake to imagine that an antiquary or geologist, who should devote a few weeks to the exploration of such a valley as that of the Somme, would himself be able to detect a single specimen.   But few tools were lying on the surface.   The rest have been exposed to view by the removal of such a volume of sand, clay, and gravel, that the price of the discovery of one of them could only be estimated by knowing how many hundred labourers have toiled at the fortifications of Abbeville, or in the sand and gravel pits near that city, and around Amiens, for road materials and other economic purposes, during the last twenty years.

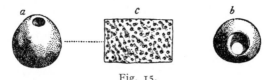

Fig. 15.

a, b Coscinopora globularis, D'Orb.   *Orbitolina concava*, Parker and Jones.
c Part of the same magnified.

In the gravel pits of St. Acheul, and in some others near Amiens, small round bodies, having a tubular cavity in the centre, occur.   They are well known as fossils of the White Chalk.   Dr. Rigollot suggested that they might have been strung together as beads, and he supposed the hole in the middle to have been artificial.   Some of these round bodies are found entire in the Chalk and in the gravel, others have naturally a hole passing through them, and sometimes one or two holes penetrating some way in from the surface, but not extending to the other side.   Others, like *b*, Fig. 15, have a large cavity, which has a very artificial aspect.   It is impossible to decide whether they have or have not served as personal ornaments, recommended by their globular form, lightness, and by being less destructible than ordinary Chalk.   Granting that there were natural cavities in the axis of some of them, it does not follow that these may not have been taken advantage of for stringing them as beads, while others may have been artificially bored through. Dr. Rigollot's argument in favour of their having been used as necklaces or bracelets, appears to me a sound one. He says he often found small heaps or groups of them in one place, all perforated, just as if, when swept into the river's bed by a flood, the bond which had united them together remained unbroken.[1]

---

[1] Rigollot, *Mémoire sur des Instruments en Silex*, etc., Amiens, 1854, p. 16.

CHAPTER VIII

PLEISTOCENE ALLUVIUM WITH FLINT IMPLEMENTS OF THE
VALLEY OF THE SOMME—*concluded*

Fluvio-marine Strata, with Flint Implements, near Abbeville—Marine
Shells in same—*Cyrena Fluminalis*—Mammalia—Entire Skeleton of
Rhinoceros—Flint Implements, why found low down in Fluviatile
Deposits—Rivers shifting their Channels—Relative Ages of higher
and lower-level Gravels—Section of Alluvium of St. Acheul—Two
Species of Elephant and Hippopotamus co-existing with Man in
France—Volume of Drift, proving Antiquity of Flint Implements—-
Absence of Human Bones in tool-bearing Alluvium, how explained—
Value of certain Kinds of Negative Evidence tested thereby—Human
Bones not found in drained Lake of Haarlem.

IN the section of the valley of the Somme given in Fig. 7, the
successive formations newer than the Chalk are numbered in
chronological order, beginning with the most modern, or the peat,
which is marked No. 1, and which has been treated of in the last
chapter. Next in the order of antiquity are the lower-level
gravels, No. 2, which we have now to describe; after which the
alluvium, No. 3, found at higher levels, or about 80 and 100 feet
above the river-plain, will remain to be considered.

I have selected, as illustrating the old alluvium of the Somme
occurring at levels slightly elevated above the present river, the
sand and gravel-pits of Menchecourt, in the northwest suburbs of
Abbeville, to which, as before stated, attention was first drawn
by M. Boucher de Perthes, in his work on Celtic antiquities.
Here, although in every adjoining pit some minor variations in
the nature and thickness of the superimposed deposits may be
seen, there is yet a general approach to uniformity in the series.
The only stratum of which the relative age is somewhat doubtful,
is the gravel marked *a*, underlying the peat, and resting on the
Chalk. It is only known by borings, and some of it may be of
the same age as No. 3; but I believe it to be for the most part
of more modern origin, consisting of the wreck of all the older
gravel, including No. 3, and formed during the last hollowing out
and deepening of the valley immediately before the commencement of the growth of peat.

The greater number of flint implements have been dug out of
No. 3, often near the bottom, and twenty-five, thirty, or even
more than thirty feet below the surface of No. 1.

A geologist will perceive by a glance at the section that the valley of the Somme must have been excavated nearly to its present depth and width when the strata of No. 3 were thrown down, and that after the deposits Nos. 3, 2, and 1 had been formed in succession, the present valley was scooped out, patches only of Nos. 3 and 2 being left. For these deposits cannot originally have ended abruptly as they now do, but must have once been continuous farther towards the centre of the valley.

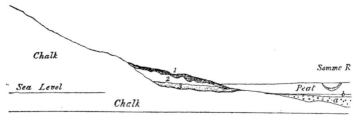

Fig. 16.

Section of fluvio-marine strata, containing flint implements and bones of extinct mammalia, at Menchecourt, Abbeville.[1]

1 Brown clay with angular flints, and occasionally Chalk rubble, unstratified, following the slope of the hill, probably of subaerial origin, of very varying thickness, from 2 to 5 feet and upwards.
2 Calcareous loam, buff-coloured, resembling loess, for the most part unstratified, in some places with slight traces of stratification, containing freshwater and land shells, with bones of elephants, etc.; thickness about 15 feet.
3 Alternations of beds of gravel, marl, and sand, with freshwater and land shells, and, in some of the lower sands, a mixture of marine shells; also bones of elephant, rhinoceros, etc., and flint implements; thickness about 12 feet.

a Gravel underlying peat, age undetermined.
b Layer of impervious clay, separating the gravel from the peat.

To begin with the oldest, No. 3, it is made up of a succession of beds, chiefly of freshwater origin, but occasionally a mixture of marine and fluviatile shells is observed in it, proving that the sea sometimes gained upon the river, whether at high tides or when the fresh water was less in quantity during the dry season, and sometimes perhaps when the land was slightly depressed in level. All these accidents might occur again and again at the mouth of any river, and give rise to alternations of fluviatile and marine strata, such as are seen at Menchecourt.

[1] For detailed sections and maps of this district, see Prestwich, *Philosophical Transactions*, 1860, p. 277.

In the lowest beds of gravel and sand in contact with the Chalk, flint hatchets, some perfect, others much rolled, have been found; and in a sandy bed in this position some workmen, whom I employed to sink a pit, found four flint knives.   Above this sand and gravel occur beds of white and siliceous sand, containing shells of the genera *Planorbis, Limnea, Paludina, Valvata, Cyclas, Cyrena, Helix*, and others, all now natives of the same part of France, except *Cyrena fluminalis* (Fig. 17), which no longer lives in Europe, but inhabits the Nile, and many parts of Asia, including Cashmere, where it abounds.   No species of Cyrena is now met with in a living state in Europe.   Mr. Prestwich first observed it fossil at Menchecourt, and it has since been found in two or three contiguous sand-pits, always in the fluvio-marine bed.   [Note 16.]

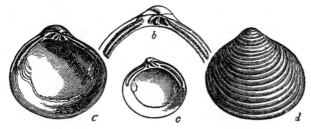

Fig. 17.

Cyrena fluminalis, O. F. Müller, sp.[1]

a Interior of left valve, from Gray's Thurrock, Essex.
b Hinge of same magnified.
c Interior of right valve of a small specimen, from Shacklewell, London.
d Outer surface of right valve, from Erith, Kent.

Living	Dates of Specific Names
Tellina fluminalis, O. F. Müller	1774
Venus fluminalis Euphratis, Chemnitz	1782
Cyclas Euphratica, Lam.	1806
Cyrena cor, Lam. (Nile)	1818
,, *consobrina*, Caillaud (Nile) . . .	1823
,, *Cashmiriensis*, Desh.	
Corbicula fluminalis, Mühlfeldt	1811
Fossil	
Cyrena trigonula, S. Woodward	1834
,, *Gemmellarii*, Philippi	1836
,, *Duchastelii*, Nyst	1838

The following marine shells occur mixed with the freshwater

[1] For synonyms, *see* S. Woodward, " Tibet Shells," *Proc. Zool. Soc.*, July 8, 1856.

species above enumerated:—*Buccinum undatum, Littorina littorea, Nassa reticulata, Purpura lapillus, Tellina solidula, Cardium edule,* and fragments of some others.   Several of these I have myself collected entire, though in a state of great decomposition, lying in the white sand called " sable aigre " by the workmen.   They are all littoral species now proper to the contiguous coast of France.   Their occurrence in a fossil state associated with freshwater shells at Menchecourt had been noticed as long ago as 1836 by MM. Ravin and Baillon, before M. Boucher de Perthes commenced the researches which have since made the locality so celebrated.[1]   The numbers since collected preclude all idea of their having been brought inland as eatable shells by the fabricators of the flint hatchets found at the bottom of the fluvio-marine sands.   From the same beds, and in marls alternating with the sands, remains of the elephant, rhinoceros, and other mammalia have been exhumed.

Above the fluvio-marine strata are those designated No. 2 in the section (Fig. 16), which are almost devoid of stratification, and probably formed of mud or sediment thrown down by the waters of the river when they overflowed the ancient alluvial plain of that day.   Some land shells, a few river shells, and bones of mammalia, some of them extinct, occur in No. 2. Its upper surface has been deeply furrowed and cut into by the action of water, at the time when the earthy matter of No. 1 was superimposed.   The materials of this uppermost deposit are arranged as if they had been the result of land floods, taking place after the formations 2 and 3 had been raised, or had become exposed to denudation.

The fluvio-marine strata and overlying loam of Menchecourt recur on the opposite or left bank of the alluvial plain of the Somme, at a distance of 2 or 3 miles.   They are found at Mautort, among other places, and I obtained there the flint hatchet shown in Fig. 9, of an oval form.   It was extracted from gravel, above which were strata containing a mixture of marine and freshwater shells, precisely like those of Menchecourt. In the alluvium of all parts of the valley, both at high and low levels, rolled bones are sometimes met with in the gravel. Some of the flint tools in the gravel of Abbeville have their angles very perfect, others have been much triturated, as if in the bed of the main river or some of its tributaries.

The mammalia most frequently cited as having been found in the deposits Nos. 2 and 3 at Menchecourt, are the following:—

[1] D'Archiac, *Histoire des Progrès*, etc., vol. ii., p. 154.

Elephas primigenius.	*Cervus somonensis,* Cuvier.
Rhinoceros tichorhinus.	*C. tarandus priscus,* Cuvier.
Equus fossilis, Owen.	*Felis spelæa.*
Bos primigenius.	*Hyæna spelæa.*

The *Ursus spelæus* has also been mentioned by some writers; but M. Lartet says he has sought in vain for it among the osteological treasures sent from Abbeville to Cuvier at Paris, and in other collections. The same palæontologist, after a close scrutiny of the bones sent formerly to the Paris Museum from the valley of the Somme, observed that some of them bore the evident marks of an instrument, agreeing well with incisions such as a rude flint-saw would produce. Among other bones mentioned as having been thus artificially cut, are those of a *Rhinoceros tichorhinus,* and the antlers of *Cervus somonensis.*[1]

The evidence obtained by naturalists that some of the extinct mammalia of Menchecourt really lived and died in this part of France, at the time of the embedding of the flint tools in fluviatile strata, is most satisfactory; and not the less so for having been put on record long before any suspicion was entertained that works of art would ever be detected in the same beds. Thus M. Baillon, writing in 1834 to M. Ravin, says: " They begin to meet with fossil bones at the depth of 10 or 12 feet in the Menchecourt sand-pits, but they find a much greater quantity at the depth of 18 and 20 feet. Some of them were evidently broken before they were embedded, others are rounded, having, without doubt, been rolled by running water. It is at the bottom of the sand-pits that the most entire bones occur. Here they lie without having undergone fracture or friction, and seem to have been articulated together at the time when they were covered up. I found in one place a whole hind limb of a rhinoceros, the bones of which were still in their true relative position. They must have been joined together by ligaments, and even surrounded by muscles at the time of their interment. The entire skeleton of the same species was lying at a short distance from the spot." [2]

If we suppose that the greater number of the flint implements occurring in the neighbourhood of Abbeville and Amiens were brought by river action into their present position, we can at once explain why so large a proportion of them are found at considerable depths from the surface, for they would naturally

---

[1] *Quart. Jour. Geol. Soc.,* vol. xvi., 1860, p. 471.
[2] *Société Roy. d'Emulation d'Abbeville,* 1834, p. 197.

be buried in gravel and not in fine sediment, or what may be
termed "inundation mud," such as No. 2 (Fig. 16), a deposit
from tranquil water, or where the stream had not sufficient force
or velocity to sweep along Chalk flints, whether wrought or un-
wrought.   Hence we have almost always to pass down through
a mass of incumbent loam with land shells, or through fine sand
with freshwater molluscs, before we get into the beds of gravel
containing hatchets.   Occasionally a weapon used as a projectile
may have fallen into quiet water, or may have dropped from a
canoe to the bottom of the river, or may have been floated by
ice, as are some stones occasionally by the Thames in severe
winters, and carried over the meadows bordering its banks; but
such cases are exceptional, though helping to explain how
isolated flint tools or pebbles and angular stones are now and
then to be seen in the midst of the finest loams.

The endless variety in the sections of the alluvium of the
valley of the Somme, may be ascribed to the frequent silting up
of the main stream and its tributaries during different stages
of the excavation of the valley, probably also during changes
in the level of the land.   As a rule, when a river attacks and
undermines one bank, it throws down gravel and sand on the
opposite side of its channel, which is growing somewhere
shallower, and is soon destined to be raised so high as to form
an addition to the alluvial plain, and to be only occasionally
inundated.   In this way, after much encroachment on cliff or
meadow at certain points, we find at the end of centuries that
the width of the channel has not been enlarged, for the new
made ground is raised after a time to the average height of
the older alluvial tract.   Sometimes an island is formed in mid-
stream, the current flowing for a while on both sides of it, and
at length scooping out a deeper channel on one side so as to
leave the other to be gradually filled up during freshets and
afterwards elevated by inundation mud, or "brick-earth."
During the levelling up of these old channels, a flood sometimes cuts into and partially removes portions of the previously
stratified matter, causing those repeated signs of furrowing and
filling up of cavities, those memorials of doing and undoing,
of which the tool-bearing sands and gravels of Abbeville and
Amiens afford such reiterated illustrations, and of which a
parallel is furnished by the ancient alluvium of the Thames
valley, where similar bones of extinct mammalia and shells,
including *Cyrena fluminalis*, are found.

Professor Noeggerath, of Bonn, informs me that, about the

year 1845, when the bed of the Rhine was deepened artificially by the blasting and removal of rock in the narrows at Bingerloch, not far from Bingen, several flint hatchets and an extraordinary number of iron weapons of the Roman period were brought up by the dredge from the bed of the great river.   The decomposition of the iron had caused much of the gravel to be cemented together into a conglomerate.   In such a case we have only to suppose the Rhine to deviate slightly from its course, changing its position, as it has often done in various parts of its plain in historical times, and then tools of the stone and iron periods would be found in gravel at the bottom with a great thickness of sand and overlying loam deposited above them.

Changes in a river plain, such as those above alluded to, give rise frequently to ponds, swamps, and marshes, marking the course of old beds or branches of the river not yet filled up, and in these depressions shells proper both to running and stagnant water may be preserved, and quadrupeds may be mired.   The latest and uppermost deposit of the series will be loam or brick-earth, with land and amphibious shells (*Helix* and *Succinea*), while below will follow strata containing freshwater shells, implying continuous submergence; and lowest of all in most sections will be the coarse gravel accumulated by a current of considerable strength and velocity.

When the St. Katharine docks were excavated at London, and similar works executed on the banks of the Mersey, old ships were dug out, as I have elsewhere noticed,[1] showing how the Thames and Mersey have in modern times been shifting their channels.   Recently, an old silted-up bed of the Thames has been discovered by boring at Shoeburyness at the mouth of the river opposite Sheerness, as I learn from Mr. Mylne. The old deserted branch is separated from the new or present channel of the Thames, by a mass of London Clay which has escaped denudation. The depth of the old branch, or the thickness of fluviatile strata with which it has been filled up, is 75 feet.   The actual channel in the neighbourhood is now 60 feet deep, but there is probably 10 or 15 feet of stratified sand and gravel at the bottom; so that, should the river deviate again from its course, its present bed might be the receptacle of a fluvio-marine formation 75 feet thick, equal to the former one of Shoeburyness, and more considerable than that of Abbeville.   It would consist both of freshwater and marine strata,

[1] *Principles of Geology*, 10th edition, vol. ii., p. 547.

as the salt water is carried by the tide far up above Sheerness; but in order that such deposits should resemble, in geological position, the Menchecourt beds, they must be raised 10 or 15 feet above their present level, and be partially eroded. Such erosion they would not fail to suffer during the process of upheaval, because the Thames would scour out its bed, and not alter its position relatively to the sea, while the land was gradually rising.

Before the canal was made at Abbeville, the tide was perceptible in the Somme for some distance above that city. It would only require, therefore, a slight subsidence to allow the salt water to reach Menchecourt, as it did in the Pleistocene period. As a stratum containing exclusively land and fresh-water shells usually underlies the fluvio-marine sands at Menchecourt, it seems that the river first prevailed there, after which the land subsided; and then there was an upheaval which raised the country to a greater height than that at which it now stands, after which there was a second sinking, indicated by the position of the peat, as already explained. All these changes happened since Man first inhabited this region.

At several places in the environs of Abbeville there are fluviatile deposits at a higher level by 50 feet than the uppermost beds at Menchecourt, resting in like manner on the Chalk. One of these occurs in the suburbs of the city at Moulin Quignon, 100 feet above the Somme and on the same side of the valley as Menchecourt, and containing flint implements of the same antique type and the bones of elephants; but no marine shells have been found there, nor in any gravel or sand at higher elevations than the Menchecourt marine shells.

It has been a matter of discussion among geologists whether the higher or the lower sands and gravels of the Somme valley are the more ancient. As a general rule, when there are alluvial formations of different ages in the same valley, those which occupy a more elevated position above the river plain are the oldest. In Auvergne and Velay, in Central France, where the bones of fossil quadrupeds occur at all heights above the present rivers from 10 to 1000 feet, we observe the terrestrial fauna to depart in character from that now living in proportion as we ascend to higher terraces and platforms. We pass from the lower alluvium, containing the mammoth, tichorhine rhinoceros, and reindeer, to various older groups of fossils, till, on a table-land 1000 feet high (near Le Puy, for example), the abrupt termination of which overlooks the present valley, we discover an old extinct river-bed covered by a current of ancient lava, showing where the lowest level was once situated. In that elevated alluvium the remains of a Tertiary mastodon and other quadrupeds of like antiquity are embedded.

If the Menchecourt beds had been first formed, and the valley, after being nearly as deep and wide as it is now, had subsided, the sea must have advanced inland, causing small delta-like accumulations at successive heights, wherever the main river and its tributaries met the sea. Such a movement, especially if it were intermittent, and interrupted occasionally by long pauses, would very well account for the accumulation of stratified débris which we encounter at certain points in the valley, especially around Abbeville and Amiens. But we are precluded from adopting this theory by the entire absence of marine shells, and the presence of freshwater and land species, and mammalian bones, in considerable abundance, in the drift both of higher and lower levels above Abbeville. Had there been a total absence of all organic remains, we might have imagined the former presence of the sea, and the destruction of such remains might have been ascribed to carbonic acid or other decomposing causes; but the Pleistocene and implement-bearing strata can be shown by their fossils to be of fluviatile origin.

Flint Implements in Gravel near Amiens. Gravel of St. Acheul

When we ascend the valley of the Somme, from Abbeville to Amiens, a distance of about 25 miles, we observe a repetition of all the same alluvial phenomena which we have seen exhibited at Menchecourt and its neighbourhood, with the single exception of the absence of marine shells and of *Cyrena fluminalis*. We find lower-level gravel, such as No. 2, Fig. 7, and higher-level alluvium, such as No. 3, the latter rising to 100 feet above the plain, which at Amiens is about 50 feet above the level of the river at Abbeville. In both the upper and lower gravels, as Dr. Rigollot stated in 1854, flint tools and the bones of extinct animals, together with river shells and land shells of living species, abound.

Immediately below Amiens, a great mass of stratified gravel, slightly elevated above the alluvial plain of the Somme, is seen at St. Roch, and half a mile farther down the valley at Montiers. Between these two places a small tributary stream, called the Celle, joins the Somme. In the gravel at Montiers,

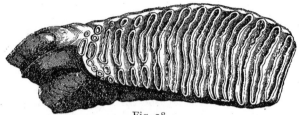

Fig. 18.

Elephas primigenius.

Penultimate molar, lower jaw, right side, one-third of natural size,
Pleistocene.   Co-existed with Man.

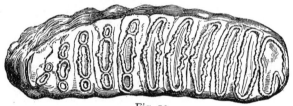

Fig. 19.

Elephas antiquus, Falconer.

Penultimate molar, lower jaw, right side, size one-third of nature,
Pleistocene and Newer Pliocene.   Co-existed with Man.

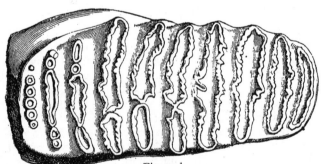

Fig. 20.[1]

Elephas meridionalis, Nesti.

Penultimate molar, lower jaw, right side, size one-third of original, Newer
Pliocene, Saint Prest, near Chartres, and Norwich Crag.   Not yet
proved to have co-existed with Man.

[1] For Fig. 20 I am indebted to M. Lartet, and Fig. 18 will be found in
his paper in *Bulletin de la Société Géologique de France*, March 1859.   Fig. 19
is from Falconer and Cautley, *Fauna Sivalensis*.

Mr. Prestwich and I found some flint knives, one of them flat on one side, but the other carefully worked, and exhibiting many fractures, clearly produced by blows skilfully applied. Some of these knives were taken from so low a level as to satisfy us that this great bed of gravel at Montiers, as well as that of the contiguous quarries of St. Roch, which seems to be a continuation of the same deposit, may be referred to the human period. Dr. Rigollot had already mentioned flint hatchets as obtained by him from St. Roch, but as none have been found there of late years, his statement was thought to require confirmation. The discovery, therefore, of these flint knives in gravel of the same age was interesting, especially as many tusks of a hippopotamus have been obtained from the gravel of St. Roch—some of these recently by Mr. Prestwich; while M. Garnier of Amiens has procured a fine elephant's molar from the same pits, which Dr. Falconer refers to *Elephas antiquus*, see Fig. 19, p. 104. Hence I infer that both these animals co-existed with Man.

The alluvial formations of Montiers are very instructive in another point of view. If, leaving the lower gravel of that place, which is topped with loam or brick-earth (of which the upper portion is about 30 feet above the level of the Somme), we ascend the Chalk slope to the height of about 80 feet, another deposit of gravel and sand, with fluviatile shells in a perfect condition, occurs, indicating most clearly an ancient river-bed, the waters of which ran habitually at that higher level before the valley had been scooped out to its present depth. This superior deposit is on the same side of the Somme, and about as high, as the lowest part of the celebrated formation of St. Acheul, 2 or 3 miles distant, to which I shall now allude.

The terrace of St. Acheul may be described as a gently sloping ledge of Chalk, covered with gravel, topped as usual with loam or fine sediment, the surface of the loam being 100 feet above the Somme, and about 150 above the sea.

Many stone coffins of the Gallo-Roman period have been dug out of the upper portion of this alluvial mass. The trenches made for burying them sometimes penetrate to the depth of 8 or 9 feet from the surface, entering the upper part of No. 3 of the sections Figs. 21 and 22. They prove that when the Romans were in Gaul they found this terrace in the same condition as it is now, or rather as it was before the removal of so much gravel, sand, clay, and loam, for repairing roads, and for making bricks and pottery.

In the annexed section, which I observed during my last visit in 1860, it will be seen that a fragment of an elephant's tooth is noticed as having been dug out of unstratified sandy loam at the point *a*, 11 feet from the surface. This was found at the time of my visit; and at a lower point, at *b*, 18 feet from

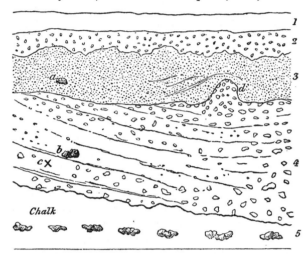

Fig. 21.

Section of a gravel pit containing flint implements at St. Acheul, near Amiens, observed in July 1860.

1 Vegetable soil and made ground, 2 to 3 feet thick.
2 Brown loam with some angular flints, in parts passing into ochreous gravel, filling up indentations on the surface of No. 3—3 feet thick.
3 White siliceous sand with layers of chalky marl, and included fragments of Chalk, for the most part unstratified—9 feet.
4 Flint-gravel, and whitish chalky sand, flints subangular, average size of fragments, 3 inches diameter, but with some large unbroken Chalk flints intermixed, cross stratification in parts. Bones of mammalia, grinder of elephant at *b*, and flint implement at *c*— 10 to 14 feet.
5 Chalk with flints.

a Part of elephant's molar, 11 feet from the surface.
b Entire molar of *E. primigenius*, 17 feet from surface.
c Position of flint hatchet, 18 feet from surface.

the surface, a large nearly entire and unrolled molar of the same species was obtained, which is now in my possession. It has been pronounced by Dr. Falconer to belong to *Elephas primigenius*.

A stone hatchet of an oval form, like that represented at

Fig. 9, was discovered at the same time, about one foot lower down, at *c*, in densely compressed gravel. The surface of the fundamental Chalk is uneven in this pit, and slopes towards the valley-plain of the Somme. In a horizontal distance of 20 feet, I found a difference in vertical height of 7 feet. In the chalky sand, sometimes occurring in interstices between the separate fragments of flint, constituting the coarse gravel No. 4, entire as well as broken freshwater shells are often met with. To some it may appear enigmatical how such fragile objects could have escaped annihilation in a river-bed, when flint tools and much gravel were shoved along the bottom; but I have seen the dredging instrument employed in the Thames, above and below London Bridge, to deepen the river, and worked by steam power, scoop up gravel and sand from the bottom, and then pour the contents pell-mell into the boat, and still many specimens of *Limnæa, Planorbis, Paludina, Cyclas*, and other shells might be taken out uninjured from the gravel.

It will be observed that the gravel No. 4 is obliquely stratified, and that its surface had undergone denudation before the white sandy loam No. 3 was superimposed. The materials of the gravel at *d* must have been cemented or frozen together into a somewhat coherent mass to allow the projecting ridge, *d*, to stand up 5 feet above the general surface, the sides being in some places perpendicular. In No. 3 we probably behold an example of a passage from river-silt to inundation mud. In some parts of it, land shells occur.

It has been ascertained by MM. Buteux, Ravin, and other observers conversant with the geology of this part of France, that in none of the alluvial deposits, ancient or modern, are there any fragments of rocks foreign to the basin of the Somme—no erratics which could only be explained by supposing them to have been brought by ice, during a general submergence of the country, from some other hydrographical basin.

But in some of the pits at St. Acheul there are seen in the beds No. 4, Fig. 21, not only well-rounded Tertiary pebbles, but great blocks of hard sandstone, of the kind called in the south of England " greywethers," some of which are 3 or 4 feet and upwards in diameter. They are usually angular, and when spherical owe their shape generally to an original concretionary structure, and not to trituration in a river's bed. These large fragments of stone abound both in the higher and lower level gravels round Amiens and at the higher level at Abbeville. They have also been traced far up the valley above

Amiens, wherever patches of the old alluvium occur. They have all been derived from the Tertiary strata which once covered the Chalk. Their dimensions are such that it is impossible to imagine a river like the present Somme, flowing through a flat country, with a gentle fall towards the sea, to have carried them for miles down its channel unless ice co-operated as a transporting power. Their angularity also favours the supposition of their having been floated by ice, or rendered so buoyant by it as to have escaped much of the wear and tear which blocks propelled along the bottom of a

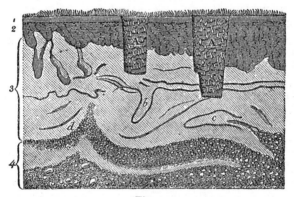

Fig. 22.

Contorted fluviatile strata at St. Acheul (Prestwich, *Phil. Trans.*, 1861, p. 299).

　1 Surface soil.
　2 Brown loam as in Fig. 21, p. 106,—thickness, 6 feet.
　3 White sand with bent and folded layers of marl—thickness, 6 feet.
　4 Gravel, as in Fig. 21, p. 106, with bones of mammalia and flint implements.

　A Graves filled with made ground and human bones.
　b and c Seams of laminated marl often bent round upon themselves.
　d Beds of gravel with sharp curves.

river channel would otherwise suffer. We must remember that the present mildness of the winters in Picardy and the northwest of Europe generally is exceptional in the northern hemisphere, and that large fragments of granite, sandstone, and limestone are now carried annually by ice down the Canadian rivers in latitudes farther south than Paris.[1]

Another sign of ice agency, of which Mr. Prestwich has given

[1] *Principles of Geology*, 9th ed., p. 220.

a good illustration in one of his published sections, and which I myself observed in several pits at St. Acheul, deserves notice. It consists in flexures and contortions of the strata of sand, marl, and gravel (as seen at *b, c,* and *d,* Fig. 22), which they have evidently undergone since their original deposition, and from which both the underlying Chalk and part of the overlying beds of sand No. 3 are usually exempt.

In my former writings I have attributed this kind of derangement to two causes; first, the pressure of ice running aground on yielding banks of mud and sand; and, secondly, the melting of masses of ice and snow of unequal thickness, on which horizontal layers of mud, sand, and other fine and coarse materials had accumulated. The late Mr. Trimmer first pointed out in what manner the unequal failure of support caused by the liquefaction of underlying or intercalated snow and ice might give rise to such complicated foldings.[1]

When "ice-jams" occur on the St. Lawrence and other Canadian rivers (latitude 46° N.), the sheets of ice, which become packed or forced under or over one another, assume in most cases a highly inclined and sometimes even a vertical position. They are often observed to be coated on one side with mud, sand, or gravel frozen on to them, derived from shallows in the river on which they rested when congelation first reached the bottom.

As often as portions of these packs melt near the margin of the river, the layers of mud, sand, and gravel, which result from their liquefaction, cannot fail to assume a very abnormal arrangement—very perplexing to a geologist who should undertake to interpret them without having the ice-clue in his mind.

Mr. Prestwich has suggested that ground-ice may have had its influence in modifying the ancient alluvium of the Somme.[2] It is certain that ice in this form plays an active part every winter in giving motion to stones and gravel in the beds of rivers in European Russia and Siberia. It appears that when in those countries the streams are reduced nearly to the freezing point, congelation begins frequently at the bottom; the reason being, according to Arago, that the current is slowest there, and the gravel and large stones, having parted with much of their heat by radiation, acquire a temperature below the average of the main body of the river. It is, therefore, when the water is

---

[1] *See* chap. xii.
[2] Prestwich, Memoir read to Royal Society, April 1862.

clear, and the sky free from clouds, that ground ice forms most readily, and oftener on pebbly than on muddy bottoms. Fragments of such ice, rising occasionally to the surface, bring up with them gravel, and even large stones.

Without dwelling longer on the various ways in which ice may affect the forms of stratification in drift, so as to cause bendings and foldings in which the underlying or over-lying strata do not participate, a subject to which I shall have occasion again to allude in the sequel, I will state in this place that such contortions, whether explicable or not, are very characteristic of glacial formations. They have also no necessary connection with the transportation of large blocks of stone, and they therefore afford, as Mr. Prestwich remarks, independent proof of ice-action in the Pleistocene gravel of the Somme.

Let us, then, suppose that, at the time when flint hatchets were embedded in great numbers in the ancient gravel which now forms the terrace of St. Acheul, the main river and its tributaries were annually frozen over for several months in winter. In that case, the primitive people may, as Mr. Prestwich hints, have resembled in their mode of life those American Indians who now inhabit the country between Hudson's Bay and the Polar Sea. The habits of those Indians have been well described by Hearne, who spent some years among them. As often as deer and other game become scarce on the land, they betake themselves to fishing in the rivers; and for this purpose, and also to obtain water for drinking, they are in the constant practice of cutting round holes in the ice, a foot or more in diameter, through which they throw baited hooks or nets. Often they pitch their tent on the ice, and then cut such holes through it, using ice-chisels of metal when they can get copper or iron, but when not, employing tools of flint or hornstone.

The great accumulation of gravel at St. Acheul has taken place in part of the valley where the tributary streams, the Noye and the Arve, now join the Somme. These tributaries, as well as the main river, must have been running at the height first of 100 feet, and afterwards at various lower levels above the present valley-plain, in those earlier times when the flint tools of the antique type were buried in successive river beds. I have said at various levels, because there are, here and there, patches of drift at heights intermediate between the higher and lower gravel, and also some deposits, showing that the river once flowed at elevations above as well as below the level of the platform of St. Acheul. As yet, however, no patch of gravel skirting the valley at heights exceeding 100 feet above the Somme has yielded flint tools or other signs of the former sojourn of Man in this region.

Possibly, in the earlier geographical condition of this country, the confluence of tributaries with the Somme afforded inducements to a hunting and fishing tribe to settle there, and some of the same natural advantages may have caused the first inhabitants of Amiens and Abbeville to fix on the same sites for their dwellings. If the early hunting and fishing tribes frequented the same spots for hundreds or thousands of years in succession, the number of the stone implements lost in the bed of the river need not surprise us. Ice-chisels, flint hatchets, and spear-heads may have slipped accidentally through holes kept constantly open, and the recovery of a lost treasure once sunk in the bed of the ice-bound stream, inevitably swept away with gravel on the breaking up of the ice in the spring, would be hopeless. During a long winter, in a country affording abundance of flint, the manufacture of tools would be continually in progress; and, if so, thousands of chips and flakes would be purposely thrown into the ice-hole, besides a great number of implements having flaws, or rejected as too unskilfully made to be worth preserving.

As to the fossil fauna of the drift, considered in relation to the climate, when, in 1859, I took a collection which I had made of all the more common species of land and freshwater shells from the Amiens and Abbeville drift, to my friend M. Deshayes at Paris, he declared them to be, without exception, the same as those now living in the basin of the Seine. This fact may seem at first sight to imply that the climate had not altered since the flint tools were fabricated; but it appears that all these species of molluscs now range as far north as Norway and Finland, and may therefore have flourished in the valley of the Somme when the river was frozen over annually in winter.[1]

In regard to the accompanying mammalia, some of them, like the mammoth and tichorhine rhinoceros, may have been able to endure the rigours of a northern winter as well as the reindeer, which we find fossil in the same gravel. It is a more difficult point to determine whether the climate of the lower gravels (those of Menchecourt, for example) was more genial than that of the higher ones. Mr. Prestwich inclines to this opinion. None of those contortions of the strata above described have as yet been observed in the lower drift. It contains large

---

[1] *See* Prestwich, Paper read to Royal Society in 1862.

blocks of Tertiary sandstone and grit, which may have required the aid of ice to convey them to their present sites; but as such blocks already abounded in the older and higher alluvium, they may simply be monuments of its destruction, having been let down successively to lower and lower levels without making much seaward progress.

The *Cyrena fluminalis* of Menchecourt and the hippopotamus of St. Roch seem to be in favour of a less severe temperature in winter; but so many of the species of mammalia, as well as of the land and freshwater shells, are common to both formations, and our information respecting the entire fauna is still so imperfect, that it would be premature to pretend to settle this question in the present state of our knowledge. We must be content with the conclusion (and it is one of no small interest), that when Man first inhabited this part of Europe, at the time that the St. Acheul drift was formed, the climate as well as the physical geography of the country differed considerably from the state of things now established there.

Among the elephant remains from St. Acheul, in M. Garnier's collection, Dr. Falconer recognised a molar of the *Elephas antiquus*, Fig. 19, the same species which has been already mentioned as having been found in the lower-level gravels of St. Roch. This species, therefore, endured while important changes took place in the geographical condition of the valley of the Somme. Assuming the lower-level gravel to be the newer, it follows that the *Elephas antiquus* and the hippopotamus of St. Roch continued to flourish long after the introduction of the mammoth, a well characterised tooth of which, as I before stated, was found at St. Acheul at the time of my visit in 1860.

As flint hatchets and knives have been discovered in the alluvial deposits both at high and low levels, we may safely affirm that Man was as old an inhabitant of this region as were any of the fossil quadrupeds above enumerated, a conclusion which is independent of any difference of opinion as to the relative age of the higher and lower gravels.

The disappearance of many large pachyderms and beasts of prey from Europe has often been attributed to the intervention of Man, and no doubt he played his part in hastening the era of their extinction; but there is good reason for suspecting that other causes co-operated to the same end. No naturalist would for a moment suppose that the extermination of the *Cyrena fluminalis* throughout the whole of Europe—a species which co-existed with our race in the valley of the Somme, and which was very abundant in the waters of the Thames at the time when the elephant, rhinoceros, and hippopotamus flourished on its banks—was accelerated by human agency. The great modification in climate and in other conditions of existence which affected this aquatic mollusc, may have mainly contributed to the gradual dying out of many of the large mammalia.

We have already seen that the peat of the valley of the Somme is a formation which, in all likelihood, took thousands of years for its growth. But no change of a marked character has occurred in the mammalian fauna since it began to accumulate. The contrast of the fauna of the ancient alluvium, whether at high or low levels, with the fauna of the oldest peat is almost as great as its contrast with the existing fauna, the memorials of Man being common to the whole series; hence we may infer that the interval of time which separated the era of the large extinct mammalia from that of the earliest peat, was of far longer duration than that of the entire growth of the peat. Yet we by no means need the evidence of the ancient fossil fauna to establish the antiquity of Man in this part of France. The mere volume of the drift at various heights would alone suffice to demonstrate a vast lapse of time during which such heaps of shingle, derived both from the Eocene and the Cretaceous rocks, were thrown down in a succession of river-channels. We observe thousands of rounded and half-rounded flints, and a vast number of angular ones, with rounded pieces of white Chalk of various sizes, testifying to a prodigious amount of mechanical action, accompanying the repeated widening and deepening of the valley, before it became the receptacle of peat; and the position of many of the flint tools leaves no doubt in the mind of the geologist that their fabrication preceded all this reiterated denudation.

On the Absence of Human Bones in the Alluvium of the Somme

It is naturally a matter of no small surprise that, after we have collected many hundred flint implements (including knives, many thousands), not a single human bone has yet been met with in the old alluvial sand and gravel of the Somme. This dearth of the mortal remains of our species holds true equally, as yet, in all other parts of Europe where the tool-bearing drift of the Pleistocene period has been investigated in valley deposits. Yet in these same formations there is no want of bones of mammalia belonging to extinct and living species. In the course of the last quarter of a century, thousands of them have been submitted to the examination of skilful osteologists, and they have been unable to detect among them one fragment of a human skeleton, not even a tooth. Yet Cuvier pointed out long ago, that the bones of Man found buried in ancient battle-fields were not more decayed than those of horses interred in the same graves. We have seen that in the Liège caverns, the skulls, jaws, and teeth, with other bones of the human race, were preserved in the same condition as those of the cave-bear, tiger, and mammoth.

That ere long, now that curiosity has been so much excited on this subject, some human remains will be detected in the older alluvium of European valleys, I confidently expect. In the meantime, the absence of all vestige of the bones which belonged to that population by which so many weapons were designed and executed, affords a most striking and instructive lesson in regard to the value of negative evidence, when adduced in proof of the non-existence of certain classes of terrestrial animals at given periods of the past. It is a new and emphatic illustration of the extreme imperfection of the geological record, of which even they who are constantly working in the field cannot easily form a just conception.

We must not forget that Dr. Schmerling, after finding extinct mammalia and *flint tools* in forty-two Belgian caverns, was only rewarded by the discovery of human bones in three or four of those rich repositories of osseous remains. In like manner, it was not till the year 1855 that the first skull of the musk ox (*Bubalus moschatus*) was detected in the fossiliferous gravel of the Thames, and not till 1860, as will be seen in the next chapter, that the same quadruped was proved to have co-existed in France with the mammoth. The same theory which will explain the comparative rarity of such species would no doubt account for the still greater scarcity of human bones, as well as for our general ignorance of the Pleistocene terrestrial fauna, with the exception of that part of it which is revealed to us by cavern researches.

In valley drift we meet commonly with the bones of quadrupeds which graze on plains bordering rivers. Carnivorous beasts, attracted to the same ground in search of their prey, sometimes leave their remains in the same deposits, but more rarely. The whole assemblage of fossil quadrupeds at present obtained from the alluvium of Picardy is obviously a mere fraction of the entire fauna which flourished contemporaneously with the primitive people by whom the flint hatchets were made.

Instead of its being part of the plan of nature to store up enduring records of a large number of the individual plants and animals which have lived on the surface, it seems to be her chief care to provide the means of disencumbering the habitable areas lying above and below the waters of those myriads of solid skeletons of animals, and those massive trunks of trees, which would otherwise soon choke up every river, and fill every valley. To prevent this inconvenience she employs the heat and moisture of the sun and atmosphere, the dissolving power of carbonic and other acids, the grinding teeth and gastric juices of quadrupeds, birds, reptiles, and fish, and the agency of many of the invertebrata. We are all familiar with the efficacy of these and other causes on the land; and as to the bottoms of seas, we have only to read the published reports of Mr. MacAndrew, the late Edward Forbes, and other experienced dredgers, who, while they failed utterly in drawing up from the deep a single human bone, declared that they scarcely ever met with a work of art even after counting tens of thousands of shells and zoophytes, collected on a coast line of several hundred miles in extent, where they often approached within less than half a mile of a land peopled by millions of human beings.

Lake of Haarlem

It is not many years since the Government of Holland re-solved to lay dry that great sheet of water formerly called the Lake of Haarlem, extending over 45,000 acres. They suc-ceeded, in 1853, in turning it into dry land, by means of powerful pumps constantly worked by steam, which raised the water and discharged it into a canal running for 20 or 30 miles round the newly-gained land. This land was depressed 13 feet beneath the mean level of the ocean. I travelled, in 1859, over part of the bed of this old lake, and found it already converted into arable land, and peopled by an agricultural population of 5000 souls. Mr. Staring, who had been for some years employed by the Dutch Government in constructing a geological map of Holland, was my companion and guide. He informed me that he and his associates had searched in vain for human bones in the deposits which had constituted for three centuries the bed of the great lake.

There had been many a shipwreck, and many a naval fight in those waters, and hundreds of Dutch and Spanish soldiers and sailors had met there with a watery grave. The population which lived on the borders of this ancient sheet of water numbered between thirty and forty thousand souls. In digging the great canal, a fine section had been laid open, about 30 miles long, of the deposits which formed the ancient bottom of the lake. Trenches, also, innumerable, several feet deep, had been freshly dug on all the farms, and their united length must have amounted to thousands of miles. In some of the sandy soil recently thrown out of the trenches, I observed specimens of freshwater and brackish-water shells, such as *Unio* and *Dreissena*, of living species; and in clay brought up from below the sand, shells of *Tellina, Lutraria,* and *Cardium,* all of species now inhabiting the adjoining sea.

As the *Dreissena* is believed by conchologists to have been introduced into Western Europe in very modern times, brought with foreign timber in the holds of vessels from the rivers flowing into the Black Sea, the layer of sand containing it in the Haarlem lake is probably not more than a hundred years old.

One or two wrecked Spanish vessels, and arms of the same period, have rewarded the antiquaries who had been watching the draining operations in the hope of a richer harvest, and who were not a little disappointed at the result. In a peaty tract on the margin of one part of the lake a few coins were dug up; but if history had been silent, and if there had been a controversy whether Man was already a denizen of this planet at the time when the area of the Haarlem lake was under water, the archæologist, in order to answer this question, must have appealed, as in the case of the valley of the Somme, not to fossil bones, but to works of art embedded in the superficial strata.

Mr. Staring, in his valuable memoir on the *Geological Map of Holland,* has attributed the general scarcity of human bones in Dutch peat, notwithstanding the many works of art preserved in it, to the power of the humic and sulphuric acids to dissolve bones, the peat in question being plentifully impregnated with such acids. His theory may be correct, but it is not applicable to the gravel of the valley of the Somme, in which the bones of fossil mammalia are frequent, nor to the uppermost freshwater strata forming the bottom of a large part of the Haarlem Lake, in which it is not pretended that such acids occur.

The primitive inhabitants of the valley of the Somme may have been too wary and sagacious to be often surprised and drowned by floods, which swept away many an incautious elephant or rhinoceros, horse and ox. But even if those rude hunters had cherished a superstitious veneration for the Somme, and had regarded it as a sacred river (as the modern Hindoos revere the Ganges), and had been in the habit of committing the bodies of their dead or dying to its waters—even had such funeral rites prevailed, it by no means follows that the bones of many individuals would have been preserved to our time.

A corpse cast into the stream first sinks, and must then be almost immediately overspread with sediment of a certain weight, or it will rise again when distended with gases, and float perhaps to the sea before it sinks again. It may then be attacked by fish of marine species, some of which are capable of digesting bones. If, before being carried into the sea and devoured, it is enveloped with fluviatile mud and sand, the next flood, if it lie in mid-channel, may tear it out again, scatter all the bones, roll some of them into pebbles, and leave others exposed to destroying agencies; and this may be repeated annually, till all vestiges of the skeleton may disappear. On the other hand, a bone washed through a rent into a subterranean cavity, even though a rarer contingency, may have a greater chance of escaping destruction, especially if there be stalactite dropping from the roof of the cave or walls of a rent, and if the cave be not constantly traversed by too strong a current of engulfed water.

CHAPTER IX

Flint Implements in Pleistocene Alluvium in the Basin of the Seine.

IN the ancient alluvium of the valleys of the Seine and its
principal tributaries, the same assemblage of fossil animals,
which has been alluded to in the last chapter as characterising
the gravel of Picardy, has long been known; but it was not
till the year 1860, and when diligent search had been expressly
made for them, that flint implements of the Amiens type were
discovered in this part of France.

In the neighbourhood of Paris deposits of drift occur answering both to those of the higher and lower levels of the basin
of the Somme before described.[1] In both are found, mingled
with the wreck of the Tertiary and Cretaceous rocks of the
vicinity, a large quantity of granitic sand and pebbles, and
occasionally large blocks of granite, from a few inches to a foot
or more in diameter. These blocks are peculiarly abundant in
the lower drift commonly called the " diluvium gris." The
granitic materials are traceable to a chain of hills called the
Morvan, where the head waters of the Yonne take their rise,
150 miles to the S.S.E. of Paris.

It was in this lowest gravel that M. H. T. Gosse, of Geneva,
found, in April 1860, in the suburbs of Paris, at La Motte
Piquet, on the left bank of the Seine, one or two well-formed
flint implements of the Amiens type, accompanied by a great
number of ruder tools or attempts at tools. I visited the spot

[1] Prestwich, *Pro. Roy. Soc.* 1862.

in 1861 with M. Hébert, and saw the stratum from which the worked flints had been extracted, 20 feet below the surface, and near the bottom of the "grey diluvium," a bed of gravel from which I have myself, in and near Paris, frequently collected the bones of the elephant, horse, and other mammalia.

More recently, M. Lartet has discovered at Clichy, in the environs of Paris, in the same lower gravel, a well-shaped flint implement of the Amiens type, together with remains both of *Elephas primigenius* and *E. antiquus*. No tools have yet been met with in any of the gravels occurring at the higher levels of the valley of the Seine; but no importance can be attached to this negative fact, as so little search has yet been made for them.

Mr. Prestwich has observed contortions indicative of ice-action, of the same kind as those near Amiens, in the higher-level drift of Charonne, near Paris; but as yet no similar derangement has been seen in the lower gravels—a fact, so far as it goes, in unison with the phenomena observed in Picardy.

In the cavern of Arcy-sur-Yonne a series of deposits have lately been investigated by the Marquis de Vibraye, who discovered human bones in the lowest of them, mixed with remains of quadrupeds of extinct and recent species. This cavern occurs in Jurassic limestone, at a slight elevation above the Cure, a small tributary of the Yonne, which last joins the Seine near Fontainebleau about 40 miles south of Paris. The lowest formation in the cavern resembles the "diluvium gris" of Paris, being composed of granitic materials, and like it derived chiefly from the waste of the crystalline rocks of the Morvan. In it have been found the two branches of a human lower jaw with teeth well-preserved, and the bones of the *Elephas primigenius*, *Rhinoceros tichorhinus*, *Ursus spelæus*, *Hyæna spelæa*, and *Cervus tarandus*, all specifically determined by M. Lartet. I have been shown this collection of fossils by M. de Vibraye, and remarked that the human and other remains were in the same condition and of the same colour.

Above the grey gravel is a bed of red alluvium, made up of fragments of Jurassic limestone, in a red argillaceous matrix, in which were embedded several flint knives, with bones of the reindeer and horse, but no extinct mammalia. Over this, in a higher bed of alluvium, were several polished hatchets of the more modern type called "celts," and above all loam or cave-mud, in which were Gallo-Roman antiquities.[1]

The French geologists have made as yet too little progress

[1] *Bulletin de la Société Géologique de France*, 1860.

in identifying the age of the successive deposits of ancient alluvium of various parts of the basin of the Seine, to enable us to speculate with confidence as to the coincidence in date of the granitic gravel with human bones of the Grotte d'Arcy and the stone-hatchets buried in " grey diluvium " of La Motte Piquet, before mentioned; but as the associated extinct mammalia are of the same species in both localities, I feel strongly inclined to believe that the stone hatchets found by M. Gosse at Paris, and the human bones discovered by M. de Vibraye, may be referable to the same period.

Valley of the Oise

A flint hatchet, of the old Abbeville and Amiens type, was found lately by M. Peigné Delacourt at Précy, near Creil, on the Oise, in gravel, resembling, in its geological position, the lower-level gravels of Montiers, near Amiens, already described. I visited these extensive gravel-pits in 1861, in company with Mr. Prestwich; but we remained there too short a time to entitle us to expect to find a flint implement, even if they had been as abundant as at St. Acheul.

In 1859, I examined, in a higher part of the same valley of the Oise, near Chauny and Noyon, some fine railway cuttings, which passed continuously through alluvium of the Pleistocene period for half a mile. All this alluvium was evidently of fluviatile origin, for, in the interstices between the pebbles, the *Ancylus fluviatilis* and other freshwater shells were abundant. My companion, the Abbé E. Lambert, had collected from the gravel a great many fossil bones, among which M. Lartet has recognised both *Elephas primigenius* and *E. antiquus*, besides a species of hippopotamus (*H. major ?*), also the reindeer, horse, and the musk ox (*Bubalus moschatus*). The latter seems never to have been seen before in the old alluvium of France.[1] Over the gravel above mentioned, near Chauny, are seen dense masses of loam like the loess of the Rhine, containing shells of the genera *Helix* and *Succinea*. We may suppose that the gravel containing the flint hatchet at Précy is of the same age as that of Chauny, with which it is continuous, and that both of them are coeval with the tool-bearing beds of Amiens, for the basins of the Oise and the Somme are only separated by a narrow water-shed, and the same fossil quadrupeds occur in both.

The alluvium of the Seine and its tributaries, like that of

[1] Lartet, *Annales des Sciences Naturelles Zoologiques*, tom. xv., p. 224.

the Somme, contains no fragments of rocks brought from any other hydrographical basin; yet the shape of the land, or fall of the river, or the climate, or all these conditions, must have been very different when the grey alluvium in which the flint tools occur at Paris was formed. The great size of some of the blocks of granite, and the distance which they have travelled, imply a power in the river which it no longer possesses. We can hardly doubt that river-ice once played a much more active part than now in the transportation of such blocks, one of which may be seen in the Museum of the Ecole des Mines at Paris, 3 or 4 feet in diameter.

Pleistocene Alluvium of England, containing Works of Art

In the ancient alluvium of the basin of the Thames, at moderate heights above the main river and its tributaries, we find fossil bones of the same species of extinct and living mammalia, accompanied by recent species of land and freshwater shells, as we have shown to be characteristic of the basins of the Somme and the Seine. We can scarcely therefore doubt that these quadrupeds, during some part of the Pleistocene period, ranged freely from the continent of Europe to England, at a time when there was an uninterrupted communication by land between the two countries. The reader will not therefore be surprised to learn that flint implements of the same antique type as those of the valley of the Somme have been detected in British alluvium.

The most marked feature of this alluvium in the Thames valley is that great bed of ochreous gravel, composed chiefly of broken and slightly worn Chalk flints, on which a great part of London is built. It extends from above Maidenhead through the metropolis to the sea, a distance from west to east of 50 miles, having a width varying from 2 to 9 miles. Its thickness ranges commonly from 5 to 15 feet.[1] Interstratified with this gravel, in many places, are beds of sand, loam, and clay, the whole containing occasionally remains of the mammoth and other extinct quadrupeds. Fine sections have been exposed to view, at different periods, at Brentford and Kew Bridge, others in London itself, and below it at Erith in Kent, on the right bank of the Thames, and at Ilford and Gray's Thurrock in Essex, on the left bank. The united thickness of the beds of sand,

---

[1] Prestwich, *Quart. Jour. Geol. Soc.*, vol. xii., 1856, p. 131.

gravel, and loam amounts sometimes to 40 or even 60 feet. They are for the most part elevated above, but in some cases they descend below, the present level of the overflowed plain of the Thames.

If the reader will refer to the section of the Pleistocene sands and gravels of Menchecourt, near Abbeville, given at p. 96, he will perfectly understand the relations of the ancient Thames alluvium to the modern channel and plain of the river, and their relation, on the other hand, to the boundary formations of older date, whether Tertiary or Cretaceous.

So far as they are known, the fossil mollusca and mammalia of the two districts also agree very closely, the *Cyrena fluminalis* being common to both, and being the only extra-European shell, this and all the species of testacea being Recent. Of this agreement with the living fauna there is a fine illustration in Essex; for the determination of which we are indebted to the late Mr. John Brown, F.G.S., who collected at Copford, in Essex, from a deposit containing bones of the mammoth, a large bear (probably *Ursus spelæus*), a beaver, stag, and aurochs, no less than sixty-nine species of land and freshwater shells. Forty-eight of these were terrestrial, and two of them, *Helix incarnata* and *H. ruderata*, no longer inhabit the British Isles, but are still living on the continent, *ruderata* in high northern latitudes.[1] The *Cyrena fluminalis* and the *Unio littoralis*, to which last I shall presently allude, were not among the number.

I long ago suggested the hypothesis, that in the basin of the Thames there are indications of a meeting in the Pleistocene period of a northern and southern fauna. To the northern group may have belonged the mammoth (*Elephas primigenius*) and the *Rhinoceros tichorhinus*, both of which Pallas found in Siberia, preserved with their flesh in the ice. With these are occasionally associated the reindeer. In 1855 the skull of the musk ox (*Bubalus moschatus*) was also found in the ochreous gravel of Maidenhead, by the Rev. C. Kingsley and Mr. Lubbock; the identification of this fossil with the living species being made by Professor Owen. A second fossil skull of the same arctic animal was afterwards found by Mr. Lubbock near Bromley, in the valley of a small tributary of the Thames; and two other skulls, those of a bull and a cow were dug up near Bath Easton

[1] *Quart. Jour. Geol. Soc.*, vol. viii., 1852, p. 190. Mr. Brown calls them extinct species, which may mislead some readers, but he merely meant extinct in England. *See also* Jeffreys, *Brit. Conch.*, p. 174.

from the gravel of the valley of the Avon by Mr. Charles Moore. Professor Owen has truly said, that " as this quadruped has a constitution fitting it at present to inhabit the high northern regions of America, we can hardly doubt that its former companions, the warmly-clad mammoth and the two-horned woolly rhinoceros (*R. tichorhinus*), were in like manner capable of supporting life in a cold climate." [1]

I have already alluded to the recent discovery of this same ox near Chauny, in the valley of the Oise, in France; and in 1856 I found a skull of it preserved in the museum at Berlin, which Professor Quenstedt, the curator, had correctly named so long ago as 1836, when the fossil was dug out of drift, in the hill called the Kreuzberg, in the southern suburbs of that city. By an account published at the time, we find that the mammalia which accompanied the musk ox were the mammoth and tichorhine rhinoceros, with the horse and ox; [2] but I can find no record of the occurrence of a hippopotamus, nor of *Elephas antiquus* or *Rhinoceros leptorhinus*, in the drift of the north of Germany, bordering the Baltic.

On the other hand, in another locality in the same drift of North Germany, Dr. Hensel, of Berlin, detected, near Quedlinburg, the Norwegian Lemming (*Myodes lemmus*), and another species of the same family called by Pallas *Myodes torquatus* (by Hensel, *Misothermus torquatus*)—a still more arctic quadruped, found by Parry in latitude 82°, and which never strays farther south than the northern borders of the woody region. Professor Beyrich also informs me that the remains of the *Rhinoceros tichorhinus* were obtained at the same place. [3]

As an example of what may possibly have constituted a more southern fauna in the valley of the Thames, I may allude to the fossil remains found in the fluviatile alluvium of Gray's Thurrock, in Essex, situated on the left bank of the river, 21 miles below London. The strata of brick-earth, loam, and gravel exposed to view in artificial excavations in that spot, are precisely such as would be formed by the silting up of an old river channel. Among the mammalia are *Elephas antiquus*, *Rhinoceros leptorhinus* (*R. megarhinus*, Christol), *Hippopotamus major*, species of horse, bear, ox, stag, etc., and, among the accompanying shells, *Cyrena fluminalis*, which is extremely abundant, instead of being scarce, as at Abbeville. It is

[1] *Quart. Jour. Geol. Soc.*, vol. xii., 1856, p. 124.
[2] *Leonhard and Bronn's Jahrbuch*, 1836, p. 215.
[3] *Zeitschrift der Deutschen Geologischen Gesellschaft*, vol. vii., 1855, p. 497, etc.

associated with *Unio littoralis* also in great numbers and with both valves united. This conspicuous freshwater mussel is no longer an inhabitant of the British Isles, but still lives in the Seine, and is still more abundant in the Loire. Another fresh-water univalve (*Paludina marginata*, Michaud), not British, but common in the south of France, likewise occurs, and a peculiar variety of *Cyclas amnica*, which by some naturalists has been regarded as a distinct species. With these, moreover, is found a peculiar variety of *Valvata piscinalis*.

If we consult Dr. Von Schrenck's account of the living mammalia of Mongolia, lying between latitude 45° and 55° North, we learn that, in that part of North-Eastern Asia recently annexed to the Russian empire, no less than thirty-four out of fifty-eight living quadrupeds are identical with European species, while some of those which do not extend their range to Europe are arctic, others tropical forms. The Bengal tiger ranges northwards occasionally to latitude 52° North, where he chiefly subsists on the flesh of the reindeer, and the same tiger abounds in latitude 48°, to which the small tailless hare or pika, a polar resident, sometimes wanders southwards.[1] We may readily conceive that the countries now drained by the Thames, the Somme, and the Seine, were, in the Pleistocene period, on the borders of two distinct zoological provinces, one lying to the north, the other to the south, in which case many species belonging to each fauna endowed with migratory habits, like the living musk-ox or the Bengal tiger, may have been ready to take advantage of any, even the slightest, change in their favour to invade the neighbouring province, whether in the summer or winter months, or permanently for a series of years, or centuries. The *Elephas antiquus* and its associated *Rhinoceros leptorhinus* may have preceded the mammoth and tichorhine rhinoceros in the valley of the Thames, or both may have alternately prevailed in the same area in the Pleistocene period.

In attempting to settle the chronology of fluviatile deposits, it is almost equally difficult to avail ourselves of the evidence of organic remains and of the superposition of the strata, for we may find two old river-beds on the same level in juxta-position, one of them perhaps many thousands of years posterior in date to the other. I have seen an example of this at Ilford, where the Thames, or a tributary stream, has at some former

[1] Mammalia of Amoorland, *Natural History Review*, vol. i., 1861, p. 12.

period cut through sands containing *Cyrena fluminalis*, and again filled up the channel with argillaceous matter, evidently derived from the waste of the Tertiary London Clay. Such shiftings of the site of the main channel of the river, the frequent removal of gravel and sand previously deposited, and the throwing down of new alluvium, the flooding of tributaries, the rising and sinking of the land, fluctuations in the cold and heat of the climate—all these changes seem to have given rise to that complexity in the fluviatile deposits of the Thames, which accounts for the small progress we have hitherto made in determining their order of succession, and that of the imbedded groups of quadrupeds. It may happen, as at Brentford and Ilford, that sand-pits in two adjoining fields may each contain distinct species of elephant and rhinoceros; and the fossil remains in both cases may occur at the same depth from the surface, yet may be severally referable to different parts of the Pleistocene epoch, separated by thousands of years.

The relation of the glacial period to alluvial deposits, such as that of Gray's Thurrock, where the *Cyrena fluminalis*, *Unio littoralis*, and the hippopotamus seem rather to imply a warmer climate, has been a matter of long and animated discussion. Patches of the northern drift, at elevations of about 200 feet above the Thames, occur in the neighbourhood of London, as at Muswell Hill, near Highgate. In this drift, blocks of granite, syenite, greenstone, Coal-measure sandstone with its fossils, and other Palæozoic rocks, and the wreck of Chalk and Oolite, occur confusedly mixed together. The same glacial formation is also found capping some of the Essex hills farther to the east, and extending some way down their southern slopes towards the valley of the Thames. Although no fragments washed out of these older and upland drifts have been found in the gravel of the Thames containing elephants' bones, it is fair to presume, as Mr. Prestwich has contended,[1] that the glacial formation is the older of the two. In short, we must suppose that the basin of the Thames and all its fluviatile deposits are post-glacial, in the modified sense of that term; *i.e.* that they were subsequent to the drift of the central and northern counties.

Having offered these general remarks on the alluvium of the Thames, I may now say something of the implements hitherto discovered in it. In the British Museum there is a flint weapon

---

[1] Prestwich, *Quart. Jour. Geol. Soc.*, vol. xi., 1855, p. 110; *ibid.* vol. xii. 1856, p. 133; *ibid.* vol. xvii., 1861, p. 446.

of the spear-headed form, such as is represented in Fig. 8, which we are told was found with an elephant's tooth at Black Mary's, near Gray's Inn Lane, London.   In a letter dated 1715, printed in Herne's edition of *Leland's Collectanea*, vol. i., p. 73, it is stated to have been found in the presence of Mr. Conyers, with the skeleton of an elephant.[1]   So many bones of the elephant, rhinoceros, and hippopotamus have been found in the gravel on which London stands, that there is no reason to doubt the statement as handed down to us.   Fossil remains of all these three genera have been dug up on the site of Waterloo Place, St. James's Square, Charing Cross, the London Docks, Limehouse, Bethnal Green, and other places within the memory of persons now living.   In the gravel and sand of Shacklewell, in the north-east district of London, I have myself collected specimens of the *Cyrena fluminalis* in great numbers (*see* Fig. 17 c), with the bones of deer and other mammalia.

In the alluvium also of the Wey, near Guildford, in a place called Pease Marsh, a wedge-shaped flint implement, resembling one brought from St. Acheul by Mr. Prestwich, and compared by some antiquaries to a sling-stone, was obtained in 1836 by Mr. Whitburn, 4 feet deep in sand and gravel, in which the teeth and tusks of elephants had been found.   The Wey flows through the gorge of the North Downs at Guildford to join the Thames.   Mr. Austen has shown that this drift is so ancient that one part of it had been disturbed and tilted before another part was thrown down.[2]

Among other places where flint tools of the antique type have been met with in the course of the last three years, I may mention one of an oval form found by Mr. Whitaker in the valley of the Darent, in Kent, and another which Mr. Evans found lying on the shore at Swalecliff, near Whitstable, in the same county, where Mr. Prestwich had previously described a freshwater deposit, resting on the London Clay, and consisting chiefly of gravel, in which an elephant's tooth and the bones of a bear were embedded.   The flint implement was deeply discoloured and of a peculiar bright light-brown colour, similar to that of the old fluviatile gravel in the cliff.

Another flint implement was found in 1860 by Mr. T. Leech, at the foot of the cliff between Herne Bay and the Reculvers, and on further search five other specimens of the spear-head pattern so common at Amiens.   Messrs. Prestwich and Evans

[1] Evans, *Archæologia*, 1860.
[2] *Quart. Jour. Geol. Soc.*, vol. vii., 1851, p. 278.

have since found three other similar tools on the beach, at the base of the same wasting cliff, which consists of sandy Eocene strata, covered by a gravelly deposit of freshwater origin, about 50 feet above the sea-level, from which the flint weapons must have been derived. Such old alluvial deposits now capping the cliffs of Kent seem to have been the river-beds of tributaries of the Thames before the sea encroached to its present position and widened its estuary. On following up one of these freshwater deposits westward of the Reculvers, Mr. Prestwich found in it, at Chislet, near Grove Ferry, the *Cyrena fluminalis* among other shells.

The changes which have taken place in the physical geography of this part of England during, or since, the Pleistocene period, have consisted partly of such encroachments of the sea on the coast as are now going on, and partly of a general subsidence of the land. Among the signs of the latter movement may be mentioned a freshwater formation at Faversham, below the level of the sea. The gravel there contains exclusively land and fluviatile shells of the same species as those of other localities of the Pleistocene alluvium before mentioned, and must have been formed when the river was at a higher level and when it extended farther east. At that era it was probably a tributary of the Rhine, as represented by Mr. Trimmer in his ideal restoration of the geography of the olden time.[1] For England was then united to the continent, and what is now the North Sea was land. It is well known that in many places, especially near the coast of Holland, elephants' tusks and other bones are often dredged up from the bed of that shallow sea, and the reader will see in the map given in Chap. XIII. how vast would be the conversion of sea into land by an upheaval of 600 feet. Vertical movements of much less than half that amount would account for the annexation of England to the continent, and the extension of the Thames and its valley far to the north-east, and the flowing of rivers from the easternmost parts of Kent and Essex into the Thames, instead of emptying themselves into its estuary.

More than a dozen flint weapons of the Amiens type have already been found in the basin of the Thames; but the geological position of no one of them has as yet been ascertained with the same accuracy as that of many of the tools dug up in the valley of the Somme.

[1] *Quart. Jour. Geol. Soc.*, vol. ix., 1853, Pl. XIII. No. 4.

Flint Implements of the Valley of the Ouse, near Bedford

The ancient fluviatile gravel of the valley of the Ouse, around Bedford, has been noted for the last thirty years for yielding to collectors a rich harvest of the bones of extinct mammalia. By observations made in 1854 and 1858, Mr. Prestwich had ascertained that the valley was bounded on both sides by Oolitic strata, capped by boulder clay, and that the gravel No. 3, Fig. 23, contained bones of the elephant, rhinoceros, hippopotamus, ox, horse, and deer, which animals he therefore inferred must have been posterior in date to the boulder clay, through which, as well as the subjacent Oolite, the valley had been excavated. Mr. Evans had found in the same gravel many land and freshwater shells, and these discoveries induced Mr. James Wyatt, of Bedford, to pay two visits to St. Acheul in order to compare the implement-bearing gravels of the Somme with the drift of the valley of the Ouse. After his return he resolved to watch carefully the excavation of the gravel-pits at Biddenham, 2 miles W.N.W. of Bedford, in the hope of finding there similar works of art. With this view he paid almost daily visits for months in succession to those pits, and was at last rewarded by the discovery of two well-formed implements, one of the spear-head and the other of the oval shape, perfect counterparts of the two prevailing French types. Both specimens were thrown out by the workmen on the same day from the lowest bed of stratified gravel and sand, 13 feet thick, containing bones of the elephant, deer, and ox, and many freshwater shells. The two implements occurred at the depth of 13 feet from the surface of the soil, and rested immediately on solid beds of Oolitic limestone, as represented in the accompanying section.

Having been invited by Mr. Wyatt to verify these facts, I went to Biddenham within a fortnight of the date of his discovery (April 1861), and, for the first time, saw evidence which satisfied me of the chronological relations of those three phenomena, the antique tools, the extinct mammalia, and the glacial formation. On that occasion I examined the pits in company with Messrs. Prestwich, Evans, and Wyatt, and we collected ten species of shells from the stratified drift No. 3, or the beds overlying the lowest gravel from which the flint implements had been exhumed. They were all of common fluviatile and land species now living in the same part of

England. Since our visit, Mr. Wyatt has added to them
Paludina marginata, Michaud (*Hydrobia* of some authors), a
species of the South of France no longer inhabiting the British
Isles. The same geologist has also found, since we were at
Biddenham, several other flint tools of corresponding type, both
there and at other localities in the valley of the Ouse, near
Bedford.

The boulder clay No. 2 extends for miles in all directions,
and was evidently once continuous from *b* to *c* before the
valley was scooped out. It is a portion of the great marine
glacial drift of the midland counties of England, and contains
blocks, some of large size, not only of the Oolite of the neighbourhood, but of Chalk and other rocks transported from still

Fig. 23.

Section across the Valley of the Ouse, two miles W.N.W. of Bedford.[1]

1 Oolitic strata.
2 Boulder clay, or marine northern drift, rising to about ninety feet
   above the Ouse.
3 Ancient gravel, with elephant bones, freshwater shells, and flint implements.
4 Modern alluvium of the Ouse.

a Biddenham gravel pits, at the bottom of which flint tools were
   found.

greater distances, such as syenite, basalt, quartz, and New Red
Sandstone. These erratic blocks of foreign origin are often
polished and striated, having undergone what is called glaciation, of which more will be said by and by. Blocks of the
same mineral character, embedded at Biddenham in the gravel
No. 3, have lost all signs of this striation by the friction to
which they were subjected in the old river bed.

The great width of the valley of the Ouse, which is sometimes 2 miles, has not been expressed in the diagram. It may
have been shaped out by the joint action of the river and the
tides when this part of England was emerging from the waters
of the glacial sea, the boulder clay being first cut through, and
then an equal thickness of underlying Oolite. After this denuda-

[1] Prestwich, *Quart. Jour. Geol. Soc.*, vol. xvii., 1861, p. 364; and Wyatt·
Geologist, 1861, p. 242.

tion, which may have accompanied the emergence of the land, the country was inhabited by the primitive people who fashioned the flint tools. The old river, aided perhaps by the continued upheaval of the whole country, or by oscillations in its level, went on widening and deepening the valley, often shifting its channel, until at length a broad area was covered by a succession of the earliest and latest deposits, which may have corresponded in age to the higher and lower gravels of the valley of the Somme, already described.

At Biddenham, and elsewhere in the same gravel, remains of *Elephas antiquus* have been discovered, and Mr. Wyatt obtained, Jan. 1863, a flint implement associated with bones and teeth of hippopotamus from gravel at Summerhouse hill, which lies east of Bedford, lower down the valley of the Ouse, and 4 miles from Biddenham.

One step at least we gain by the Bedford sections, which those of Amiens and Abbeville had not enabled us to make. They teach us that the fabricators of the antique tools, and the extinct mammalia coeval with them, were all post-glacial.

Flint Implements in a Freshwater Deposit at Hoxne in Suffolk [Note 17]

So early as the first year of the nineteenth century, a remarkable paper was communicated to the Society of Antiquaries by Mr. John Frere, in which he gave a clear description of the discovery at Hoxne, near Diss, in Suffolk, of flint tools of the type since found at Amiens, adding at the same time good geological reasons for presuming that their antiquity was very great, or, as he expressed it, beyond that of the present world, meaning the actual state of the physical geography of that region. "The flints," he said, "were evidently weapons of war, fabricated and used by a people who had not the use of metals. They lay in great numbers at the depth of about 12 feet in a stratified soil which was dug into for the purpose of raising clay for bricks. Under a foot and a half of vegetable earth was clay 7½ feet thick, and beneath this one foot of sand with shells, and under this 2 feet of gravel, in which the shaped flints were found generally at the rate of 5 or 6 in a square yard. In the sandy beds with shells were found the jawbone and teeth of an enormous unknown animal. The manner in which the flint weapons lay would lead to the persuasion that it was a place of their manufacture, and not of their accidental deposit.

Their numbers were so great that the man who carried on the brick-work told me that before he was aware of their being objects of curiosity, he had emptied baskets full of them into the ruts of the adjoining road."

Mr. Frere then goes on to explain that the strata in which the flints occur are disposed horizontally, and do not lie at the foot of any higher ground, so that portions of them must have been removed when the adjoining valley was hollowed out. If the author had not mistaken the freshwater shells associated with the tools for marine species, there would have been nothing to correct in his account of the geology of the district, for he distinctly perceived that the strata in which the implements were embedded had, since that time, undergone very extensive denudation.[1]   Specimens of the flint spear-heads, sent to London by Mr. Frere, are still preserved in the British Museum, and others are in the collection of the Society of Antiquaries.

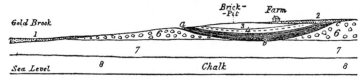

Fig. 24.

Section showing the position of the flint weapons at Hoxne, near Diss, Suffolk. *See* Prestwich, *Philosophical Transactions*, Pl. 11, 1860.

1 Gravel of Gold Brook, a tributary of the Waveney.
2 Higher-level gravel overlying the freshwater deposit.
3 and 4 Sand and gravel, with freshwater shells, and flint implements, and bones of mammalia.
5 Peaty and clayey beds, with same fossils.
6 Boulder clay or glacial drift.
7 Sand and gravel below boulder clay.
8 Chalk with flints.

Mr. Prestwich's attention was called by Mr. Evans to these weapons, as well as to Mr. Frere's memoir after his return from Amiens in 1859, and he lost no time in visiting Hoxne, a village five miles eastward of Diss.   It is not a little remarkable that he should have found, after a lapse of sixty years, that the extraction of clay was still going on in the same brick-pit. Only a few months before his arrival, two flint instruments had been dug out of the clay, one from a depth of 7 and the other of 10 feet from the surface.   Others have since been disinterred from undisturbed beds of gravel in the same pit.   Mr. Amyot

[1] Frere, *Archæologia*, vol. xiii., 1800, p. 206.

of Diss has also obtained from the underlying freshwater strata the astragalus of an elephant, and bones of the deer and horse; but although many of the old implements have recently been discovered *in situ* in regular strata and preserved by Sir Edward Kerrison, no bones of extinct mammalia seem as yet to have been actually seen in the same stratum with one of the tools.

By reference to the annexed section, the geologist will see that the basin-shaped hollow *a, b, c* has been filled up gradually with the freshwater strata 3, 4, 5, after the same cavity *a, b, c* had been previously excavated out of the more ancient boulder clay No. 6. The relative position of these formations will be better understood when I have described in the twelfth chapter the structure of Norfolk and Suffolk as laid open in the sea-cliffs at Mundesley, about 30 miles distant from Hoxne, in a N.N.E. direction.

I examined the deposits at Hoxne in 1860, when I had the advantage of being accompanied by the Rev. J. Gunn and the Rev. S. W. King. In the loamy beds 3 and 4, Fig. 24, we observed the common river shell *Valvata piscinalis* in great numbers. With it, but much more rare, were *Limnæa palustris, Planorbis albus, P. spirorbis, Succinea putris, Bithynia tentaculata, Cyclas cornea;* and Mr. Prestwich mentions *Cyclas amnica* and fragments of a *Unio,* besides several land shells. In the black peaty mass No. 5, fragments of wood of the oak, yew, and fir have been recognised. The flint weapons which I have seen from Hoxne are so much more perfect, and have their cutting edge so much sharper than those from the valley of the Somme, that they seem neither to have been used by Man, nor to have been rolled in the bed of a river. The opinion of Mr. Frere, therefore, that there may have been a manufactory of weapons on the spot, appears probable.

Flint Implements at Icklingham in Suffolk

In another part of Suffolk, at Icklingham, in the valley of the Lark, below Bury St. Edmund's, there is a bed of gravel, in which teeth of *Elephas primigenius* and several flint tools, chiefly of a lance-head form, have been found. I have twice visited the spot, which has been correctly described by Mr. Prestwich.[1]

The section of the Bedford tool-bearing alluvium, given in

[1] *Quart. Jour. Geol. Soc.,* vol. xvii., 1861, p. 364.

Fig. 23, p. 129, may serve to illustrate that of Icklingham, if we substitute Chalk for Oolite, and the river Lark for the Ouse.   In both cases, the present bed of the river is about 30 feet below the level of the old gravel, and the Chalk hill, which bounds the valley of the Lark on the right side, is capped like the Oolite of Biddenham by boulder clay, which rises to the height of 100 feet above the Lark.   About twelve years ago, a large erratic block, above 4 feet in diameter, was dug out of the boulder clay at Icklingham, which I found to consist of a hard siliceous schist, which must have come from a remote region. The tool-bearing gravel here, as in the case to which it has been compared near Bedford, is proved to be newer than the glacial drift, by containing pebbles of basalt and other rocks derived from that formation.

CHAPTER X

*Works of Art associated with extinct Mammalia in a
Cavern in Somersetshire*

THE only British cave from which implements resembling those
of Amiens have been obtained, since the attention of geologists
has been awakened to the importance of minutely observing
the position of such relics relatively to the associated fossil
mammalia, is that recently opened near Wells in Somersetshire.
It occurs near the cave of Wookey Hole, from the mouth of
which the river Axe issues on the southern flanks of the Mendips.
No one had suspected that on the left side of the ravine, through
which the river flows after escaping from its subterranean
channel, there were other caves and fissures concealed beneath
the green sward of the steep sloping bank. About ten years
ago, a canal was made, several hundred yards in length, for the
purpose of leading the waters of the Axe to a paper-mill, now
occupying the middle of the ravine. In carrying out this work,
about 12 feet of the left bank was cut away, and a cavernous
fissure, choked up to the roof with ossiferous loam, was then,
for the first time, exposed to view. This great cavity, originally 9 feet high and 36 wide, traversed the Dolomitic Conglomerate; and fragments of that rock, some angular and others
water-worn, were scattered through the red mud of the cave,
in which fossil remains were abundant. For an account of
them and the position they occupied we are indebted to Mr.
Dawkins, F.G.S., who, in company with Mr. Williamson, explored the cavern in 1859, and obtained from it the bones of the

Hyæna spelæa in such numbers as to lead him to conclude that the cavern had for a long time been a hyæna's den. Among the accompanying animals found fossil in the same bone-earth, were observed *Elephas primigenius, Rhinoceros tichorhinus, Ursus spelæus, Bos primigenius, Megaceros hibernicus, Cervus tarandus* (and other species of *Cervus*), *Felis spelæa, Canis lupus, Canis vulpes*, and teeth and bones of the genus *Equus* in great numbers.

Intermixed with the above fossil bones were some arrowheads, made of bone, and many chipped flints, and chipped pieces of chert, a white or bleached flint weapon of the spearhead Amiens type, which was taken out of the undisturbed matrix by Mr. Williamson himself, together with a hyæna's tooth, showing that Man had either been contemporaneous with or had preceded the extinct fauna. After penetrating 34 feet from the entrance, Mr. Dawkins found the cave bifurcating into two branches, one of which was vertical. By this rent, perhaps, some part of the contents of the cave may have been introduced.[1]

When I examined the spot in 1860, after I had been shown some remains of the hyæna collected there, I felt convinced that a complete revolution must have taken place in the topography of the district since the time of the extinct quadrupeds. I was not aware at the time that flint tools had been met with in the same bone-deposit.

Caves of Gower in Glamorganshire, South Wales

The ossiferous caves of the peninsula of Gower in Glamorganshire have been diligently explored of late years by Dr. Falconer and Lieutenant-Colonel E. R. Wood, who have thoroughly investigated the contents of many which were previously unknown. Among these Dr. Falconer's skilled eye has recognised the remains of almost every quadruped which he had elsewhere found fossil in British caves: in some places the *Elephas primigenius*, accompanied by its usual companion, the *Rhinoceros tichorhinus*, in others *Elephas antiquus*, associated with *Rhinoceros hemitœchus*, Falconer; the extinct animals being often embedded, as in the Belgian caves, in the same matrix with species now living in Europe, such as the common badger (*Meles taxus*), the common wolf, and the fox.

In a cavernous fissure called the Raven's Cliff, teeth of several individuals of *Hippopotamus major*, both young and old, were found; and this in a district where there is now scarce a rill

[1] Boyd Dawkins, *Proc. Geol. Soc.*, January 1862.

of running water, much less a river in which such quadrupeds could swim. In one of the caves, called Spritsail Tor, bones of the elephants above named were observed, with a great many other quadrupeds of Recent and extinct species.

From one fissure, called Bosco's Den, no less than one thousand antlers of the reindeer, chiefly of the variety called *Cervus Guettardi*, were extracted by the persevering exertions of Colonel Wood, who estimated that several hundred more still remained in the bone-earth of the same rent.

They were mostly shed horns, and of young animals; and had been washed into the rent with other bones, and with angular fragments of limestone, and all enveloped in the same ochreous mud. Among the other bones, which were not numerous, were those of the cave-bear, wolf, fox, ox, stag, and field-mouse.

But the discovery of most importance, as bearing on the subject of the present work, is the occurrence in a newly-discovered cave, called Long Hole, by Colonel Wood, in 1861, of the remains of two species of rhinoceros, *R. tichorhinus* and *R. hemitœchus*, Falconer, in an undisturbed deposit, in the lower part of which were some well-shaped flint knives, evidently of human workmanship It is clear from their position that Man was coeval with these two species. We have elsewhere independent proofs of his co-existence with every other species of the cave-fauna of Glamorganshire; but this is the first well-authenticated example of the occurrence of *R. hemitœchus* in connection with human implements.

In the fossil fauna of the valley of the Thames, *Rhinoceros leptorhinus* was mentioned as occurring at Gray's Thurrock with *Elephas antiquus*. Dr. Falconer, in a memoir which he is now preparing for the press on the European Pliocene and Pleistocene species of the genus *Rhinoceros*, has shown that, under the above name of *R. leptorhinus*, three distinct species have been confounded by Cuvier, Owen, and other palæontologists:—

1. *R. megarhinus*, Christol, being the original and typical *R. leptorhinus* of Cuvier, founded on Cortesi's Monte Zago cranium, and the *only* Pliocene, or Pleistocene European species, that had not a nasal septum.—Gray's Thurrock, etc.

2. *R. hemitœchus*, Falconer, in which the ossification of the septum dividing the nostrils is incomplete in the middle, besides other cranial and dental characters distinguishing it from *R. tichorhinus*, accompanies *Elephas antiquus* in most of the oldest British bone-caves, such as Kirkdale, Cefn, Durdham Down,

Minchin Hole, and other Gower caverns—also found at Clacton, in Essex, and in Northamptonshire.

3. *R. etruscus*, Falconer, a comparatively slight and slender form, also with an incomplete bony septum,[1] occurs deep in the Val d'Arno deposits, and in the " Forest bed," and super-imposed blue clays, with lignite, of the Norfolk coast, but nowhere as yet found in the ossiferous caves in Britain.

Dr. Falconer announced in 1860 his opinion that the filling up of the Gower caves in South Wales took place after the deposition of the marine boulder clay,[2] an opinion in harmony with what we have since learnt from the section of the gravels near Bedford, given above (Fig. 23), where a fauna correspond-ing to that of the Welsh caves characterises the ancient alluvium, and is shown to be clearly post-glacial, in the sense of being posterior in date to the boulder-clay of the midland counties. In the same sense the late Edward Forbes declared, in 1846, his conviction that not only the *Cervus megaceros*, but also the mammoth and other extinct pachyderms and carnivora, had lived in Britain in post-glacial times.[3] The Gower caves in general have their floors strewed over with sand, containing marine shells, all of living species; and there are raised beaches on the adjoining coast, and other geological signs of great alteration in the relative level of land and sea, since that country was inhabited by the extinct mammalia, some of which, as we have seen, were certainly coeval with Man.

Ossiferous Caves in the North of Sicily

Geologists have long been familiar with the fact that on the northern coast of Sicily, between Termini on the east, and Trapani on the west, there are several caves containing the bones of extinct animals. These caves are situated in rocks of Hippurite limestone, a member of the Cretaceous series, and some of them may be seen on both sides of the Bay of Palermo. If in the neighbourhood of that city we proceed from the sea inland, ascending a sloping terrace, composed of the marine Newer Pliocene strata, we reach about a mile from the shore, and at the height of about 180 feet above it a precipice of limestone, at the base of which appear the entrances of several caves. In that of San Ciro, on the east side of the bay, we find at the bottom sand with marine shells, forty species of

[1] Falconer, *Quart. Jour. Geol. Soc.*, vol. xv., 1859, p. 602.
[2] *Ibid.*, vol. xvi., 1860, p. 491.   [3] *Mem. Geol. Survey*, pp. 394-397.

which have been examined, and found almost all to agree specifically with mollusca now inhabiting the Mediterranean. Higher in position, and resting on the sand, is a breccia, composed of pieces of limestone, quartz, and schist in a matrix of brown marl, through which land shells are dispersed, together with bones of two species of hippopotamus, as determined by Dr. Falconer. Certain bones of the skeleton were counted in such numbers as to prove that they must have belonged to several hundred individuals. With these were associated the remains of *Elephas antiquus*, and bones of the genera *Bos,* *Cervus, Sus, Ursus, Canis,* and a large *Felis.* Some of these bones have been rolled as if partially subjected to the action of water, and may have been introduced by streams through rents in the Hippurite limestone; but there is now no running water in the neighbourhood, no river such as the hippopotamus might frequent, not even a small brook, so that the physical geography of the district must have been altogether changed since the time when such remains were swept into fissures, or into the channels of engulfed rivers.

No proofs seem yet to have been found of the existence of Man at the period when the hippopotamus and *Elephas antiquus* flourished at San Ciro. But there is another cave called the Grotto di Maccagnone, which much resembles it in geological position, on the opposite or west side of the Bay of Palermo, near Carini. In the bottom of this cave a bone deposit like that of San Ciro occurs, and above it other materials reaching to the roof, and evidently washed in from above, through crevices in the limestone. In this upper and newer breccia Dr. Falconer discovered flint knives, bone splinters, bits of charcoal, burnt clay, and other objects indicating human intervention, mingled with entire land shells, teeth of horses, coprolites of hyænas, and other bones, the whole agglutinated to one another and to the roof by the infiltration of water holding lime in solution. The perfect condition of the large fragile helices (*Helix vermiculata*) afforded satisfactory evidence, says Dr. Falconer, that the various articles were carried into the cave by the tranquil agency of water, and not by any tumultuous action. At a subsequent period other geographical changes took place, so that the cave, after it had been filled, was washed out again, or emptied of its contents with the exception of those patches of breccia which, being cemented together by stalactite, still adhere to the roof.[1]

[1] *Quart. Jour. Geol. Soc.*, vol. xvi., 1860, p. 105.

Baron Anca, following up these investigations, explored, in 1859, another cave at Mondello, west of Palermo, and north of Mount Gallo, where he discovered molars of the living African elephant, and afterwards additional specimens of the same species in the neighbouring grotto of Olivella. In reference to this elephant, Dr. Falconer has reminded us that the distance between the nearest part of Sicily and the coast of Africa, between Marsala and Cape Bon, is not more than 80 miles, and Admiral Smyth, in his Memoir on the Mediterranean, states (p. 499) that there is a subaqueous plateau, named by him Adventure Bank, uniting Sicily to Africa by a succession of ridges which are not more than from 40 to 50 fathoms under water.[1] Sicily therefore might be re-united to Africa by movements of upheaval not greater than those which are already known to have taken place within the human period on the borders of the Mediterranean, of which I shall now proceed to cite a well-authenticated example, observed in Sardinia.

Rise of the Bed of the Sea to the Height of 300 Feet, in the Human Period, in Sardinia

Count Albert de la Marmora, in his description of the geology of Sardinia,[2] has shown that on the southern coast of that island, at Cagliari and in the neighbourhood, an ancient bed of the sea, containing marine shells of living species, and numerous fragments of antique pottery, has been elevated to the height of from 230 to 324 feet above the present level of the Mediterranean. Oysters and other shells, of which a careful list has been published, including the common mussel (*Mytilus edulis*), many of them having both valves united, occur, embedded in a breccia in which fragments of limestone abound. The mussels are often in such numbers as to impart, when they have decomposed, a violet colour to the marine stratum. Besides pieces of coarse pottery, a flattened ball of baked earthenware, with a hole through its axis, was found in the midst of the marine shells. It is supposed to have been used for weighting a fishing net. Of this and of one of the fragments of ancient pottery Count de la Marmora has given figures.

The upraised bed of the sea probably belongs in this instance to the Pleistocene period, for in a bone breccia, filling fissures in the rocks around Cagliari, the remains of extinct mammalia

---

[1] Cited by Horner, Pres. Address to Geol. Soc., 1861, p. 42.
[2] *Partie Géologique*, vol. i., pp. 382, 387.

have been detected; among which is a new genus of carnivorous quadruped, named *Cynotherium* by M. Studiati, and figured by Count de la Marmora in his Atlas (Pl. vii.), also an extinct species of *Lagomys*, determined by Cuvier in 1825. Embedded in the same bone-breccia, and enveloped with red earth like the mammalian remains, were detected shells of the *Mytilus edulis* before mentioned, implying that the marine formation containing shells and pottery had been already upheaved and exposed to denudation before the remains of quadrupeds were washed into these rents and included in the red earth. In the vegetable soil covering the upraised marine stratum, fragments of Roman pottery occur.

If we assume the average rate of upheaval to have been, as before hinted, $2\frac{1}{2}$ feet in a century, 300 feet would give an antiquity of 12,000 years to the Cagliari pottery, even if we simply confine our estimate to the upheaval above the sea-level, without allowing for the original depth of water in which the mollusca lived. Even then our calculation would merely embrace the period during which the upward movement was going on; and we can form at present no conjecture as to the probable era of its commencement or termination.

I learn from Captain Spratt, R.N., that the island of Crete or Candia, about 135 miles in length, has been raised at its western extremity about 25 feet; so that ancient ports are now high and dry above the sea, while at its eastern end it has sunk so much that the ruins of old towns are seen under water. Revolutions like these in the physical geography of the countries bordering the Mediterranean, may well help us to understand the phenomena of the Palermo caves, and the presence in Sicily of African species of mammalia.

Climate and Habits of the Hippopotamus

As I have alluded more than once in this chapter to the occurrence of the remains of the hippopotamus in places where there are now no rivers, not even a rill of water, and as other bones of the same genus have been met with in the lower-level gravels of the Somme where large blocks of sandstone seem to imply that ice once played a part in their transportation, it may be well to consider, before proceeding farther, what geographical and climatal conditions are indicated by the presence of these fossil pachyderms.

It is now very generally conceded that the mammoth and

tichorhine rhinoceros were fitted to inhabit northern regions, and it is therefore natural to begin by asking whether the extinct hippopotamus may not in like manner have flourished in a cold climate. In answer to this inquiry, it has been remarked that the living hippopotami, anatomically speaking so closely allied to the extinct species, are so aquatic and fluviatile in their habits as to make it difficult to conceive that their congeners could have thriven all the year round in regions where, during winter, the rivers were frozen over for months. Moreover, I have been unable to learn that, in any instance, bones of the hippopotamus have been found in the drift of northern Germany associated with the remains of the mammoth, tichorhine rhinoceros, musk-ox, reindeer, lemming, and other arctic quadrupeds before alluded to; yet, though not proved to have ever made a part of such a fauna, the presence of the fossil hippopotamus north of the fiftieth parallel of latitude naturally tempts us to speculate on the migratory powers and instincts of some of the extinct species of the genus. They may have resembled, in this respect, the living musk-ox, herds of which pass for hundreds of miles over the ice to the rich pastures of Melville Island, and then return again to southern latitudes before the ice breaks up.

We are indebted to Sir Andrew Smith,[1] an experienced zoologist, for having given us an account of the migratory habits of the living hippopotamus of Southern Africa (*H. amphibius*, Linn.).

He states that, when the Dutch first colonised the Cape of Good Hope, this animal abounded in all the great rivers, as far south as the land extends; whereas, in 1849, they had all disappeared, scarcely one remaining even within a moderate distance of the colony. He also tells us that this species evinces great sagacity in changing its quarters whenever danger threatens, quitting every district invaded by settlers bearing fire-arms. Bulky as they are, they can travel speedily for miles over land from one pool of a dried-up river to another; but it is by water that their powers of locomotion are surpassingly great, not only in rivers, but in the sea, for they are far from confining themselves to fresh water. Indeed, Sir A. Smith finds it " difficult to decide whether, during the daytime and when not feeding, they prefer the pools of rivers or the waters of the ocean for their abode." In districts where they have been disturbed by Man, they feed almost entirely in the night, chiefly on certain kinds of grass, but also on brushwood. Sir A. Smith

---

[1] *Illustrations of the Zoology of South Africa*: art. " Hippopotamus."

relates that, in an expedition which he made north of Port Natal, he found them swarming in all the rivers about the tropic of Capricorn. Here they were often seen to have left their footprints on the sands, entering or coming out of the salt water; and on one occasion Smith's party tried in vain to intercept a female with her young as she was making her way to the sea. Another female, which they had wounded on her precipitate retreat to the sea, was afterwards shot in that element.

The geologist, therefore, may freely speculate on the time when herds of hippopotami issued from North African rivers, such as the Nile, and swam northwards in summer along the coasts of the Mediterranean, or even occasionally visited islands near the shore. Here and there they may have landed to graze or browse, tarrying awhile and afterwards continuing their course northwards. Others may have swum in a few summer days from rivers in the south of Spain or France to the Somme, Thames, or Severn, making timely retreat to the south before the snow and ice set in.

Burial-place at Aurignac, in the South of France, of Pleistocene Date

I have alluded in the beginning of the fourth chapter to a custom prevalent among rude nations of consigning to the tomb works of art, once the property of the dead, or objects of their affection, and even of storing up, in many cases, animal food destined for the manes of the defunct in a future life. I also cited M. Desnoyers' comments on the absence among the bones of wild and domestic animals found in old Gaulish tombs of all intermixture of extinct species of quadrupeds, as proving that the oldest sepulchral monuments then known in France (1845) had no claims to high antiquity founded on palæontological data.

M. Lartet, however, has recently published a circumstantial account of what seems clearly to have been a sepulchral vault of the Pleistocene period, near Aurignac, not far from the foot of the Pyrenees. I have had the advantage of inspecting the fossil bones and works of art obtained by him from that grotto, and of conversing and corresponding with him on the subject, and can see no grounds for doubting the soundness of his conclusions.[1]

The town of Aurignac is situated in the department of the

[1] See Lartet, Annales des Sci. Nat., 4me Ser., Zoologie, vol. xv., p. 177, translated in Natural History Review, London, January 1862.

Haute Garonne, near a spur of the Pyrenees; adjoining it is the small flat-topped hill of Fajoles, about 60 feet above the brook called Rodes, which flows at its foot on one side. It consists of Nummulitic limestone, presenting a steep escarpment towards the north-west, on which side in the face of the rock, about 45 feet above the brook, is now visible the entrance of a grotto *a*, Fig. 25, which opened originally on the terrace *h, c, k*; which slopes gently towards the valley.

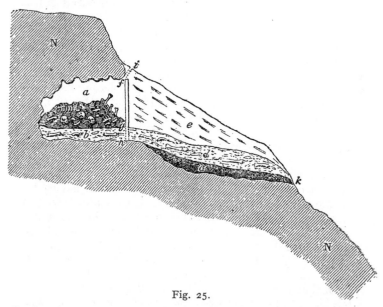

Fig. 25.

Section of part of the hill of Fajoles passing through the sepulchral grotto of Aurignac (E. Lartet).

a Part of the vault in which the remains of seventeen human skeletons were found.
b Layer of made ground, two feet thick, inside the grotto in which a few human bones, with entire bones of extinct and living species of animals, and many works of art were embedded.
c Layers of ashes and charcoal, six inches thick, with broken, burnt, and gnawed bones of extinct and Recent mammalia; also hearth-stones and works of art; no human bones.
d Deposit with similar contents and a few scattered cinders.
e Talus of rubbish washed down from the hill above.
f, g Slab of rock which closed the vault, not ascertained whether it extended to *h*.
f, i Rabbit burrow which led to the discovery of the grotto.
h, k Original terrace on which the grotto opened.
N Nummulitic limestone of hill of Fajoles.

Until the year 1852, the opening into this grotto was masked by a talus of small fragments of limestone and earthy matter *e*, such as the rain may have washed down the slope of the hill. In that year a labourer named Bonnemaison, employed in repairing the roads, observed that rabbits, when hotly pursued by the sportsman, ran into a hole which they had burrowed in the talus, at *i f*, Fig. 25. On reaching as far into the opening as the length of his arm, he drew out, to his surprise, one of the long bones of the human skeleton; and his curiosity being excited, and having a suspicion that the hole communicated with a subterranean cavity, he commenced digging a trench through the midde of the talus, and in a few hours found himself opposite a large heavy slab of rock *f h*, placed vertically against the entrance. Having removed this, he discovered on the other side of it an arched cavity *a*, 7 or 8 feet in its greatest height, 10 in width, and 7 in horizontal depth. It was almost filled with bones, among which were two entire skulls, which he recognised at once as human. The people of Aurignac, astonished to hear of the occurrence of so many human relics in so lonely a spot, flocked to the cave, and Dr. Amiel, the Mayor, ordered all the bones to be taken out and reinterred in the parish cemetery. But before this was done, having as a medical man a knowledge of anatomy, he ascertained by counting the homologous bones that they must have formed parts of no less than seventeen skeletons of both sexes, and all ages; some so young that the ossification of some of the bones was incomplete. Unfortunately the skulls were injured in the transfer; and what is worse, after the lapse of eight years, when M. Lartet visited Aurignac, the village sexton was unable to tell him in what exact place the trench was dug, into which the skeletons had been thrown, so that this rich harvest of ethnological knowledge seems for ever lost to the antiquary and geologist.

M. Lartet having been shown, in 1860, the remains of some extinct animals and works of art, found in digging the original trench made by Bonnemaison through the bed *d* under the talus, and some others brought out from the interior of the grotto, determined to investigate systematically what remained intact of the deposits outside and inside the vault, those inside, underlying the human skeletons, being supposed to consist entirely of made ground. Having obtained the assistance of some intelligent workmen, he personally superintended their labours, and found outside the grotto, resting on the sloping terrace *h k*, the layer of ashes and charcoal c, about 6 inches thick, extending over an area of 6 or 7 square yards, and going as far as the entrance of the grotto and no farther, there being no cinders or charcoal in the interior. Among the cinders outside the vault were fragments of fissile sandstone, reddened by heat, which were observed to rest on a levelled surface of Nummulitic limestone and to have formed a hearth. The nearest place from whence such slabs of sandstone could have been brought was the opposite side of the valley.

Among the ashes, and in some overlying earthy layers, d, separating the ashes from the talus e, were a great variety of bones and implements; amongst the latter not fewer than a hundred flint articles—knives, projectiles, sling stones, and chips, and among them one of those siliceous cores or nuclei with numerous facets, from which flint flakes or knives had been struck off, seeming to prove that some instruments were occasionally manufactured on the very spot.

Among other articles outside the entrance was found a stone of a circular form, and flattened on two sides, with a central depression, composed of a tough rock which does not belong to that region of the Pyrenees. This instrument is supposed by the Danish antiquaries to have been used for removing by skilful blows the edges of flint knives, the fingers and thumb being placed in the two opposite depressions during the operation. Among the bone instruments were arrows without barbs, and other tools made of reindeer horn, and a bodkin formed out of the more compact horn of the roedeer. This instrument was well shaped, and sharply pointed, and in so good a state of preservation that it might still be used for piercing the tough skins of animals.

Scattered through the same ashes and earth were the bones of the various species of animals enumerated in the subjoined lists, with the exception of two, marked with an asterisk, which only occurred in the interior of the grotto:—

1. CARNIVORA

Number of individuals.

1. *Ursus spelæus* (cave-bear)	5 — 6
2. *Ursus arctos?* (brown bear)	1
3. *Meles taxus* (badger)	1 — 2
4. *Putorius vulgaris* (polecat)	1
5.*Felis spelæa (cave-lion)	1
6. *Felis catus ferus* (wild cat)	1
7. *Hyæna spelæa* (cave-hyæna)	5 — 6
8. *Canis lupus* (wolf)	3
9. *Canis vulpes* (fox)	18 —20

2. Herbivora

Number of individuals.

1. *Elephas primigenius* (mammoth, two molars).
2. *Rhinoceros tichorhinus* (Siberian rhinoceros) . 1
3. *Equus caballus* (horse) . . . . . 12 —15
4. *Equus asinus ?* (ass) 1
5. *Sus scrofa (pig, two incisors).
6. *Cervus elaphus* (stag) 1
7. *Megaceros hibernicus* (gigantic Irish deer). . 1
8. *C. capreolus* (roebuck) . . . . . 3 — 4
9. *C. tarandus* (reindeer) . . . . . 10 —12
10. *Bison europæus* (aurochs) . . . . 12 —15

The bones of the herbivora were the most numerous, and all those on the outside of the grotto which had contained marrow were invariably split open, as if for its extraction, many of them being also burnt. The spongy parts, moreover, were wanting, having been eaten off and gnawed after they were broken, the work, according to M. Lartet, of hyænas, the bones and coprolites of which were mixed with the cinders, and dispersed through the overlying soil *d*. These beasts of prey are supposed to have prowled about the spot and fed on such relics of the funeral feasts as remained after the retreat of the human visitors, or during the intervals between successive funeral ceremonies which accompanied the interment of the corpses within the sepulchre. Many of the bones were also streaked, as if the flesh had been scraped off by a flint instrument.

Among the various proofs that the bones were fresh when brought to the spot, it is remarked that those of the herbivora not only bore the marks of having had the marrow extracted and having afterwards been gnawed and in part devoured as if by carnivorous beasts, but that they had also been acted upon by fire (and this was especially noticed in one case of a cave-bear's bone), in such a manner as to show that they retained in them at the time all their animal matter.

Among other quadrupeds which appear to have been eaten at the funeral feasts, and of which the bones occurred among the ashes, were those of a young *Rhinoceros tichorhinus,* the bones of which had been, like those of the accompanying herbivora, broken and gnawed by a beast of prey at both extremities.

Outside of the great slab of stone forming the door, not one human bone occurred; inside of it there were found, mixed with loose soil, the remains of as many as seventeen human individuals, besides some works of art and bones of animals. We know nothing of the arrangement of these bones when they were first broken into. M. Lartet inferred at first that the bodies were bent down upon themselves in a squatting attitude, a posture known to have been adopted in most of the sepulchres of primitive times; and he has so represented them in his restoration of the cave: but this opinion he has since retracted. His artist also has inadvertently, in the same drawing, delineated the arched grotto as if it were shaped very regularly and smoothly, like a finished piece of masonry, whereas the surface was in truth as uneven and irregular as are the roofs of all natural grottos.

There was no stalagmite in the grotto, and M. Lartet, an experienced investigator of ossiferous caverns in the south of France, came to the conclusion that all the bones and soil found in the inside were artificially introduced. The substratum *b*, Fig. 25, which remained after the skeletons had been removed, was about 2 feet thick. In it were found about ten detached human bones, including a molar tooth; and M. Delesse ascertained by careful analysis of one of these, as well as of the bones of a rhinoceros, bear, and some other extinct animals, that they all contained precisely the same proportion of nitrogen, or had lost an equal amount of their animal matter. My friend Mr. Evans, before cited, has suggested to me that such a fact, taken alone, may not be conclusive in favour of the equal antiquity of the human and other remains. No doubt, had the human skeletons been found to contain more gelatine than those of the extinct mammalia, it would have shown that they were the more modern of the two; but it is possible that after a bone has gone on losing its animal matter up to a certain point, it may then part with no more so long as it continues enveloped in the same matrix. If this be so, it follows that bones of very different degrees of antiquity, after they have lain for many thousands of years in a particular soil, may all have reached long ago the maximum of decomposition attainable in such a matrix. In the present case, however, the proof of the contemporaneousness of Man and the extinct animals does not depend simply on the identity of their mineral condition. The chemical analysis of M. Delesse is only a fact in corroboration of a great mass of other evidence.

Mixed with the human bones inside the grotto first removed by Bonnemaison, were eighteen small, round, and flat plates of a white shelly substance, made of some species of cockle (*Cardium*), pierced through the middle as if for being strung into a bracelet. In the substratum also in the interior examined by M. Lartet was found the tusk of a young *Ursus spelæus*, the crown of which had been stripped of its enamel, and which had been carved perhaps in imitation of the head of a bird. It was perforated lengthwise as if for suspension as an ornament or amulet. A flint knife also was found in the interior which had evidently never been used; in this respect, unlike the numerous worn specimens found outside, so that it is conjectured that it may, like other associated works of art, have been placed there as part of the funeral ceremonies.

A few teeth of the cave-lion, *Felis spelæa*, and two tusks of the wild boar, also found in the interior, were memorials perhaps of the chase. No remains of the same animals were met with among the external relics.

On the whole, the bones of animals inside the vault offer a remarkable contrast to those of the exterior, being all entire and uninjured, none of them broken, gnawed, half-eaten, scraped, or burnt like those lying among the ashes on the other side of the great slab which formed the portal. The bones of the interior seem to have been clothed with their flesh when buried in the layer of loose soil strewed over the floor. In confirmation of this idea, many bones of the skeleton were often observed to be in juxtaposition, and in one spot all the bones of the leg of an *Ursus spelæus* were lying together uninjured. Add to this, the entire absence in the interior of cinders and charcoal, and we can scarcely doubt that we have here an example of an ancient place of sepulture, closed at the opening so effectually against the hyænas or other carnivora that no marks of their teeth appear on any of the bones, whether human or brute.

John Carver, in his travels in the interior of North America in 1766-68 (chap. xv.), gave a minute account of the funeral rites of an Indian tribe which inhabited the country now called Iowa, at the junction of the St. Peter's River with the Mississippi; and Schiller, in his famous *Nadowessische Todtenklage*, has faithfully embodied in a poetic dirge all the characteristic features of the ceremonies so graphically described by the English traveller, not omitting the many funeral gifts which, we are told, were placed " in a cave " with the bodies of the dead. The lines beginning, " Bringet her die letzten Gaben," have been thus translated, truthfully, and with all the spirit of the original, by Sir E. L. Bulwer[1] —

> " Here bring the last gifts!—and with these
> The last lament be said;
> Let all that pleased, and yet may please,
> Be buried with the dead.

[1] *Poems and Ballads of Schiller.*

" Beneath his head the hatchet hide,
   , That he so stoutly swung;
And place the bear's fat haunch beside—
   The journey hence is long!

" And let the knife new sharpened be
   That on the battle-day
Shore with quick strokes—he took but three—
   The foeman's scalp away!

" The paints that warriors love to use,
   Place here within his hand,
That he may shine with ruddy hues
   Amidst the spirit-land."

If we accept M. Lartet's interpretation of the ossiferous deposits of Aurignac, both inside and outside the grotto, they add nothing to the palæontological evidence in favour of Man's antiquity, for we have seen all the same mammalia associated elsewhere with flint implements, and some species, such as the *Elephas antiquus, Rhinoceros hemitœchus,* and *Hippopotamus major,* missing here, have been met with in other places. An argument, however, having an opposite leaning may perhaps be founded on the phenomena of Aurignac. It may—indeed it has been said, that they imply that some of the extinct mammalia survived nearly to our times:

First—Because of the modern style of the works of art at Aurignac.

Secondly—Because of the absence of any signs of change in the physical geography of the country since the cave was used for a place of sepulture.

In reference to the first of these propositions, the utensils, it is said, of bone and stone indicate a more advanced state of the arts than the flint implements of Abbeville and Amiens. M. Lartet, however, is of opinion that they do not, and thinks that we have no right to assume that the fabricators of the various spear-headed and other tools of the Valley of the Somme possessed no bone instruments or ornaments resembling those discovered at Aurignac. These last, moreover, he regards as extremely rude in comparison with others of the stone period in France, which can be proved palæontologically, at least by strong negative evidence, to be of subsequent date. Thus, for example, at Savigné, near Civray, in the department of Vienne, there is a cave in which there are no extinct mammalia, but where remains of the reindeer abound. The works of art of the stone period found there indicate considerable progress in skill beyond that attested by the objects found in the Aurignac grotto. Among the Savigné articles, there is the bone of a stag, on which figures of two animals, apparently meant for deer, are engraved in outline, as if by a sharp-pointed flint. In another cave, that of Massat, in the department of Ariége, which M. Lartet ascribes to the period of the aurochs, a quadruped which survived the reindeer in the south of France, there are bone instruments of a still more advanced state of the arts, as, for example, barbed arrows with a small canal in each, believed to have served for the insertion of poison; also a needle of bird's bone, finely shaped, with an eye or perforation at one end, and a stag's horn, on which is carved a representation of a bear's head, and a hole at one end as if for suspending it. In this figure we see, says M. Lartet, what may perhaps be the earliest known example of lines used to express shading.

The fauna of the aurochs (*Bison europæus*) agrees with that of the earlier lake dwellings in Switzerland, in which hitherto the reindeer is wanting; whereas the reindeer has been found in a Swiss cave, in Mont Salève, supposed by Lartet to be more ancient than the lake dwellings.

According to this view, the mammalian fauna has undergone at least two fluctuations since the remains of some extinct quadrupeds were eaten, and others buried as funeral gifts in the sepulchral vault of Aurignac.

As to the absence of any marked changes in the physical configuration of the district since the same grotto was a place of sepulture, we must remember that it is the normal state of the earth's surface to be undergoing great alterations in one place, while other areas, often in close proximity, remain for ages without any modification. In one region, rivers are deepening and widening their channels, or the waves of the sea are undermining cliffs, or the land is sinking beneath or rising above the waters, century after century, or the volcano is pouring forth torrents of lava or showers of ashes; while, in tracts hard by, the ancient forest, or extensive heath, or the splendid city continue scatheless and motionless. Had the talus which concealed from view the ancient hearth with its cinders and the massive stone portal of the Aurignac grotto escaped all human interference for thousands of years to come, there is no reason to suppose that the small stream at the foot of the hill of Fajoles would have undermined it. At the end of a long period the only alteration might have been the thickening of the talus which protected the loose cinders and bones from waste. We behold in many a valley of Auvergne, within 50 feet of the present river channel, a volcanic cone of loose ashes, with a crater at its summit, from which powerful currents of basaltic lava have poured, usurping the ancient bed of the torrent. By the action of the stream, in the course of ages, vast masses of the hard columnar basalt have been removed, pillar after pillar, and much vesicular lava, as in the case, for example, of the Puy Rouge, near Chalucet, and of the Puy de Tartaret, near Nechers.[1] The rivers have even in some cases, as the Sioule, near Chalucet, cut through not only the basalt which dispossessed them of their ancient channels, but have actually eaten 50 feet into the subjacent gneiss; yet the cone, an incoherent heap of scoriæ and spongy ejectamenta, stands unmolested. Had the waters once risen, even for a day, so high as to reach the level of the base of one of these cones—had there been a single flood 50 or 60 feet in height since the last eruption occurred, a great part of these volcanoes must inevitably have been swept away as readily as all traces of the layer of cinders; and the accompanying bones would have been obliterated by the Rodes near Aurignac, had it risen, since the days of the mammoth, rhinoceros, and cave-bear, 50 feet above its present level.

The Aurignac cave adds no new species to the list of extinct quadrupeds, which we have elsewhere, and by independent evidence, ascertained to have once flourished contemporaneously with Man. But if the fossil memorials have been correctly interpreted—if we have here before us at the northern base of the Pyrenees a sepulchral vault with skeletons of human beings, consigned by friends and relatives to their last resting-place—if we have also at the portal of the tomb the relics of funeral feasts, and within it indications of viands destined for the use of the departed on their way to a land of spirits; while among the funeral gifts are weapons wherewith in other fields to chase the gigantic deer, the cave-lion, the cave-bear, and woolly rhinoceros—we have at last succeeded in tracing back the sacred rites of burial, and, more interesting still, a belief in a future state, to times long anterior to those of history and tradition. Rude and superstitious as may have been the savage of that remote era, he still deserved, by cherishing hopes of a hereafter, the epithet of " noble," which Dryden gave to what he seems to have pictured to himself as the primitive condition of our race,

" as Nature first made Man
When wild in woods the noble savage ran." [2]

[1] Scrope's *Volcanoes of Central France*, 1858, p. 97.
[2] *Siege of Granada*, Part I., Act I., Scene i.

CHAPTER XI

AGE OF HUMAN FOSSILS OF LE PUY IN CENTRAL FRANCE AND
OF NATCHEZ ON THE MISSISSIPPI DISCUSSED

Question as to the Authenticity of the Fossil Man of Denise, near Le Puy-
en-Velay, considered—Antiquity of the Human Race implied by that
Fossil—Successive Periods of Volcanic Action in Central France—
With what Changes in the Mammalian Fauna they correspond—The
Elephas Meridionalis anterior in Time to the implement-bearing
Gravel of St. Acheul—Authenticity of the Human Fossil of Natchez
on the Mississippi discussed—The Natchez Deposit, containing Bones
of *Mastodon* and *Megalonyx*, probably not older than the Flint
Implements of St. Acheul.

AMONG the fossil remains of the human species supposed to
have claims to high antiquity, and which have for many years
attracted attention, two of the most prominent examples are—

First—" The fossil man of Denise," comprising the remains
of more than one skeleton, found in a volcanic breccia near the
town of Le Puy-en-Velay, in Central France.

Secondly—The fossil human bone of Natchez, on the Mississippi, supposed to have been derived from a deposit containing remains of *Mastodon* and *Megalonyx*. Having carefully
examined the sites of both of these celebrated fossils, I shall
consider in this chapter the nature of the evidence on which
the remote date of their entombment is inferred.

Fossil Man of Denise

An account of the fossil remains, so called, was first published
in 1844 by M. Aymard of Le Puy, a writer of deservedly high
authority both as a palæontologist and archæologist.[1] M. Pictet,
after visiting Le Puy and investigating the site of the alleged
discovery, was satisfied that the fossil bones belonged to the
period of the last volcanic eruptions of Velay; but expressly
stated in his important treatise on palæontology that this conclusion, though it might imply that Man had co-existed with
the extinct elephant, did not draw with it the admission
that the human race was anterior in date to the filling of

[1] *Bulletin de la Société Géologique de France*, 1844, 1845, 1847.

the caverns of France and Belgium with the bones of extinct mammalia.[1]

At a meeting of the " Scientific Congress " of France, held at Le Puy in 1856, the question of the age of the Denise fossil bones was fully gone into, and in the report of their proceedings published in that year, the opinions of some of the most skilful osteologists respecting the point in controversy are recorded. The late Abbé Croizet, a most experienced collector of fossil bones in the volcanic regions of Central France, and an able naturalist, and the late M. Laurillard, of Paris, who assisted Cuvier in modelling many fossil bones, and in the arrangement of the museum of the Jardin, declared their opinion that the specimen preserved in the museum of Le Puy is no counterfeit. They believed the human bones to have been enveloped by natural causes in the tufaceous matrix in which we now see them.

In the year 1859, Professor Hébert and M. Lartet visited Le Puy, expressly to investigate the same specimen, and to inquire into the authenticity of the bones and their geological age. Later in the same year, I went myself to Le Puy, having the same object in view, and had the good fortune to meet there my friend Mr. Poulett Scrope, with whom I examined the Montagne de Denise, where a peasant related to us how he had dug out the specimen with his own hands and in his own vineyard, not far from the summit of the volcano. I employed a labourer to make under his directions some fresh excavations, following up those which had been made a month earlier by MM. Hébert and Lartet, in the hope of verifying the true position of the fossils, but all of us without success. We failed even to find *in situ* any exact counterpart of the stone of the Le Puy Museum.

The osseous remains of that specimen consist of a frontal and some other parts of the skull, including the upper jaw with teeth, both of an adult and young individual; also a radius, some lumbar vertebræ, and some metatarsal bones. They are all embedded in a light porous tuff, resembling in colour and mineral composition the ejectamenta of several of the latest eruptions of Denise. But none of the bones penetrate into another part of the same specimen, which consists of a more compact rock thickly laminated. Nevertheless, I agree with the Abbé Croizet and M. Aymard, that it is not conceivable even that the less coherent part of the museum specimen which

---

[1] *Traité de Paléontologie*, vol. i., 1853, p. 152.

envelopes the human bones should have been artificially put together, whatever may have been the origin of certain other slabs of tuff which were afterwards sold as coming from the same place, and which also contained human remains. Whether some of these were spurious or not is a question more difficult to decide. One of them, now in the possession of M. Pichot-Dumazel, an advocate of Le Puy, is suspected of having had some plaster of Paris introduced into it to bind the bones more firmly together in the loose volcanic tuff. I was assured that a dealer in objects of natural history at Le Puy had been in the habit of occasionally securing the cohesion in that manner of fragments of broken bones, and the juxtaposition of un-injured ones found free and detachable in loose volcanic tuffs. From this to the fabrication of a factitious human fossil was, it is suggested, but a short step. But in reference to M. Pichot's specimen, an expert anatomist remarked to me that it would far exceed the skill, whether of the peasant who owned the vineyard or of the dealer above mentioned, to put together in their true position all the thirty-eight bones of the hand and fingers, or the sixteen of the wrist, without making any mistake, and especially without mixing those of the right with the homolo-gous bones of the left hand, assuming that they had brought bones, from some other spot, and then artificially introduced them into a mixture of volcanic tuff and plaster of Paris.

Granting, however, that the high prices given for " human fossils " at Le Puy may have led to the perpetration of some frauds, it is still an interesting question to consider whether the admission of the genuineness of a single fossil, such as that now in the museum at Le Puy, would lead us to assign a higher antiquity to the existence of Man in France than is deducible from many other facts explained in the last seven chapters. In reference to this point, I may observe that although I was not able to fix with precision the exact bed in the volcanic mountain from which the rock containing the human bones was taken, M. Felix Robert has, nevertheless, after studying " the volcanic alluviums " of Denise, ascertained that, on the side of Cheyrac and the village of Malouteyre, blocks of tuff frequently occur exactly like the one in the museum. That tuff he considers a product of the latest eruption of the volcano. In it have been found the remains of *Hyæna spelæa* and *Hippopotamus major*. The eruptions of steam and gaseous matter which burst forth from the crater of Denise broke through laminated Tertiary clays, small pieces of which, some of them scarcely altered, others half converted into scoriæ, were cast out in abundance, while other portions must have been in a state of argillaceous mud. Showers of such materials would be styled by the Neapolitans " aqueous lava " or " lava d'aqua," and we may well suppose that some human individuals, if any existed, would, together with wild animals, be occasionally overwhelmed in these tuffs. From near the place on the mountain whence the block with human bones now in the museum is said to have come, a stream of lava, well marked by its tabular structure, flowed down the flanks of the hill, within a few feet of the alluvial plain of the Borne, a small tributary of the Loire, on the opposite bank of which stands the town of Le Puy. Its continuous extension to so low a level clearly shows that the valley had already been deepened to within a few feet of its present depth at the time of the flowing of the lava.

We know that the alluvium of the same district, having a similar relation to the present geographical outline of the valleys, is of Pleistocene date, for it contains around Le Puy the bones of *Elephas primigenius* and *Rhinoceros tichorhinus*; and this affords us a palæontological test of the age of the human skeleton of Denise, if the latter be assumed to be coeval with the lava stream above referred to.

It is important to dwell on this point, because some geologists have felt disinclined to believe in the genuineness of the " fossil man of Denise," on the ground that, if conceded, it would imply that the human race was contemporary with an older fauna, or that of the *Elephas meridionalis*. Such a fauna is found fossil in another layer of tuff covering the slope of Denise, opposite to that where the museum specimen was exhumed. The quadrupeds obtained from that more ancient tuff comprise *Elephas meridionalis*, *Hippopotamus major*, *Rhinoceros megarhinus*, *Antilope torticornis*, *Hyæna brevirostris*, and twelve others of the genera horse, ox, stag, goat, tiger, etc., all supposed to be of extinct species. This tuff, found between Malouteyre and Polignac, M. Robert regards as the product of a much older eruption, and referable to the neighbouring Montagne de St. Anne, a volcano in a much more wasted and denuded state than Denise, and classed by M. Bertrand de Doue as of intermediate age between the ancient and modern cones of Velay.

The fauna to which *Elephas meridionalis* and its associates belong, can be shown to be of anterior date, in the north of

France, to the flint implements of St. Acheul, by the following train of reasoning. The valley of the Seine is not only geographically contiguous to the valley of the Somme, but its ancient alluvium contains the same mammoth and other fossil species. The Eure, one of the tributaries of the Seine, in its way to join that river, flows in a valley which follows a line of fault in the Chalk; and this valley is seen to be comparatively modern, because it intersects at St. Prest, 4 miles below Chartres, an older valley belonging to an anterior system of drainage, which has been filled by a more ancient fluviatile alluvium, consisting of sand and gravel, 90 feet thick. I have examined the site of this older drift, and the fossils have been determined by Dr. Falconer. They comprise *Elephas meridionalis*, a species of rhinoceros (not *R. tichorhinus*), and other mammalia differing from those of the implement-bearing gravels of the Seine and Somme. The latter, belonging to the period of the mammoth, might very well have been contemporary with the modern volcanic eruptions of Central France; and we may presume, even without the aid of the Denise fossil, that Man may have witnessed these. But the tuffs and gravels in which the *Elephas meridionalis* are embedded were synchronous with an older epoch of volcanic action, to which the cone of St. Anne, near Le Puy, and many other mountains of M. Bertrand de Doue's middle period belong, having cones and craters, which have undergone much waste by aqueous erosion. We have as yet no proof that Man witnessed the origin of these hills of lava and scoriæ of the middle phase of volcanic action.

Some surprise was expressed in 1856, by several of the assembled naturalists at Le Puy, that the skull of the "fossil man of Denise," although contemporary with the mammoth, and coeval with the last eruptions of the Le Puy volcanoes [Note 18], should be of the ordinary Caucasian or European type; but the observations of Professor Huxley on the Engis skull, cited in the fifth chapter, showing the near approach of that ancient cranium to the European standard, will help to remove this source of perplexity.

Human Fossil of Natchez on the Mississippi

I have already alluded to Dr. Dowler's attempt to calculate, in years, the antiquity of the human skeleton said to have been buried under four cypress forests in the delta of the Mississippi, near New Orleans (*see* page 34). In that case no remains of extinct animals were found associated with those of Man: but in another part of the basin of the Mississippi, a human bone, accompanied by bones of *Mastodon* and *Megalonyx*, is supposed to have been washed out of a more ancient alluvial deposit.

After visiting the spot in 1846, I described the geological position of the bones, and discussed their probable age, with a stronger bias, I must confess, as to the antecedent improbability of the contemporaneous entombment of Man and the mastodon than any geologist would now be justified in entertaining.

In the latitude of Vicksburg, 32° 50′ N., the broad, flat, alluvial plain of the Mississippi, *a b*, Fig. 26, is bounded on its eastern side by a table-land *d e*, about 200 feet higher than the river, and extending 12 miles eastward with a gentle upward slope. This elevated platform ends abruptly at *d*, in a line of perpendicular cliffs or bluffs, the base of which is continually undermined by the great river.

Fig. 26.

1 Modern alluvium of the Mississippi.	3, *f* Eocene.
2 Loam or loess.	4 Cretaceous.

The table-land *d e* consists at Vicksburg, through which the annexed section, Fig. 26, passes, of loam, overlying the Tertiary strata *f f*. Between the loam and the Tertiary formation there is usually a deposit of stratified sand and gravel, containing large fragments of silicified corals and the wreck of older Palæozoic rocks. The age of this underlying drift, which is 140 feet thick at Natchez, has not yet been determined; but it may possibly belong to the glacial period. Natchez is about 80 miles in a straight line south of Vicksburg, on the same left bank of the Mississippi. Here there is a bluff, the upper 60 feet of which consists of a continuous portion of the same calcareous loam as at Vicksburg, equally resembling the Rhenish loess in mineral character and in being sometimes barren of fossils, sometimes so full of them that bleached land-shells stand out conspicuously in relief in the vertical and weathered face of cliffs which form the banks of streams, everywhere intersecting the loam.

So numerous are the shells that I was able to collect at

Natchez, in a few hours, in 1846, no less than twenty species of the genera *Helix, Helicina, Pupa, Cyclostoma, Achatina,* and *Succinea,* all identical with shells now living in the same country; and in one place I observed (as happens also occasionally in the valley of the Rhine) a passage of the loam with land-shells into an underlying marly deposit of subaqueous origin, in which shells of the genera *Limnæa, Planorbis, Paludina, Physa,* and *Cyclas* were embedded, also consisting of recent American species. Such deposits, more distinctly stratified than the loam containing land-shells, are produced, as before stated, in all great alluvial plains, where the river shifts its position, and where marshes, ponds, and lakes are formed in its old deserted channels. In this part of America, however, it may have happened that some of these lakes were caused by partial subsidences, such as were witnessed, during the earthquakes of 1811-12, around New Madrid, in the valley of the Mississippi.

Owing to the destructible nature of the yellow loam, *d e,* Fig. 26, every streamlet flowing over the platform has cut for itself, in its way to the Mississippi, a deep gully or ravine; and this erosion has of late years, especially since 1812, proceeded with accelerated speed, ascribable in some degree to the partial clearing of the native forest, but partly also to the effects of the earthquake of 1811-12. By that convulsion the region around Natchez was rudely shaken and much fissured. One of the narrow valleys near Natchez, due to this fissuring, is now called the Mammoth Ravine. Though no less than 7 miles long, and in some parts 60 feet deep, I was assured by a resident proprietor, Colonel Wiley, that it had no existence before 1812. With its numerous ramifications, it is said to have been entirely formed since the earthquake at New Madrid. Before that event, Colonel Wiley had ploughed some of the land exactly over a spot now traversed by part of this water-course.

I satisfied myself that the ravine had been considerably enlarged and lengthened a short time before my visit, and it was then freshly undermined and undergoing constant waste. From a clayey deposit immediately below the yellow loam, bones of the *Mastodon ohioticus,* a species of *Megalonyx,* bones of the genera *Equus, Bos,* and others, some of extinct and others presumed to be of living species, had been detached, and had fallen to the base of the cliffs. Mingled with the rest, the pelvic bone of a man, *os innominatum,* was obtained by Dr. Dickeson of Natchez, in whose collection I saw it. It appeared to be quite in the same state of preservation, and was of the same black colour as the other fossils, and was believed to have come like them from a depth of about 30 feet from the surface. In my *Second Visit to America*, in 1846, I suggested, as a possible explanation of this association of a human bone with remains of *Mastodon* and *Megalonyx*, that the former may possibly have been derived from the vegetable soil at the top of the cliff, whereas the remains of extinct mammalia were dislodged from a lower position, and both may have fallen into the same heap or talus at the bottom of the ravine. The pelvic bone might, I conceived, have acquired its black colour by having lain for years or centuries in a dark superficial peaty soil, common in that region. I was informed that there were many human bones, in old Indian graves in the same district, stained of as black a dye. On suggesting this hypothesis to Colonel Wiley of Natchez, I found that the same idea had already occurred to his mind. No doubt, had the pelvic bone belonged to any recent mammifer other than Man, such a theory would never have been resorted to; but so long as we have only one isolated case, and are without the testimony of a geologist who was present to behold the bone when still engaged in the matrix, and to extract it with his own hands, it is allowable to suspend our judgment as to the high antiquity of the fossil.

If, however, I am asked whether I consider the Natchez loam, with land-shells and the bones of *Mastodon* and *Megalonyx*, to be more ancient than the alluvium of the Somme containing flint implements and the remains of the mammoth and hyæna, I must declare that I do not. Both in Europe and America the land and freshwater shells accompanying the extinct pachyderms are of living species, and I could detect no shell in the Natchez loam so foreign to the basin of the Mississippi as is the *Cyrena fluminalis* to the rivers of modern Euorpe. If, therefore, the relative ages of the Picardy and Natchez alluvium were to be decided on conchological data alone, the fluviomarine beds of Abbeville might rank as a shade older than the loess of Natchez. My reluctance in 1846 to regard the fossil human bone as of Pleistocene date arose in part from the reflection that the ancient loess of Natchez is anterior in time to the whole modern delta of the Mississippi. The table-land, *d e*, Fig. 26, p. 157, was, I believe, once a part of the original alluvial plain or delta of the great river before it was upraised. It has now risen more than 200 feet above its pristine level. After the upheaval, or during it, the Mississippi cut through the old fluviatile formation of which its bluffs are now formed, just as the Rhine has in many parts of its valley excavated a passage through the ancient loess. If I was right in calculating that the present delta of the Mississippi must have required many tens of thousands of years for its growth, and if the claims of the Natchez man to have co-existed with the *mastodon* are admitted, it would follow that North America was peopled by the human race many tens of thousands of years before our time. But even were that true, we could not presume, reasoning from ascertained geological data, that the Natchez bone was anterior in date to the antique flint hatchets of St. Acheul. When we ascend the Mississippi from Natchez to Vicksburg, and then enter the Ohio, we are accompanied everywhere by a continuous fringe of terraces of sand and gravel at a certain height above the alluvial plain, first of the great river, and then of its tributary. We also find that the older alluvium contains the remains of *Mastodon* everywhere, and in some places, as at Evansville, those of the *Megalonyx*. As in the valley of the Somme in Europe, those old Pleistocene gravels often occur at more than one level, and the ancient mounds of the Ohio, with their works of art, are newer than the old terraces of the *mastodon* period, just as the Gallo-Roman tombs of St. Acheul or the Celtic weapons of the Abbeville peat are more modern than the tools of the mammoth-bearing alluvium.

In the first place, I may remind the reader that the vertical movement of 250 feet, required to elevate the loess of Natchez to its present height, is exceeded by the upheaval which the marine stratum of Cagliari, containing pottery, has been ascertained by Count de la Marmora to have experienced. Such changes of level, therefore, have actually occurred in Europe in the human epoch, and may therefore have happened in America. In the second place, I may observe that if, since the Natchez *mastodon* was embedded in clay, the delta of the Mississippi has been formed, so, since the mammoth and rhinoceros of Abbeville and Amiens we e enveloped in fluviatile mud and gravel, together with flint tools, a great thickness of peat has accumulated in the valley of the Somme ; and antecedently to the first growth of peat, there had been time for the extinction of a great many mammalia, requiring, perhaps, a lapse of ages many times greater than that demanded for the formation of 30 feet of peat, for since the earliest growth of the latter there has been no change in the species of mammalia in Europe.

Should future researches, therefore, confirm the opinion that the Natchez man co-existed with the *mastodon*, it would not enhance the value of the geological evidence in favour of Man's antiquity, but merely render the delta of the Mississippi available as a chronometer, by which the lapse of Pleistocene time could be measured somewhat less vaguely than by any means of measuring which have as yet been discovered or rendered available in Europe.

CHAPTER XII

ANTIQUITY OF MAN RELATIVELY TO THE GLACIAL PERIOD AND TO
THE EXISTING FAUNA AND FLORA

Chronological Relation of the Glacial Period, and the earliest known Signs
of Man's Appearance in Europe—Series of Tertiary Deposits in Nor-
folk and Suffolk immediately antecedent to the Glacial Period—
Gradual Refrigeration of Climate proved by the Marine Shells of
successive Groups—Marine Newer Pliocene Shells of northern Character near Woodbridge—Section of the Norfolk Cliffs—Norwich
Crag—Forest Bed and fluvio-marine Strata—Fossil Plants and
Mammalia of the same—Overlying Boulder Clay and contorted Drift
—Newer freshwater Formation of Mundesley compared to that of
Hoxne—Great Oscillations of Level implied by the Series of Strata
in the Norfolk Cliffs—Earliest known Date of Man long subsequent
to the existing Fauna and Flora.

FREQUENT allusions have been made in the preceding pages
to a period called the glacial, to which no reference is made
in the Chronological Table of Formations given at p. 5. It
comprises a long series of ages, during which the power of cold,
whether exerted by glaciers on the land, or by floating ice on
the sea, was greater in the northern hemisphere, and extended
to more southern latitudes than now. [Note 19.]

It often happens that when in any given region we have
pushed back our geological investigations as far as we can in
search of evidence of the first appearance of Man in Europe,
we are stopped by arriving at what is called the "boulder
clay" or "northern drift." This formation is usually quite
destitute of organic remains, so that the thread of our inquiry
into the history of the animate creation, as well as of man, is
abruptly cut short. The interruption, however, is by no means
encountered at the same point of time in every district. In
the case of the Danish peat, for example, we get no farther back
than the Recent period of our Chronologic l Table (p. 5), and
then meet with the boulder clay; and it is the same in the
valley of the Clyde, where the marine strata contain the ancient
canoes before described (p. 37), and where nothing intervenes
between that Recent formation and the glacial drift. But we
have seen that, in the neighbourhood of Bedford the memorials
of Man can be traced much farther back into the past, namely,
into the Pleistocene epoch, when the human race was contemporary with the mammoth and many other species of mam-

malia now extinct. Nevertheless, in Bedfordshire as in Denmark, the formation next antecedent in date to that containing the human implements is still a member of the glacial drift, with its erratic blocks.

If the reader remembers what was stated in the eighth chapter as to the absence or extreme scarcity of human bones and works of art in all strata, whether marine or freshwater, even in those formed in the immediate proximity of land inhabited by millions of human beings, he will be prepared for the general dearth of human memorials in glacial formations, whether Recent, Pleistocene, or of more ancient date. If there were a few wanderers over lands covered with glaciers, or over seas infested with ice-bergs, and if a few of them left their bones or weapons in moraines or in marine drift, the chances, after the lapse of thousands of years, of a geologist meeting with one of them must be infinitesimally small.

It is natural, therefore, to encounter a gap in the regular sequence of geological monuments bearing on the past history of Man, wherever we have proofs of glacial action having prevailed with intensity, as it has done over large parts of Europe and North America, in the Pleistocene period. As we advance into more southern latitudes approaching the 50th parallel of latitude in Europe, and the 40th in North America, this disturbing cause ceases to oppose a bar to our inquiries; but even then, in consequence of the fragmentary nature of all geological annals, our progress is inevitably slow in constructing anything like a connected chain of history, which can only be effected by bringing the links of the chain found in one area to supply the information which is wanting in another.

The least interrupted series of consecutive documents to which we can refer in the British Islands, when we desire to connect the Pliocene with the Pleistocene periods, are found in the counties of Norfolk, Suffolk, and Essex; and I shall speak of them in this chapter, as they have a direct bearing on the relations of the human and glacial periods, which will be the subject of several of the following chapters. The fossil shells of the deposits in question clearly point to a gradual refrigeration of climate, from a temperature somewhat warmer than that now prevailing in our latitudes to one of intense cold; and the successive steps which have marked the coming on of the increasing cold are matters of no small geological interest. [Note 20.]

It will be seen in the table at p. 5, that next before the

Pleistocene period stands the Pliocene. The shelly and sandy beds representing these periods in Norfolk and Suffolk are termed provincially Crag, having under the name been long used in agriculture to fertilise soils deficient in calcareous matter, or to render them less stiff and impervious. In Suffolk, the older Pliocene strata called Crag are divisible into the Coralline and the Red Crags, the former being the older of the two. In Norfolk, a more modern formation, commonly termed the "Norwich," or sometimes the "mammaliferous" Crag, which is referable to the newer Pliocene period, occupies large areas.

We are indebted to Mr. Searles Wood, F.G.S., for an admirable monograph on the fossil shells of these British Pliocene formations. He has not himself given us an analysis of the results of his treatise, but the following tables have been drawn up for me by Mr. S. P. Woodward, the well-known author of the *Manual of the Mollusca, Recent and Fossil* (London, 1851-56), in order to illustrate some of the general conclusions to which Mr. Wood's careful examination of 442 species of mollusca has led.

Number of known Species of Marine Testacea in the three English Pliocene Deposits, called the Norwich, the Red, and the Coralline Crags

Brachiopoda	6
Lamellibranchia	206
Gasteropoda	230
Total . . .	442

Distribution of the above Marine Testacea

Number of Species.		Species common to the	
Norwich Crag . .	81	Norwich and Red Crag (not in Cor.)	33
Red Crag . . .	225	Norwich and Coralline (not in Red)	4
Coralline Crag . .	327	Red and Coralline (not in Norwich)	116
		Norwich, Red, and Coralline	19[1]

Proportion of Recent to Extinct Species

	Recent.	Extinct.	Percentage of Recent.
Norwich Crag . . .	69	12	85
Red Crag . . .	130	95	57
Coralline Crag . . .	168	159	51

Recent Species not living now in British Seas

	Northern Species.	Southern.
Norwich Crag . . .	12	0
Red Crag	8	16
Coralline Crag . . .	2	27

[1] These 19 species must be added to the numbers 33, 4, and 116 respectively, in order to obtain the full amount of *common* species in each of those cases.

In the above list I have not included the shells of the glacial beds of the Clyde and of several other British deposits of newer origin than the Norwich Crag, in which nearly all—perhaps all—the species are Recent. The land and freshwater shells, thirty-two in number, have also been purposely omitted, as well as three species of London Clay shells, suspected by Mr. Wood himself to be spurious.

By far the greater number of the living marine species included in these tables are still inhabitants of the British seas; but even these differ considerably in their relative abundance, some of the commonest of the Crag shells being now extremely scarce; as, for example, *Buccinopsis Dalei;* and others, rarely met with in a fossil state, being now very common, as *Murex erinaceus* and *Cardium echinatum.*

The last table throws light on a marked alteration in the climate of the three successive periods. It will be seen that in the Coralline Crag there are twenty-seven southern shells, including twenty-six Mediterranean, and one West Indian species (*Erato Maugeriæ*). Of these only thirteen occur in the Red Crag, associated with three new southern species, while the whole of them disappear from the Norwich beds. On the other hand, the Coralline Crag contains only two shells closely related to arctic forms of the genera *Admete* and *Limopsis.* The Red Crag contains, as stated in the table, eight northern species, all of which recur in the Norwich Crag, with the addition of four others, also inhabitants of the arctic regions; so that there is good evidence of a continual refrigeration of climate during the Pliocene period in Britain. The presence of these northern shells cannot be explained away by supposing that they were inhabitants of the deep parts of the sea; for some of them, such as *Tellina calcarea* and *Astarte borealis,* occur plentifully, and sometimes, with the valves united by their ligament, in company with other littoral shells, such as *Mya arenaria* and *Littorina rudis,* and evidently not thrown up from deep water. Yet the northern character of the Norwich Crag is not fully shown by simply saying that it contains twelve northern species. It is the predominance of certain genera and species, such as *Tellina calcarea, Astarte borealis, Scalaria groenlandica,* and *Fusus carinatus,* which satisfies the mind of a conchologist as to the arctic character of the Norwich Crag. In like manner, it is the presence of such genera as *Pyrula, Columbella, Terebra, Cassidaria, Pholadomya, Lingula, Discina,* and others which give a southern aspect to the Coralline Crag shells.

The cold, which had gone on increasing from the time of the Coralline to that of the Norwich Crag, continued, though not perhaps without some oscillations of temperature, to become more and more severe after the accumulation of the Norwich Crag, until it reached its maximum in what has been called the glacial epoch. The marine fauna of this last period contains, both in Ireland and Scotland, Recent species of mollusca now living in Greenland and other seas far north of the areas where we find their remains in a fossil state.

The refrigeration of climate from the time of the older to that of the newer Pliocene strata is not now announced for the first time, as it was inferred from a study of the Crag shells in 1846 by the late Edward Forbes.[1]

The most southern point to which the marine beds of the Norwich Crag have yet been traced is at Chillesford, near Woodbridge, in Suffolk, about 80 miles N.E. of London, where, as Messrs. Prestwich and Searles Wood have pointed out,[2] they exhibit decided marks of having been deposited in a sea of a much lower temperature than that now prevailing in the same latitude. Out of twenty-three shells obtained in that locality from argillaceous strata 20 feet thick, two only, namely, *Nucula Cobboldiæ* and *Tellina obliqua*, are extinct, and not a few of the other species, such as *Leda lanceolata*, *Cardium groenlandicum*, *Lucina borealis*, *Cyprina islandica*, *Panopæa norvegica*, and *Mya truncata*, betray a northern, and some of them an arctic character.

These Chillesford beds are supposed to be somewhat more modern than any of the purely marine strata of the Norwich Crag exhibited by the sections of the Norfolk cliffs N.W. of Cromer, which I am about to describe. Yet they probably preceded in date the " Forest Bed " and fluvio-marine deposits of those same cliffs. They are, therefore, of no small importance in reference to the chronology of the glacial period, since they afford evidence of an assemblage of fossil shells with a proportion of between eight and nine in a hundred of extinct species occurring so far south as latitude 53° N., and indicating so cold a climate as to imply that the glacial period commenced before the close of the Pliocene era.

The annexed section will give a general idea of the ordinary succession of the Pliocene and Pleistocene strata which rest upon the Chalk in the Norfolk and Suffolk cliffs. These cliffs vary in height from fifty to above three hundred feet. At the north-

---

[1] *Memoirs of Geological Survey*, London, 1846, p. 391.
[2] *Quart. Jour. Geol. Soc.*, vol. v., 1849, p. 345.

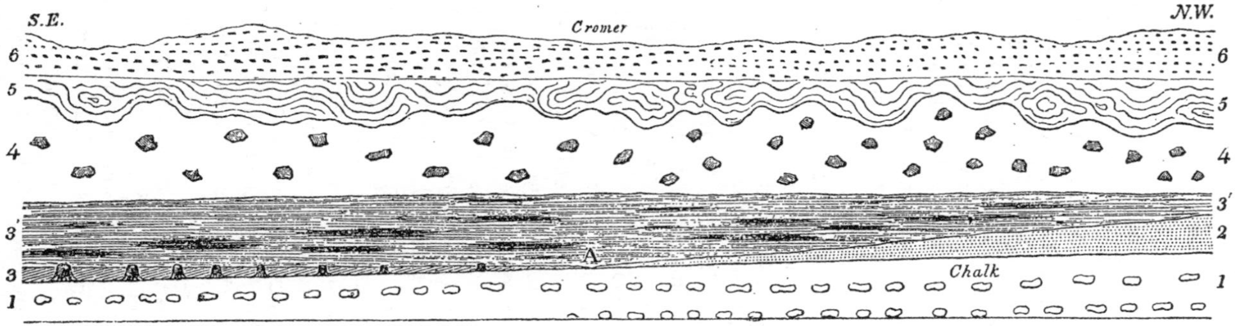

Fig. 27.

DIAGRAM TO ILLUSTRATE THE GENERAL SUCCESSION OF THE STRATA IN THE NORFOLK CLIFFS,
EXTENDING SEVERAL MILES N.W. AND S.E. OF CROMER

A Site of Cromer Jetty.

1 Upper Chalk with flints in regular stratification.

2 Norwich Crag, rising from low water at Cromer to the top of the cliffs at Weybourn, seven miles distant.

3 "Forest Bed," with stumps of trees *in situ* and remains of *Elephas meridionalis*, *E. primigenius*, *E. antiquus*, *Rhinoceros etruscus*, etc. This bed increases in depth and thickness eastward. No Crag (No. 2) known east of Cromer Jetty.

3' Fluvio-marine series. At Cromer and eastward, with abundant lignite beds and mammalian remains, and

with cones of the Scotch and spruce firs and wood. At Runton, north-west of Cromer, expanding into a thick freshwater deposit, with overlying marine strata, elsewhere consisting of alternating sands and clays, tranquilly deposited, some with marine, others with freshwater shells.

4 Boulder clay of glacial period, with far transported erratics, some of them polished and scratched, 20 to 80 feet in thickness.

5 Contorted drift.

6 Superficial gravel and sand with covering of vegetable soil.

western extremity of the section at Weybourn (beyond the limits of the annexed diagram), and from thence to Cromer, a distance of 7 miles, the Norwich Crag, a marine deposit, reposes immediately upon the Chalk. A vast majority of its shells are of living species such as *Cardium edule*, *Cyprina islandica*, *Scalaria groenlandica*, and *Fusus antiquus*, and some few extinct, as *Tellina obliqua*, and *Nucula Cobboldiæ*. At Cromer jetty this formation thins out, as expressed in the diagram at A; and to the south we find No. 3, or what is commonly called the " Forest Bed," reposing immediately upon the Chalk, and occupying, as it were, the place previously held by the marine Crag No. 2. This buried forest has been traced for more than 40 miles, being exposed at certain seasons and states of the beach between high and low water mark. It extends from Cromer to near Kessingland, and consists of the stumps of numerous trees standing erect, with their roots attached to them, and penetrating in all directions into the loam or ancient vegetable soil on which they grew. They mark the site of a forest which existed there for a long time, since, besides the erect trunks of trees, some of them 2 and 3 feet in diameter, there is a vast accumulation of vegetable matter in the immediately overlying clays. Thirty years ago, when I first examined this bed, I saw many trees, with their roots in the old soil, laid open at the base of the cliff near Happisburgh; and long before my visit, other observers, and among them the late Mr. J. C. Taylor, had noticed the buried forest. Of late years it has been repeatedly seen at many points by Mr. Gunn, and, after the great storms of the autumn of 1861, by Mr. King. In order to expose the stumps to view, a vast body of sand and shingle must be cleared away by the force of the waves. [Note 21.]

As the sea is always gaining on the land, new sets of trees are brought to light from time to time, so that the breadth as well as length of the area of ancient forest land seems to have been considerable. Next above No. 3, we find a series of sands and clays with lignite (No. 3'), sometimes 10 feet thick, and containing alternations of fluviatile and marine strata, implying that the old forest land, which may at first have been considerably elevated above the level of the sea, had sunk down so as to be occasionally overflowed by a river, and at other times by the salt waters of an estuary. There were probably several oscillations of level which assisted in bringing about these changes, during which trees were often uprooted and laid prostrate, giving rise to layers of lignite. Occasionally marshes were formed and peaty matter accumulated, after which salt water again predominated, so that species of *Mytilus, Mya, Leda,* and other marine genera, lived in the same area where the *Unio, Cyclas,* and *Paludina* had flourished for a time. That the marine shells lived and died on the spot, and were not thrown up by the waves during a storm, is proved, as Mr. King has remarked, by the fact that at West Runton, N.W. of Cromer, the *Mya truncata* and *Leda myalis* are found with both valves united and erect in the loam, all with their posterior or siphuncular extremities uppermost. This attitude affords as good evidence to the conchologist that those mollusca lived and died on the spot as the upright position of the trees proves to the botanist that there was a forest over the Chalk east of Cromer.

Between the stumps of the buried forest, and in the lignite above them, are many well-preserved cones of the Scotch and spruce firs, *Pinus sylvestris,* and *Pinus abies.* The specific names of these fossils were determined for me in 1840, by a botanist of no less authority than the late Robert Brown; and Professor Heer has lately examined a large collection from the same stratum, and recognised among the cones of the spruce some which had only the central part or axis remaining, the rest having been bitten off, precisely in the same manner as when in our woods the squirrel has been feeding on the seeds. There is also in the forest-bed a great quantity of resin in lumps, resembling that gathered for use, according to Professor Heer, in Switzerland, from beneath spruce firs.

The following is a list of some of the plants and seeds which were collected by the Rev. S. W. King, in 1861, from the forest bed at Happisburgh, and named by Professor Heer:—

Plants and Seeds of the Forest and Lignite Beds below the Glacial Drift of the Norfolk Cliffs

Pinus sylvestris, Scotch fir.	*Prunus spinosa,* common sloe.
Pinus abies, spruce fir.	*Menyanthes trifoliata,* buckbean.
Taxus baccata, yew.	*Nymphæa alba,* white water-lily.
Nuphar luteum, yellow water-lily.	*Alnus,* alder.
Ceratophyllum demersum, hornwort.	*Quercus,* oak.
Potamogeton, pondweed.	*Betula,* birch.

The insects, so far as they are known, including several species of *Donacia,* are, like the plants and freshwater shells, of living species. It may be remarked, however, that the Scotch fir has been confined in historical times to the northern parts of the British Isles, and the spruce fir is nowhere indigenous in Great Britain.   The other plants are such as might now be found in Norfolk, and many of them indicate fenny or marshy ground.[1]

When we consider the familiar aspect of the flora, the accompanying mammalia are certainly most extraordinary.   There are no less than three elephants, a rhinoceros and hippopotamus, a large extinct beaver, and several large estuarine and marine mammalia, such as the walrus, the narwhal, and the whale.

The following is a list of some of the species of which the bones have been collected by Messrs. Gunn and King.

Those marked * have been recorded by Professor Owen in his British Fossil Mammalia.   Those marked † have been recognised by the same authority in the cabinets of Messrs. Gunn and King, or in the Norwich Museum; the other three are given on the authority of Dr. Falconer.

Mammalia of the Forest and Lignite Beds below the Glacial Drift of the Norfolk Cliffs

Elephas meridionalis.	**Cervus capreolus.*
*Elephas primigenius.	*†Cervus tarandus.*
Elephas antiquus.	*†Cervus Sedgwickii.*
Rhinoceros etruscus.	**Arvicola amphibia.*
*Hippopotamus (major ?).	**Castor (Trogontherium) Cuvieri.*
*Sus scrofa.	**Castor europæus.*
*Equus (fossilis ?).	**Palæospalax magnus.*
*Ursus (sp.?).	*†Trichecus rosmarus,* Walrus.
†Canis lupus.	*†Monodon monoceros,* Narwhal.
†Bison priscus.	*†Balænoptera.*
†Megaceros hibernicus.	

Mr. Gunn informs me that the vertebræ of two distinct whales were found in the fluvio-marine beds at Bacton, and that one of them, shown to Professor Owen, is said by him to imply that the animal was 60 feet long.   A narwhal's tusk was discovered by Mr. King near Cromer, and the remains of a walrus.   No less than three species of elephant, as determined by Dr. Falconer, have been obtained from the strata 3 and 3′, of which, according to Mr. King, *E. meridionalis* is the most common, the mammoth next in abundance, and the third, *E. antiquus,* comparatively rare.

The freshwater shells accompanying the fossil quadrupeds,

[1] Mr. King discovered in 1863, in the forest bed, several rhizomes of the large British fern *Osmunda regalis,* of such dimensions as they are known to attain in marshy places.   They are distinguishable from those of other British ferns by the peculiar arrangement of the vessels, as seen under the microscope in a cross section.

above enumerated, are such as now inhabit rivers and ponds in England; but among them, as at Runton, between the "forest bed" and the glacial deposits, a remarkable variety of the *Cyclas amnica* occurs (Fig. 28), identical with that which accompanies the *Elephas antiquus* at Ilford and Grays in the valley of the Thames.

All the freshwater shells of the beds intervening between the Forest-bed No. 3, and the glacial formation 4, Fig. 27, are of Recent species. As to the small number of marine shells occurring in the same fluvio-marine series, I have seen none which belonged to extinct species, although one or two have been cited by authors. I am in doubt, therefore, whether to class the forest bed and overlying strata as Pleistocene, or to consider them as beds of passage between the Pliocene and Pleistocene periods. The fluvio-marine series usually terminates upwards in finely laminated sands and clays without fossils, on which reposes the boulder clay.

Fig. 28.

Cyclas (Pisidium) amnica var.?

The two middle figures are of the natural size.

This formation, No. 4, is of very varying thickness. Its glacial character is shown, not only by the absence of stratification, and the great size and angularity of some of the included blocks of distant origin, but also by the polished and scratched surfaces of such of them as are hard enough to retain any markings.

Near Cromer, blocks of granite from 6 to 8 feet in diameter have been met with, and smaller ones of syenite, porphyry, and trap, besides the wreck of the London Clay, Chalk, Oolite, and Lias, mixed with more ancient fossiliferous rocks. Erratics of Scandinavian origin occur chiefly in the lower portions of the till. I came to the conclusion in 1834, that they had really come from Norway and Sweden, after having in that year traced the course of a continuous stream of such blocks from those countries to Denmark, and across the Elbe, through Westphalia, to the borders of Holland. It is not surprising that they should then reappear on our eastern coast between the Tweed and the Thames, regions not half so remote from parts of Norway as are many Russian erratics from the sources whence they came. [Note 22.]

According to the observations of the Rev. J. Gunn and the late Mr. Trimmer, the glacial drift in the cliffs at Lowestoft consists of two divisions, the lower of which abounds in the Scandinavian blocks, supposed to have come from the north-east; while the upper, probably brought by a current from the north-west, contains chiefly fragments of Oolitic rocks, more rolled than those of the lower deposit. The united thickness

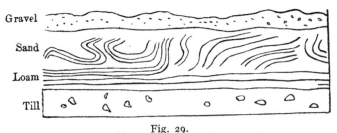

Fig. 29.

Cliff 50 feet high between Bacton Gap and Mundesley.

of the two divisions, without reckoning some interposed laminated beds, is 80 feet, but it probably exceeds 100 feet near Happisburgh.[1] Although these subdivisions of the drift may be only of local importance, they help to show the changes of currents and other conditions, and the great lapse of time which the accumulation of so varied a series of deposits must have required.

The lowest part of the glacial till, resting on the laminated clays before mentioned, is very even and regular, while its upper surface is remarkable for the unevenness of its outline, owing partly, in all likelihood, to denudation, but still more to other causes presently to be discussed.

The overlying strata of sand and gravel, No. 5, p. 167, often display a most singular derangement in their stratification, which in many places seems to have a very intimate relation to the irregularities of outline in the subjacent till. There are some cases, however, where the upper strata are much bent, while the lower beds of the same series have continued horizontal. Thus the annexed section (Fig. 29) represents a cliff about 50

[1] *Quart. Jour. Geol. Soc.*, vol. vii., 1851, p. 21.

feet high, at the bottom of which is till, or unstratified clay, containing boulders, having an even horizontal surface, on which repose conformably beds of laminated clay and sand about 5 feet thick, which, in their turn, are succeeded by vertical, bent, and contorted layers of sand and loam 20 feet thick, the whole being covered by flint gravel. The curves of the variously coloured beds of loose sand, loam, and pebbles, are so complicated that not only may we sometimes find portions of them which maintain their verticality to a height of 10 or 15 feet, but they have also been folded upon themselves in such a manner that continuous layers might be thrice pierced in one perpendicular boring.

At some points there is an apparent folding of the beds round a central nucleus, as at *a*, Fig. 30, where the strata seem bent round a small mass of Chalk, or, as in Fig. 31, where the

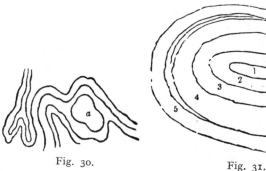

Fig. 30.

Folding of the strata between East and West Runton.

Fig. 31.

Section of concentric beds west of Cromer.

1 Blue clay.   3 Yellow Sand.
2 White sand.   4 Striped loam and clay.
      5 Laminated blue clay.

blue clay No. 1 is in the centre; and where the other strata 2, 3, 4, 5 are coiled round it; the entire mass being 20 feet in perpendicular height. This appearance of concentric arrangement around a nucleus is, nevertheless, delusive, being produced by the intersection of beds bent into a convex shape; and that which seems the nucleus being, in fact, the innermost bed of the series, which has become partially visible by the removal of the protuberant portions of the outer layers.

To the north of Cromer are other fine illustrations of contorted drift reposing on a floor of Chalk horizontally stratified and having a level surface. These phenomena, in themselves sufficiently difficult of explanation, are rendered still more anomalous by the occasional enclosure in the drift of huge fragments of Chalk many yards in diameter. One striking instance occurs west of Sheringham, where an enormous pinnacle of Chalk, between 70 and 80 feet in height, is flanked on both sides by vertical layers of loam, clay, and gravel (Fig. 32).

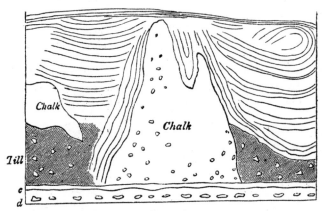

Fig. 32.

Included pinnacle of Chalk at Old Hythe point, west of Sheringham.

d Chalk with regular layers of flints.
c Layer called " the pan," of Chalk, flints, and marine shells of Recent species, cemented by oxide of iron.

This chalky fragment is only one of many detached masses which have been included in the drift, and forced along with it into their present position. The level surface of the Chalk *in situ* (*d*) may be traced for miles along the coast, where it has escaped the violent movements to which the incumbent drift has been exposed.[1]

We are called upon, then, to explain how any force can have been exerted against the upper masses, so as to produce movements in which the subjacent strata have not participated. It may be answered that, if we conceive the till and its boulders to have been drifted to their present place by ice, the lateral pressure may have been supplied by the stranding of ice-islands. We learn, from the observations of Messrs. Dease and Simpson

[1] For a full account of the drift of East Norfolk, see a paper by the author, *Philosophical Magazine*, No. 104, May 1840.

in the polar regions, that such islands, when they run aground, push before them large mounds of shingle and sand. It is therefore probable that they often cause great alterations in the arrangement of pliant and incoherent strata forming the upper part of shoals or submerged banks, the inferior portions of the same remaining unmoved. Or many of the complicated curvatures of these layers of loose sand and gravel may have been due to another cause, the melting on the spot of ice-bergs and coast ice in which successive deposits of pebbles, sand, ice, snow, and mud, together with huge masses of rock fallen from cliffs, may have become interstratified. Ice-islands so constituted often capsize when afloat, and gravel once horizontal may have assumed, before the associated ice was melted, an inclined or vertical position. The packing of ice forced up on a coast may lead to a similar derangement in a frozen conglomerate of sand or shingle, and, as Mr. Trimmer has suggested,[1] alternate layers of earthy matter may have sunk down slowly during the liquefaction of the intercalated ice so as to assume the most fantastic and anomalous positions, while the strata below, and those afterwards thrown down above, may be perfectly horizontal (*see above*).

In most cases where the principal contortions of the layers of gravel and sand have a decided correspondence with deep indentations in the underlying till, the hypothesis of the melting of large lumps and masses of ice once mixed up with the till affords the most natural explanation of the phenomena. The quantity of ice now seen in the cliffs near Behring's Straits, in which the remains of fossil elephants are common, and the huge fragments of solid ice which Meyendorf discovered in Siberia, after piercing through a considerable thickness of incumbent soil, free from ice, is in favour of such an hypothesis, the partial failure of support necessarily giving rise to foldings in the overlying and previously horizontal layers, as in the case of creeps in coal mines.[2]

In the diagram of the cliffs at p. 167, the bent and contorted beds No. 5, last alluded to, are represented as covered by undisturbed beds of gravel and sand No. 6. These are usually destitute of organic remains; but at some points marine shells of Recent species are said to have been found in them. They afford evidence at many points of repeated denudation and redeposition, and may be the monuments of a long series of ages.

[1] *Quart. Jour. Geol. Soc.*, vol. vii., 1851, pp. 22, 30.
[2] See *Manual of Geology*, by the author, p. 51.

Mundesley Post-glacial Freshwater Formation

In the range of cliffs above described at Mundesley, about 8 miles S.E. of Cromer, a fine example is seen of a freshwater formation, newer than all those already mentioned, a deposit which has filled up a depression hollowed out of all the older beds 3, 4, and 5 of the section Fig. 27.

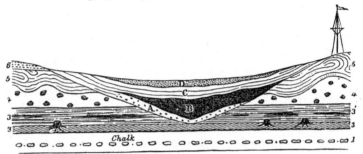

Fig. 33.

Section of the newer freshwater formation in the cliffs at Mundesley, eight miles S.E. of Cromer, drawn up by the Rev. S. W. King.

Height of cliff where lowest, 35 feet above high water.

Older Series.

1 Fundamental Chalk, below the beach line.
3 Forest bed, with elephant, rhinoceros, stag, etc., and with tree roots and stumps, also below the beach line.
3′ Finely laminated sands and clays, with thin layer of lignite, and shells of *Cyclas* and *Valvata*, and with *Mytilus* in some beds.
4 Glacial boulder till.
5 Contorted drift.
6 Gravel overlying contorted drift.

N.B.—No. 2 of the section, Fig. 27, at p. 167, is wanting here.

Newer Freshwater Beds.

A Coarse river gravel, in layers inclined against the till and laminated sands.
B Black peaty deposit, with shells of *Anodon, Valvata, Cyclas, Succinea, Limnæa, Paludina*, etc., seeds of *Ceratophyllum demersum, Nuphar lutea*, scales and bones of pike, perch, salmon, etc., elytra of *Donacia, Copris, Harpalus*, and other beetles.
C Yellow sands.
D Drift gravel.

When I examined this line of coast in 1839, the section alluded to was not so clearly laid open to view as it has been of late years, and finding at that period not a few of the fossils in the lignite beds No. 3′ above the forest bed, identical in species with those from the post-glacial deposits B C, I supposed the whole to have been of contemporaneous origin, and so described them in my paper on the Norfolk cliffs.[1]

Mr. Gunn was the first to perceive this mistake, which he explained to me on the spot when I revisited Mundesley in the autumn of 1859 in company with Dr. Hooker and Mr. King. The last-named geologist has had the kindness to draw up for me the annexed diagram of the various beds which he has recently studied in detail.[2]

The formations 3, 4, and 5 already described, Fig. 27, were evidently once continuous, for they may be followed for miles N.W. and S.E. without a break, and always in the same order. A valley or river channel was cut through them, probably during the gradual upheaval of the country, and the hollow became afterwards the receptacle of the comparatively modern fresh-water beds A, B, C, and D. They may well represent a silted up river-channel, which remained for a time in the state of a

Fig. 34.

Paludina marginata, Michaud (*P. minuta,* Strickland).
Hydrobia marginata.[3]

The middle figure is of the natural size.

lake or mere, and in which the black peaty mass B accumulated by a very slow growth over the gravel of the river-bed A. In B we find remains of some of the same plants which were enumerated as common in the ancient lignite in 3', such as the yellow water-lily and hornwort, together with some freshwater shells which occur in the same fluvio-marine series 3'.

The only shell which I found not referable to a British species is the minute *Paludina,* Fig. 34, already alluded to.

[1] *Philosophical Magazine,* vol. xvi., 1840, p. 345.
[2] Mr. Prestwich has given a correct account of this section in a paper read to the British Association, Oxford, 1860. See *The Geologist,* vol. iv., 1861.
[3] This shell is said to have a sub-spiral operculum (not a concentric one, as in *Paludina*), and therefore to be referable to the *Hydrobia,* a sub-genus of *Rissoa.* But this species is always associated with freshwater shells, while the *Rissoæ* frequent marine and brackish waters.

When I showed the scales and teeth of the pike, perch, roach, and salmon, which I obtained from this formation, to M. Agassiz, he thought they varied so much from their nearest living representatives that they might rank as distinct species; but Mr. Yarrell doubted the propriety of so distinguishing them. The insects, like the shells and plants, are identical, so far as they are known, with living British species. No progress has yet been made at Mundesley in discovering the contemporary mammalia.

By referring to the description and section before given of the freshwater deposit at Hoxne, the reader will at once perceive the striking analogy of the Mundesley and Hoxne deposits, the latter so productive of flint implements of the Amiens type. Both of them, like the Bedford gravel with flint tools and the bones of extinct mammalia, are post-glacial. It will also be seen that a long series of events, accompanied by changes in physical geography, intervened between the "forest bed," No. 3, Fig. 27, when the *Elephas meridionalis* flourished, and the period of the Mundesley fluviatile beds A, B, C; just as in France I have shown that the same *E. meridionalis* belonged to a system of drainage different from and anterior to that with which the flint implements of the old alluvium of the Somme and the Seine were connected.

Before the growth of the ancient forest, No. 3, Fig. 33, the *Mastodon arvernensis*, a large proboscidian, characteristic of the Norwich Crag, appears to have died out, or to have become scarce, as no remains of it have yet been found in the Norfolk cliffs. There was, no doubt, time for other modifications in the mammalian fauna between the era of the marine beds, No. 2, Fig. 27 (the shells of which imply permanent submergence beneath the sea), and the accumulation of the uppermost of the fluvio-marine, and lignite beds, No. 3', which overlie both Nos. 3 and 2, or the buried forest and the Crag. In the interval we must suppose repeated oscillations of level, during which land covered with trees, an estuary with its freshwater shells, and the sea with its *Mya truncata* and other mollusca still retaining their erect position, gained by turns the ascendency. These changes were accompanied by some denudation followed by a grand submergence of several hundred feet, probably brought about slowly, and when floating ice aided in transporting erratic blocks from great distances. The glacial till No. 4 then originated, and the gravel and sands No. 5 were afterwards superimposed on the boulder clay, first in horizontal beds, which became subsequently contorted. These were covered in their turn by other layers of gravel and sand, No. 6, Figs. 27 and 33, the downward movement still continuing.

The entire thickness of the beds above the Chalk at some points near the coast, and the height at which they now are raised, are such as to show that the subsidence of the country after the growth of the forest bed exceeded 400 feet. The re-elevation must have amounted to nearly as many feet, as the site of the ancient forest, originally subaërial, has been brought up again to within a few feet of high-water mark. Lastly, after all these events, and probably during the final process of emergence, the valley was scooped out in which the newer fresh-water strata of Mundesley, Fig. 33, were gradually deposited.

Throughout the whole of this succession of geographical changes, the flora and invertebrate fauna of Europe appear to have undergone no important revolution in their specific characters. The plants of the forest bed belonged already to what has been called the Germanic flora. The mollusca, the insects, and even some of the mammalia, such as the European beaver and roebuck, were the same as those now co-existing with Man. Yet the oldest memorials of our species at present discovered in Great Britain are post-glacial, or posterior in date to the boulder clay, No. 4, Figs. 27 and 33. The position of the Hoxne flint implements corresponds with that of the Mundesley beds, from A to D, Fig. 33, and the most likely stratum in which to find hereafter flint tools is no doubt the gravel A of that section, which has all the appearance of an old river-bed. No flint tools have yet been observed there, but had the old alluvium of Amiens or Abbeville occurred in the Norfolk cliffs instead of the valley of the Somme, and had we depended on the waves of the sea instead of the labour of many hundred workmen continued for twenty years, for exposing the flint implements to view, we might have remained ignorant to this day of the fossil relics brought to light by M. Boucher de Perthes and those who have followed up his researches.

Neither need we despair of one day meeting with the signs of Man's existence in the forest bed No. 3, or in the overlying strata 3', on the ground of any uncongeniality in the climate or incongruity in the state of the animate creation with the well-being of our species. For the present we must be content to wait and consider that we have made no investigations which entitle us to wonder that the bones or stone weapons of the era of the *Elephas meridionalis* have failed to come to light. If any such lie hid in those strata, and should hereafter be revealed to us, they would carry back the antiquity of Man to a distance of time probably more than twice as great as that which separates our era from that of the most ancient of the tool-bearing gravels yet discovered in Picardy, or elsewhere. But even then the reader will perceive that the age of Man, though preglacial, would be so modern in the great geological calendar, as given in Chap. I., that he would scarcely date so far back as the commencement of the Pleistocene period.

CHAPTER XIII

CHRONOLOGICAL RELATIONS OF THE GLACIAL PERIOD AND THE EARLIEST SIGNS OF MAN'S APPEARANCE IN EUROPE

Chronological Relations of the Close of the Glacial Period and the earliest geological Signs of the Appearance of Man—Effects of Glaciers and Icebergs in polishing and scoring Rocks—Scandinavia once encrusted with Ice like Greenland—Outward Movement of Continental Ice in Greenland—Mild Climate of Greenland in the Miocene Period—Erratics of Recent Period in Sweden—Glacial State of Sweden in the Pleistocene Period—Scotland formerly encrusted with Ice—Its subsequent Submergence and Re-elevation—Latest Changes produced by Glaciers in Scotland—Remains of the Mammoth and Reindeer in Scotch Boulder Clay—Parallel Roads of Glen Roy formed in Glacier Lakes—Comparatively modern Date of these Shelves.

THE chronological relations of the human and glacial periods were frequently alluded to in the last chapter, and the sections obtained near Bedford, and at Hoxne, in Suffolk, and a general view of the Norfolk cliffs, have taught us that the earliest signs of Man's appearance in the British isles, hitherto detected, are of post-glacial date. We may now therefore inquire whether the peopling of Europe by the human race and by the mammoth and other mammalia now extinct, was brought about during the concluding phases of the glacial epoch.

Although it may be impossible in the present state of our knowledge to come to a positive conclusion on this head, I know of no inquiry better fitted to clear up our views respecting the geological state of the northern hemisphere at the time when the fabricators of the flint implements of the Amiens type flourished. I shall therefore now proceed to consider the chronological relations of that ancient people with the final retreat of the glaciers from the mountains of Scandinavia, Scotland, Wales, and Switzerland.

Superficial Markings and Deposits left by Glaciers and Icebergs

In order fully to discuss this question, I must begin by referring to some of the newest theoretical opinions entertained on the glacial question. When treating of this subject in the *Principles of Geology*, chap. xv., and in the *Manual (or*

Elements) *of Geology,* chap. xi., I have stated that the whole
mass of the ice in a glacier is in constant motion, and that the
blocks of stone detached from boundary precipices, and the
mud and sand swept down by avalanches of snow, or by rain
from the surrounding heights, are lodged upon the surface and
slowly borne along in lengthened mounds, called in Switzerland
moraines.   These accumulations of rocky fragments and detrital
matter are left at the termination of the glacier, where it melts
in a confused heap called the " terminal moraine," which is
unstratified, because all the blocks, large and small, as well as
the sand and the finest mud, are carried to equal distances and
quietly deposited in a confused mass without being subjected
to the sorting power of running water, which would convey the
finer materials farther than the coarser ones, and would produce,
as the strength of the current varied from time to time in the
same place, a stratified arrangement.

In those regions where glaciers reach the sea, and where large
masses of ice break off and float away, moraines, such as I
have just alluded to, may be transported to indefinite distances,
and may be deposited on the bottom of the sea wherever the
ice happens to melt.   If the liquefaction take place when the
berg has run aground and is stationary, and if there be no
current, the heap of angular and rounded stones, mixed with
sand and mud, may fall to the bottom in an unstratified form
called " till " in Scotland, and which has been shown in the
last chapter to abound in the Norfolk cliffs; but should the
action of a current intervene at certain points or at certain
seasons, then the materials will be sorted as they fall, and
arranged in layers according to their relative weight and size.
Hence there will be passages from till to stratified clay, gravel,
and sand.

Some of the blocks of stone with which the surfaces of glaciers
are loaded, falling occasionally through fissures in the ice, get
fixed and frozen into the bottom of the moving mass, and are
pushed along under it.   In this position, being subjected to
great pressure, they scoop out long rectilinear furrows or grooves
parallel to each other on the subjacent solid rock.   Smaller
scratches and striæ are made on the polished surface by crystals
or projecting edges of the hardest minerals, just as a diamond
cuts glass.

In all countries the fundamental rock on which the boulder
formation reposes, if it consists of granite, gneiss, marble, or
other hard stone capable of permanently retaining any super-

ficial markings which may have been imprinted upon it, is smoothed or polished, and exhibits parallel striæ and furrows having a determinate direction. This prevailing direction, both in Europe and North America, is evidently connected with the course taken by the erratic blocks in the same district, and is very commonly from north to south, or if it be twenty or thirty or more degrees to the east or west of north, still always corresponds to the direction in which the large angular and rounded stones have travelled. These stones themselves also are often furrowed and scratched on more than one side, like those already spoken of as occurring in the glacial drift of Bedford, and in that of Norfolk.

When we contemplate the area which is now exposed to the abrading action of ice, or which is the receptacle of moraine matter thrown down from melting glaciers or bergs, we at once perceive that the submarine area is the most extensive of the two. The number of large icebergs which float annually to great distances in the northern and southern hemispheres is extremely great, and the quantity of stone and mud which they carry about with them enormous. Some floating islands of ice have been met with from 2 to 5 miles in length, and from 100 to 225 feet in height above water, the submerged portion, according to the weight of ice relatively to sea water, being from six to eight times more considerable than the part which is visible. Such masses, when they run aground on the bottom of the sea, must exert a prodigious mechanical power, and may polish and groove the subjacent rocks after the manner of glaciers on the land. Hence there will often be no small difficulty in distinguishing between the effects of the submarine and supramarine agency of ice.

Scandinavia once covered with Ice, and a Centre of Dispersion of Erratics

In the north of Europe, along the borders of the Baltic, where the boulder formation is continuous for hundreds of miles east and west, it has been long known that the erratic blocks, often of very large size, are of northern origin. Some of them have come from Norway and Sweden, others from Finland, and their present distribution implies that they were carried southwards, for a part at least of their way, by floating ice, at a time when much of the area over which they are scattered was under water. But it appears from the observations of Boetlingk, in

1840, and those of more recent inquirers, that while many blocks have travelled to the south, others have been carried northwards, or to the shores of the Polar Sea, and others northeastward, or to those of the White Sea. In fact, they have wandered towards all points of the compass, from the mountains of Scandinavia as a centre, and the rectilinear furrows imprinted by them on the polished surfaces of the mountains where the rocks are hard enough to retain such markings, radiate in all directions, or point outwards from the highest land, in a manner corresponding to the course of the erratics above mentioned.[1]

Before the glacial theory was adopted, the Swedish and Norwegian geologists speculated on a great flood, or the sudden rush of an enormous body of water charged with mud and stones, descending from the central heights or watershed into the adjoining lower lands. The erratic blocks were supposed in their downward passage to have smoothed and striated the rock surfaces over which they were forced along.

It would be a waste of time, in the present state of science, to controvert this hypothesis, as it is now admitted that even if the rush of a diluvial current, invented for the occasion and wholly without analogy in the known course of nature, be granted, it would be inadequate to explain the uniformity, parallelism, persistency, and rectilinearity of the so-called glacial furrows. It is moreover ascertained that heavy masses of rock, not fixed in ice, and moving as freely as they do when simply swept along by a muddy current, do not give rise to such scratches and furrows.

M. Kjerulf of Christiania, in a paper lately communicated to the Geological Society of Berlin,[2] has objected, and perhaps with reason, to what he considers the undue extent to which I have, in some of my writings, supposed the mountains of northern Europe, to have been submerged during the glacial period. He remarks that the signs of glacial action on the Scandinavian mountains ascend as high as 6000 feet, whereas fossil marine shells of the same period never reach elevations exceeding 600 feet. The land, he says, may have been much higher than it now is, but it has evidently not been much lower since the commencement of the glacial period, or marine shells

---

[1] Sir R. I. Murchison, in his *Russia and the Ural Mountains* (1845), has indicated on a map, not only the southern limits of the Scandinavian drift, but by arrows the direction in which " it proceeded eccentrically from a common central region."

[2] *Zeitschrift der Deutschen Geologischen Gesellschaft*, Berlin, 1860.

would be traceable to more elevated points. In regard to the absence of marine shells, I shall point out in the sequel how small is the dependence we can place on this kind of negative evidence, if we desire to test by it the extent to which the land has been submerged. I cannot therefore consent to limit the probable depression and re-elevation of Scandinavia to 600 feet. But that the larger part of the glaciation of that country has been supramarine, I am willing to concede. In support of this view M. Kjerulf observes that the direction of the furrows and striæ, produced by glacial abrasion, neither conforms to a general movement of floating ice from the Polar regions, nor to the shape of the existing valleys, as it would do if it had been caused by independent glaciers generated in the higher valleys after the land had acquired its actual shape. Their general arrangement and apparent irregularities are, he contends, much more in accordance with the hypothesis of there having been at one time a universal covering of ice over the whole of Norway and Sweden, like that now existing in Greenland, which, being annually re-cruited by fresh falls of snow, was continually pressing outwards and downwards to the coast and lower regions, after crossing many of the lower ridges, and having no relation to the minor depressions, which were all choked up with ice and reduced to one uniform level.

Continental Ice of Greenland

In support of this view, he appeals to the admirable description of the continental ice of Greenland, lately published by Dr. H. Rink of Copenhagen,[1] who resided three or four years in the Danish settlements in Baffin's Bay, on the west coast of Greenland, between latitudes 69° and 73° N. " In that country, the land," says Dr. Rink, " may be divided into two regions, the ' inland ' and the ' outskirts.' The ' inland,' which is 800 miles from west to east, and of much greater length from north to south, is a vast unknown continent, buried under one continuous and colossal mass of permanent ice, which is always moving seaward, but a small proportion only of it in an easterly direction, since nearly the whole descends towards Baffin's Bay." At the heads of the fjords which intersect the coast, the ice is seen to rise somewhat abruptly from the level of the sea to the height of 2000 feet, beyond which the ice of the interior rises continuously as far as the eye can reach, and to an unknown

[1] *Journal of Royal Geographical Society,* vol. xxiii., 1853, p. 145.

altitude. All minor ridges and valleys are levelled and concealed, but here and there steep mountains protrude abruptly from the icy slope, and a few superficial lines of stones or moraines are visible at seasons when no recent snow has fallen. [Note 23.]

Although all the ice is moving seaward, the greatest quantity is discharged at the heads of certain large fjords, usually about 4 miles wide, which, if the climate were milder, would be the outlet of as many great rivers. Through these the ice is now protruded in huge blocks, several miles wide, and from 1000 to 1500 feet in height or thickness. When these masses reach the fjords, they do not melt or break up into fragments, but continue their course in a solid form in the salt water, grating along the rocky bottom, which they must polish and score at depths of hundreds and even of more than 1000 feet. At length, when there is water enough to float them, huge portions, having broken off, fill Baffin's Bay with icebergs of a size exceeding any which could be produced by ordinary valley glaciers. Stones, sand, and mud are sometimes included in these bergs which float down Baffin's Bay. At some points, where the ice of the interior of Greenland reaches the coast, Dr. Rink saw mighty springs of clayey water issuing from under the edge of the ice even in winter, showing the grinding action of the glacial mass mixed with sand on the subjacent surface of the rocks.

The "outskirts," where the Danish colonies are stationed, consist of numerous islands, of which Disco island is the largest in latitude 70° N., and of many peninsulas, with fjords from 50 to 100 miles long, running into the land, and through which the ice above alluded to passes on its way to the bay. This area is 30,000 square miles in extent, and contains in it some mountains 4000 feet to 5000 feet high. The perpetual snow usually begins at the height of 2000 feet, below which level the land is for the most part free from snow between June and August, and supports a vegetation of several hundred species of flowering plants, which ripen their seeds before the winter. There are even some places where phanerogamous plants have been found at an elevation of 4500 feet; a fact which, when we reflect on the immediate vicinity of so large and lofty a region of continental ice in the same latitude, well deserves the attention of the geologist, who should also bear in mind, that while the Danes are settled to the west in the "outskirts," there exists, due east of the most southern portion of this ice-covered continent, at the distance of about 1200 miles, the home of the

Laplanders with their reindeer, bears, wolves, seals, walruses, and whales. If, therefore, there are geological grounds for suspecting that Scandinavia or Scotland or Wales was ever in the same glacial condition as Greenland now is, we must not imagine that the contemporaneous fauna and flora were everywhere poor and stunted, or that they may not, especially at the distance of a few hundred miles in a *southward* direction, have been very luxuriant. [Note 24.]

Another series of observations made by Captain Graah, during a survey of Greenland between 1823 and 1829, and by Dr. Pingel in 1830-32, adds not a little to the geological interest of the " outskirts," in their bearing on glacial phenomena of ancient date. Those Danish investigators, with one of whom, Dr. Pingel, I conversed at Copenhagen. in 1834, ascertained that the whole coast from latitude 60° to about 70° N. has been subsiding for the last four centuries, so that some ancient piles driven into the beach to support the boats of the settlers have been gradually submerged, and wooden buildings have had to be repeatedly shifted farther inland.[1]

In Norway and Sweden, instead of such a subsiding movement, the land is slowly rising; but we have only to suppose that formerly, when it was covered like Greenland with continental ice, it sank at the rate of several feet in a century, and we shall be able to explain why marine deposits are found above the level of the sea, and why these generally overlie polished and striated surfaces of rock.

We know that Greenland was not always covered with snow and ice, for when we examine the Tertiary strata of Disco Island (of the Upper Miocene period) we discover there a multitude of fossil plants, which demonstrate that, like many other parts of the arctic regions, it formerly enjoyed a mild and genial climate. Among the fossils brought from that island, latitude 70° N., Professor Heer has recognised *Sequoia Langsdorfii*, a coniferous species which flourished throughout a great part of Europe in the Miocene period, and is very closely allied to the living *Sequoia sempervirens* of California. The same plant has been found fossil by Sir John Richardson within the arctic circle, far to the west on the Mackenzie River, near the entrance of Bear River, also by some Danish naturalists in Iceland to the east. The Icelandic surturbrand, or lignite, of this age has also yielded a rich harvest of plants, more than thirty-one of them, according to Steenstrup and Heer, in a good state of

[1] *Principles of Geology*, ch. xxx.

preservation, and no less than fifteen specifically identical with Miocene plants of Europe. Thirteen of the number are arborescent; and amongst others is a tulip-tree (*Liriodendron*), with its fruit and characteristic leaves, a plane (*Platanus*), a walnut, and a vine, affording unmistakable evidence of a climate in the parallel of the arctic circle which precludes the supposition of glaciers then existing in the neighbourhood, still less any general crust of continental ice, like that of Greenland.[1]

As the older Pliocene flora of the Tertiary strata of Italy, like the shells of the Coralline Crag, before adverted to, p. 165, indicate a temperature milder than that now prevailing in Europe, though not so warm as that of the Upper Miocene period, it is probable that the accumulation of snow and glaciers on the mountains and valleys of Greenland did not begin till after the commencement of the Pliocene period, and may not have reached its maximum until the close of that period.

Norway and Sweden appear to have passed through all the successive phases of glaciation which Greenland has experienced, and others which that country will one day undergo, if the climate which it formerly enjoyed should ever be restored to it. There must have been first a period of separate glaciers in Scandinavia, then a Greenlandic state of continental ice, and thirdly, when that diminished, a second period of enormous separate glaciers filling many a valley now wooded with fir and birch. Lastly, under the influence of the Gulf Stream, and various changes in the height and extent of land in the arctic circle, a melting of nearly all the permanent ice between latitudes 60° and 70° N., corresponding to the parallels of the continental ice of Greenland, has occurred, so that we have now to go farther north than latitude 70° before we encounter any glacier coming down to the sea coast. Among other signs of the last retreat of the extinct glaciers, Kjerulf and other authors describe large transverse moraines left in many of the Norwegian and Swedish glens.

Chronological Relations of the Human and Glacial Periods in Sweden

We may now consider whether any, and what part, of these changes in Scandinavia may have been witnessed by Man. In Sweden, in the immediate neighbourhood of Upsala, I observed, in 1834, a ridge of stratified sand and gravel, in the midst

[1] Heer, *Recherches sur la Végétation du Pays tertiaire*, etc., 1861, p. 178.

of which occurs a layer of marl, evidently formed originally at
the bottom of the Baltic, by the slow growth of the mussel,
cockle, and other marine shells of living species intermixed with
some proper to fresh water. The marine shells are all of
dwarfish size, like those now inhabiting the brackish waters of
the Baltic; and the marl, in which myriads of them are
embedded, is now raised more than 100 feet above the level of
the Gulf of Bothnia. Upon the top of this ridge (one of those
called ösars in Sweden) repose several huge erratics consisting
of gneiss, for the most part unrounded, from 9 to 16 feet in
diameter, and which must have been brought into their present
position since the time when the neighbouring gulf was already
characterised by its peculiar fauna. Here, therefore, we have
proof that the transport of erratics continued to take place,
not merely when the sea was inhabited by the existing Testacea,
but when the north of Europe had already assumed that remarkable feature of its physical geography, which separates the
Baltic from the North Sea, and causes the Gulf of Bothnia to
have only one-fourth of the saltness belonging to the ocean.

I cannot doubt that these large erratics of Upsala were
brought into their present position during the Recent period,
not only because of their moderate elevation above the sea-
level in a country where the land is now rising every century,
but because I observed signs of a great oscillation of level
which had taken place at Södertelje, south of Stockholm (about
45 miles distant from Upsala), after the country had been inhabited by Man. I described, in the *Philosophical Transactions*
for 1835, the section there laid open in digging a level in 1819,
which showed that a subsidence followed by a re-elevation of
land, each movement amounting to more than 60 feet, had
occurred since the time when a rude hut had been built on the
ancient shore. The wooden frame of the hut, with a ring of
hearthstones on the floor, and much charcoal, were found, and
over them marine strata, more than 60 feet thick, containing
the dwarf variety of *Mytilus edulis,* and other brackish-water
shells of the Bothnian Gulf. Some vessels put together with
wooden pegs, of anterior date to the use of metals, were also
embedded in parts of the same marine formation, which has
since been raised, so that the upper beds are more than 60 feet
above the sea-level, the hut being thus restored to about its
original position relatively to the sea.

We have seen in the account of the Danish kitchen-middens
of the Recent period that even at the comparatively late period

of their origin the waters of the Baltic had been rendered more salt than they are now. The Upsala erratics may belong to nearly the same era as these. But were we to go back to a long antecedent epoch, or to that of the Belgian and British caves with their extinct animals, and the signs they afford of a state of physical geography departing widely from the present, or to the era of the implement-bearing alluvium of St. Acheul, we might expect to find Scandinavia overwhelmed with glaciers, and the country uninhabitable by Man. At a much remoter period the same country was in the state in which Greenland now is, overspread with one uninterrupted coating of continental ice, which has left its peculiar markings on the highest mountains. This period, probably anterior to the earliest traces yet brought to light of the human race, may have coincided with the submergence of England, and the accumulation of the boulder-clay of Norfolk, Suffolk, and Bedfordshire, before mentioned. It has already been stated that the syenite and some other rocks of the Norfolk till seem to have come from Scandinavia, and there is no era when icebergs are so likely to have floated them so far south as when the whole of Sweden and Norway were enveloped in a massive crust of ice; a state of things the existence of which is deduced from the direction of the glacial furrows, and their frequent unconformity to the shape of the minor valleys.

Glacial Period in Scotland [Note 25]

Professor Agassiz, after his tour in Scotland in 1840, announced the opinion that erratic blocks had been dispersed from the Scottish mountains as from an independent centre, and that the capping of ice had been of extraordinary thickness.[1]

Mr. Robert Chambers, after visiting Norway and Sweden, and comparing the signs of glacial action observed there with similar appearances in the Grampians, came to the conclusion that the Highlands both of Scandinavia and Scotland had once been " moulded in ice," and that the outward and downward movement and pressure of the frozen mass had not only smoothed, polished, and scratched the rocks, but had, in the course of ages, deepened and widened the valleys, and produced much of that denudation which has commonly been ascribed exclusively to aqueous action. The glaciation of the Scotch

[1] Agassiz, *Proc. Geol. Soc.*, 1840, and *Edinb. Phil. Journ.*, xlix., p. 79.

mountains was traced by him to the height of at least 3000 feet.[1]

Mr. T. F. Jamieson, of Ellon, in Aberdeenshire, has recently brought forward an additional body of facts in support of this theory. According to him the Grampians were at the period of extreme cold enveloped " in one great winding sheet of snow and ice," which reached everywhere to the coast-line, the land being then more elevated than it is now. He describes the glacial furrows sculptured on the solid rocks as pointing in Aberdeenshire to the south-east, those of the valley of the Forth at Edinburgh, from west to east, and higher up the same valley at Stirling, from north-west to south-east, as they should. do if the ice had followed the lines of what is now the principal drainage. The observations of Sir James Hall, Mr. Maclaren, Mr. Chambers, and Dr. Fleming, are cited by him in confirmation of this arrangement of the glacial markings, while in Sutherland and Ross-shire he shows that the glacial furrows along the north coast point northwards, and in Argyleshire westwards, always in accordance with the direction of the principal glens and fjords.

Another argument is also adduced by him in proof of the ice having exerted its mechanical force in a direction from the higher and more inland country to the lower region and sea-coast. Isolated hills and minor prominences of rock are often polished and striated on the land side, while they remain rough and jagged on the side fronting the sea. This may be seen both on the east and west coast. Mention is also made of blocks of granite which have travelled from south to north in Aberdeenshire, of which there would have been no examples had the erratics been all brought by floating ice from the arctic regions when Scotland was submerged. It is also urged against the doctrine of attributing the general glaciation to submergence, that the glacial grooves, instead of radiating as they do from a centre, would, if they had been due to ice coming from the north, have been parallel to the coast-line, to which they are now often almost at right angles. The argument, moreover, which formerly had most weight in favour of floating ice, namely, that it explained why so many of the stones did not conform to the contour and direction of the minor hills and valleys, is now brought forward, and with no small effect, in favour of the doctrine of continental ice on the Greenlandic

[1] *Ancient Sea Margins*, Edinburgh, 1848. Glacial Phenomena, *Edinburgh New Philosophical Journal*, April 1853 and January 1855.

scale, which, after levelling up the lesser inequalities, would occasionally flow in mighty ice-currents, in directions often at a high angle to the smaller ridges and glens.

The application to Scandinavia and Scotland of this theory makes it necessary to reconsider the validity of the proofs formerly relied on as establishing the submergence of a great part of Scotland beneath the sea, at some period subsequent to the commencement of the glacial period. In all cases where marine shells overlie till, or rest on polished and striated surfaces of rock, the evidence of the land having been under water, and having been since upheaved, remains unshaken; but this special proof rarely extends to heights exceeding 500 feet. In the basin of the Clyde we have already seen that Recent strata occur 25 feet above the sea-level, with existing species of marine testacea, and with buried canoes, and other works of art. At the higher level of 50 feet occurs the well-known raised beach of the western coast, which, according to Mr. Jamieson, contains, near Fort William and on Loch Fyne and elsewhere, an assemblage of shells implying a colder climate than that of the 25-foot terrace, or that of the present sea; just as, in the valley of the Somme, the higher level gravels are supposed to belong to a colder period than the lower ones, and still more decidedly than that of the present era. At still greater elevations, older beds containing a still more arctic group of shells have been observed at Airdrie, 14 miles south-east of Glasgow, 524 feet above the level of the sea. They were embedded in stratified clays, with the unstratified boulder till both above and below them, and in the overlying unstratified drift were some boulders of granite which must have come from distances of 60 miles at the least.[1] The presence of *Tellina calcarea*, and several other northern shells, implies a climate colder than that of the present Scottish seas. In the north of Scotland, marine shells have been found in deposits of the same age in Caithness and in Aberdeenshire at heights of 250 feet, and on the shores of the Moray Firth, as at Gamrie in Banff, at an elevation of 350 feet; and the stratified sands and beds of pebbles which belong to the same formation ascend still higher—to heights of 500 feet at least.[2]

At much greater heights, stratified masses of drift occur in which hitherto no organic remains, whether of marine or freshwater animals, have ever been found. It is still an undecided

[1] Smith of Jordanhill, *Quart. Jour. Geol. Soc.*, vol. vi., 1850, p. 387.
[2] Prestwich, *Proc. Geol. Soc.*, vol. ii., p. 545; Jamieson, *Quart. Jour. Geol. Soc.*, vol. xvi., 1860.

question whether the origin of all such deposits in the Grampians can be explained without the intervention of the sea. One of the most conspicuous examples has been described by Mr. Jamieson as resting on the flank of a hill called Meal Uaine, in Perthshire, on the east side of the valley of the Tummel, just below Killiecrankie. It consists of perfectly horizontal strata, the lowest portion of them 300 feet above the river and 600 feet above the sea. From this elevation to an altitude of nearly 1200 feet the same series of strata is traceable, continuously, up the slope of the mountain, and some patches are seen here and there even as high as 1550 feet above the sea. They are made up in great part of finely laminated silt, alternating with coarser materials, through which stones from 4 to 5 feet in length are scattered. These large boulders, and some smaller ones, are polished on one or more sides, and marked with glacial striæ. The subjacent rocks, also, of gneiss, mica slate, and quartz, are everywhere grooved and polished as if by the passage of a glacier.[1]

At one spot a vertical thickness of 130 feet of this series of strata is exposed to view by a mountain torrent, and in all more than 2000 layers of clay, sand, and gravel were counted, the whole evidently accumulated under water. Some beds consist of an impalpable mud, like putty, apparently derived from the grinding down of felspar, and resembling the mud produced by the grinding action of modern glaciers.

Mr. Jamieson, when he first gave an account of this drift, inferred, in spite of the absence of marine shells, that it implied the submergence of Scotland beneath the ocean after the commencement of the glacial period, or after the era of continental ice indicated by the subjacent floor of polished and grooved rock. This conclusion would require a submergence of the land as far up as 1550 feet above the present sea-level, after which a great re-upheaval must have occurred. But the same author, having lately revisited the valley of the Tummel, suggests another possible, and I think probable, explanation of the same phenomena. The stratified drift in question is situated in a deep depression between two buttresses of rock, and if an enormous glacier be supposed to have once filled the valley of the Tummel to the height of the stratified drift, it may have dammed up the mouth of a mountain torrent by a transverse barrier, giving rise to a deep pond, in which beds of clay and sand brought down by the waters of the torrent were

[1] Jamieson, *Quart. Jour. Geol. Soc.*, vol. xvi., 1860, p. 360.

deposited. Charpentier in his work on the Swiss glaciers has described many such receptacles of stratified matter now in progress, and due to such blockages, and he has pointed out the remnants of ancient and similar formations left by extinct glaciers of an earlier epoch. He specially notices that angular stones of various dimensions, often polished and striated, which rest on the glacier and are let fall when the torrent undermines the side of the moving ice, descend into the small lake and become interstratified with the gravel and fine sediment brought down by the torrent into the same.[1]

The evidence of the former sojourn of the sea upon the land after the commencement of the glacial period was formerly inferred from the height to which erratic blocks derived from distant regions could be traced, besides the want of conformity in the glacial furrows to the present contours of many of the valleys. Some of these phenomena may now, as we have seen, be accounted for by assuming that there was once a crust of ice resembling that now covering Greenland.

The Grampians in Forfarshire and in Perthshire are from 3000 to 4000 feet high. To the southward lies the broad and deep valley of Strathmore, and to the south of this again rise the Sidlaw Hills to the height of 1500 feet and upwards. On the highest summits of this chain, formed of sandstone and shale, and at various elevations, I have observed huge angular fragments of mica-schist, some 3 and others 15 feet in diameter, which have been conveyed for a distance of at least 15 miles from the nearest Grampian rocks from which they could have been detached. Others have been left strewed over the bottom of the large intervening vale of Strathmore.[2]

It may be argued that the transportation of such blocks may have been due not to floating ice, but to a period when Strathmore was filled up with land ice, a current of which extended from the Perthshire Highlands to the summit of the Sidlaw Hills, and the total absence of marine or freshwater shells from all deposits, stratified or unstratified, which have any connection with these erratics in Forfarshire and Perthshire may be thought to favour such a theory.

But the same mode of transport can scarcely be imagined for those fragments of mica-schist, one of them weighing from 8 to 10 tons, which were observed much farther south by Mr. Maclaren on the Pentland Hills, near Edinburgh, at the height

[1] Charpentier, *Essai sur les Glaciers*, p. 63, 1841.
[2] *Proceedings of the Geological Society*, vol. iii., p. 344.

of 1100 feet above the sea, the nearest mountain composed of this formation being 50 miles distant.[1]   On the same hills, also, at all elevations, stratified gravels occur which, although devoid of shells, it seems hardly possible to refer to any but a marine origin.[1]

Although I am willing, therefore, to concede that the glaciation of the Scotch mountains, at elevations exceeding 2000 feet, may be explained by land ice, it seems difficult not to embrace the conclusion that a subsidence took place not merely of 500 or 600 feet, as demonstrated by the marine shells, but to a much greater amount, as shown by the present position of erratics and some patches of stratified drift.   The absence of marine shells at greater heights than 525 feet above the sea, will be treated of in a future chapter.   It may in part, perhaps, be ascribed to the action of glaciers, which swept out marine strata from all the higher valleys, after the re-emergence of the land.

Latest Changes produced by Glaciers in Scotland

We may next consider the state of Scotland after its emergence from the glacial sea, when we cannot fail to be approaching the time when Man co-existed with the mammoth and other mammalia now extinct.   In a paper which I published in 1840, on the ancient glaciers of Forfarshire, I endeavoured to show that some of these existed after the mountains and glens had acquired precisely their present shape,[2] and had left moraines even in the minor valleys, just where they would now leave them were the snow and ice again to gain ground.   I described also one remarkable transverse mound, evidently the terminal moraine of a retreating glacier, which crosses the valley of the South Esk, a few miles above the point where it issues from the Grampians, and about 6 miles below the Kirktown of Clova. Its central part, at a place called Glenarm, is 800 feet above the level of the sea.   The valley is about half a mile broad, and is bounded by steep and lofty mountains, but immediately above the transverse barrier it expands into a wide alluvial plain, several miles broad, which has evidently once been a lake.   The barrier itself, about 150 feet high, consists in its lower part of till with boulders, 50 feet thick, precisely resembling the moraine of a Swiss glacier, above which there is a mass

[1] Maclaren, *Geology of Fife*, etc., p. 220.
[2] *Proceedings of the Geological Society*, vol. iii., p. 337.

of stratified sand, varying in thickness from 50 to 100 feet, which has the appearance of consisting of the materials of the moraine rearranged in a stratified form, possibly by the waters of a glacier lake. The structure of the barrier has been laid open by the Esk, which has cut through it a deep passage about 400 yards wide.

I have also given an account of another striking feature in the physical geography of Perthshire and Forfarshire, which I consider to belong to the same period; namely, a continuous zone of boulder clay, forming ridges and mounds from 50 to 70 feet high (the upper part of the mounds usually stratified), enclosing numerous lakes, some of them several miles long, and many ponds and swamps filled with shell-marl and peat. This band of till, with Grampian boulders and associated river-gravel, may be traced continuously for a distance of 34 miles, with a width of 3½ miles, from near Dunkeld, by Coupar, to the south of Blairgowrie, then through the lowest part of Strathmore, and afterwards in a straight line through the greatest depression in the Sidlaw Hills, from Forfar to Lunan Bay.

Although no great river now takes its course through this line of ancient lakes, moraines, and river gravel, yet it evidently marks an ancient line by which, first, a great glacier descended from the mountains to the sea, and by which, secondly, at a later period, the principal water drainage of this country was effected. The subsequent modification in geography is comparable in amount to that which has taken place since the higher level gravels of the valley of the Somme were formed, or since the Belgian caves were filled with mud and bone-breccia.

Fig. 35.

Mr. Jamieson has remarked, in reference to this and some other extinct river-channels of corresponding date, that we have the means of ascertaining the direction in which the waters flowed by observing the arrangement of the oval and flattish pebbles in their deserted channels; for in the bed of a fast-flowing river such pebbles are seen to dip towards the current, as represented in Fig. 35, such being the position of greatest resistance to the stream.[1] If this be admitted, it

[1] Jamieson, *Quart. Jour. Geol. Soc.*, vol. xvi., 1860, p. 349.

follows that the higher or mountainous country bore the same relation to the lower lands, at the time when a great river passed through this chain of lakes, as it does at present.

We also seem to have a test of the comparatively modern origin of the mounds of till which surround the above-mentioned chain of lakes (of which that of Forfar is one), in the species of organic remains contained in the shell-marl deposited at their bottom. All the mammalia as well as shells are of recent species. Unfortunately, we have no information as to the fauna which inhabited the country at the time when the till itself was formed. There seem to be only three or four instances as yet known in all Scotland of mammalia having been discovered in boulder clay.

Mr. R. Bald has recorded the circumstances under which a single elephant's tusk was found in the unstratified drift of the valley of the Forth, with the minuteness which such a discovery from its rarity well deserved. He distinguishes the boulder clay, under the name of " the old alluvial cover," from that more modern alluvium, in which the whales of Airthrie, described in Chap. III., were found. This cover he says is sometimes 160 feet thick. Having never observed any organic remains in it, he watched with curiosity and care the digging of the Union Canal between Edinburgh and Falkirk, which passed for no less than 28 miles almost continuously through it. Mr. Baird, the engineer who superintended the works, assisted in the inquiry, and at one place only in this long section did they meet with a fossil, namely, at Cliftonhall, in the valley of the Almond. It lay at a depth of between 15 and 20 feet from the surface, in very stiff clay, and consisted of an elephant's tusk, 39 inches long and 13 in circumference, in so fresh a state that an ivory turner purchased it and turned part of it into chessmen before it was rescued from destruction. The remainder is still preserved in the museum at Edinburgh, but by exposure to the air it has shrunk considerably.[1] In 1817, two other tusks and some bones of the elephant, as we learn from the same authorty (Mr. Bald), were met with, 3½ feet long and 13 inches in circumference, lying in an horizontal position, 17 feet deep in clay, with marine shells, at Kilmaurs, in Ayrshire. The species of shells are not given.[2]

In another excavation through the Scotch boulder clay, made in digging the Clyde and Forth Junction Railway, the antlers

[1] *Memoirs of the Wernerian Society*, Edinburgh, vol. iv., p. 58.
[2] *Ibid.*, vol. iv., p. 63.

of a reindeer were found at Croftamie, in Dumbartonshire, in the basin of the river Endrick, which flows into Loch Lomond. They had cut through 12 feet of till with angular and rounded stones, some of large size, and then through 6 feet of underlying clay, when they came upon the deer's horns, 18 feet from the surface, and within a foot of the sandstone on which the till rested. At the distance of a few yards, and in the same position, but a foot or two deeper, were observed marine shells, *Cyprina islandica, Astarte elliptica, A. compressa, Fusus antiquus, Littorina littorea,* and a *Balanus*. The height above the level of the sea was between 100 and 103 feet. The reindeer's horn was seen by Professor Owen, who considered it to be that of a young female of the large variety, called by the Hudson's Bay trappers the caribou.

The remains of elephants, now in the museums of Glasgow and Edinburgh, purporting to come from the superficial deposits of Scotland have been referred to *Elephas primigenius*. In cases where tusks alone have been found unaccompanied by molar teeth, such specific determinations may be uncertain; but if any one specimen be correctly named, the occurrence of the mammoth and reindeer in the Scotch boulder-clay, as both these quadrupeds are known to have been contemporary with Man, favours the idea which I have already expressed, that the close of the glacial period in the Grampians may have coincided in time with the existence of Man in those parts of Europe where the climate was less severe, as, for example, in the basins of the Thames, Somme, and Seine, in which the bones of many extinct mammalia are associated with flint implements of the antique type.

Parallel Roads of Glen Roy in Scotland

Perhaps no portion of the superficial drift of Scotland can lay claim to so modern an origin on the score of the freshness of its aspect, as that which forms what are called the Parallel Roads of Glen Roy. If they do not belong to the Recent epoch, they are at least posterior in date to the present outline of mountain and glen, and to the time when every one of the smaller burns ran in their present channels, though some of them have since been slightly deepened. The almost perfect horizontality, moreover, of the roads, one of which is continuous for about 20 miles from east to west, and 12 miles from north to south, shows that since the era of their formation no change

Plate II.

VIEW OF THE MOUTHS OF GLEN ROY AND GLEN SPEAN, BY SIR T. DICK LAUDER.

℣℣ Hill of Bohuntine,   ℣℣℣ Glen Roy.   ℣ Mealderry.   ℣ Entrance of Glen Spean

Λ

℣℣ Point of division between Glens Roy and Spean.

has taken place in the relative levels of different parts of the district.

Glen Roy is situated in the Western Highlands, about 10 miles E.N.E. of Fort William, near the western end of the great glen of Scotland, or Caledonian Canal, and near the foot of the highest of the Grampians, Ben Nevis. (*See* map, Fig. 36.) Throughout nearly its whole length, a distance of more than 10 miles, three parallel roads or shelves are traced along the steep sides of the mountains, as represented in the annexed view, Plate II., by the late Sir T. Dick Lauder, each maintaining a perfect horizontality, and continuing at exactly the same level on the opposite sides of the glen. Seen at a distance, they appear like ledges, or roads, cut artificially out of the sides of the hills; but when we are upon them, we can scarcely recognise their existence, so uneven is their surface, and so covered with boulders. They are from 10 to 60 feet broad, and merely differ from the side of the mountain by being somewhat less steep.

On closer inspection, we find that these terraces are stratified in the ordinary manner of alluvial or littoral deposits, as may be seen at those points where ravines have been excavated by torrents. The parallel shelves, therefore, have not been caused by denudation, but by the deposition of detritus, precisely similar to that which is dispersed in smaller quantities over the declivities of the hills above. These hills consist of clay-slate, mica schist, and granite, which rocks have been worn away and laid bare at a few points immediately above the parallel roads. The lowest of these roads is about 850 feet above the level of the sea, the next about 212 feet higher, and the third 82 feet above the second. There is a fourth shelf, which occurs only in a contiguous valley called Glen Gluoy, which is 12 feet above the highest of all the Glen Roy roads, and consequently about 1156 feet above the level of the sea.[1] One only, the lowest of the three roads of Glen Roy, is continued throughout Glen Spean, a large valley with which Glen Roy unites. (*See* Plate II. and map, Fig. 36.) As the shelves, having no slope towards the sea like ordinary river terraces, are always at the same absolute height, they become continually more elevated above the river in proportion as we descend each valley; and they at length terminate very abruptly, without any obvious cause, or any change either in the shape of the ground or in the composition or hardness of the rocks.

I should exceed the limits of this work, were I to attempt to

[1] Another detached shelf also occurs at Kilfinnan. (*See* Map, Fig. 36.)

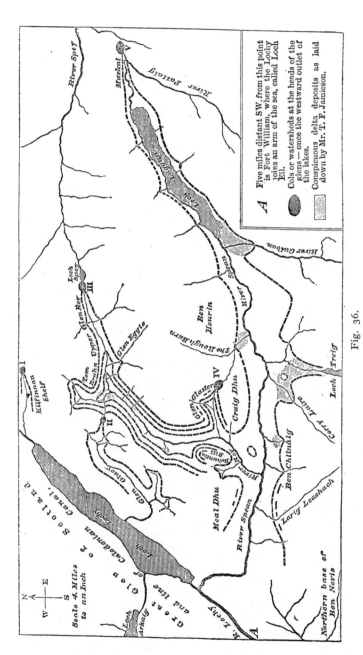

Fig. 36.

MAP OF THE PARALLEL ROADS OF GLEN ROY OR LOCHABER.

give a full description of all the geographical circumstances attending these singular terraces, or to discuss the ingenious theories which have been severally proposed to account for them by Dr. Macculloch, Sir T. Lauder, and Messrs. Darwin, Agassiz, Milne, and Chambers. There is one point, however, on which all are agreed, namely, that these shelves are ancient beaches, or littoral formations, accumulated round the edges of one or more sheets of water which once stood for a long time successively at the level of the several shelves.

It is well known, that wherever a lake or marine fjord exists surrounded by steep mountains subject to disintegration by frost or the action of torrents, some loose matter is washed down annually, especially during the melting of snow, and a check is given to the descent of this detritus at the point where it reaches the waters of the lake. The waves then spread out the materials along the shore, and throw some of them upon the beach; their dispersing power being aided by the ice, which often adheres to pebbles during the winter months, and gives buoyancy to them. The annexed diagram illustrates the manner in which Dr. Macculloch and Mr. Darwin suppose " the roads " to constitute mere excrescences of the superficial alluvial coating which rests upon the hillside, and consists chiefly of clay and sharp unrounded stones.

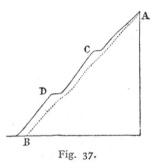

Fig. 37.

A B Supposed original surface of rock.
C D Roads or shelves in the outer alluvial covering of the hill.

Among other proofs that the parallel roads have really been formed along the margin of a sheet of water, it may be mentioned, that wherever an isolated hill rises in the middle of the glen above the level of any particular shelf, as in Mealderry, Plate II., a corresponding shelf is seen at the same level passing round the hill, as would have happened if it had once formed an island in a lake or fjord. Another very remarkable peculiarity in these terraces is this; each of them comes in some portion of its course to a *col*, or parting ridge, between the heads of glens, the explanation of which will be considered in the sequel.

Those writers who first advocated the doctrine that the roads were the ancient beaches of freshwater lakes, were unable to offer any probable hypothesis respecting the formation and subsequent removal of barriers of sufficient height and solidity to dam up the water. To introduce any violent convulsion for their removal was inconsistent with the uninterrupted horizontality of the roads, and with the undisturbed aspect of those parts of the glens where the shelves come suddenly to an end.

Mr. Agassiz and Dr. Buckland, desirous, like the defenders of the lake theory, to account for the limitation of the shelves to certain glens, and their absence in contiguous glens, where the rocks are of the same composition, and the slope and inclination of the ground very similar, first started the theory that these valleys were once blocked up by enormous glaciers descending from Ben Nevis, giving rise to what are called, in Switzerland and in the Tyrol, glacier-lakes. In corroboration of this view, they contended that the alluvium of Glen Roy, as well as of other parts of Scotland, agrees in character with the moraines of glaciers seen in the Alpine valleys of Switzerland. It will readily be conceded that this hypothesis was preferable to any previous lacustrine theory, by accounting more easily for the temporary existence and entire disappearance of lofty transverse barriers, although the height required for the supposed dams of ice appeared very enormous.

Before the idea of glacier-lakes had been suggested by Agassiz, Mr. Darwin examined Glen Roy, and came to the opinion that the shelves were formed when the glens were still arms of the sea, and, consequently, that there never were any seaward barriers. According to him, the land emerged during a slow and uniform upward movement, like that now experienced throughout a large part of Sweden and Finland; but there were certain pauses in the upheaving process, at which times the waters of the sea remained stationary for so many centuries as to allow of the accumulation of an extraordinary quantity of detrital matter, and the excavation, at many points immediately above the sea-level, of deep notches and bare cliffs in the hard and solid rock.

This theory I adopted in 1841 (*Elements*, 2nd ed.), as appearing to me less objectionable than any other then proposed. The phenomena most difficult to reconcile with it are, first, the abrupt cessation of the roads at certain points in the different glens; secondly, their unequal number in different valleys connecting with each other, there being three, for example, in Glen Roy, and only one in Glen Spean; thirdly, the precise horizontality of level maintained by the same shelf over a space many leagues in length, requiring us to assume, that during a rise of 1156 feet no one portion of the land was raised even a few yards above another; fourthly, the coincidence of level already alluded to of each shelf with a *col*, or the point forming the head of two glens, from which the rain-waters flow in opposite directions. This last-mentioned feature in the physical geography of Lochaber Mr. Darwin endeavoured to explain in the following manner. He called these *cols* " land-straits," and regarding them as having been anciently sounds or channels between islands, he pointed out that there is a tendency in such sounds to be silted up, and always the more so in proportion to their narrowness. In a chart of the Falkland Islands, by Captain Sulivan, R.N., it appears that there are several examples there of straits where the soundings diminish regularly towards the narrowest part. One is so nearly dry that it can be walked over at low water, and another, no longer covered by the sea, is supposed to have recently dried up in consequence of a small alteration in the relative level of sea and land. " Similar straits," observes Mr. Chambers, " hovering, in character, between sea and land, and which may be called fords, are met with in the Hebrides. Such, for example, is the passage dividing the islands of Lewis and Harris, and that between North Uist and Benbecula, both of which would undoubtedly appear as *cols*, coinciding with a terrace or raised beach, all round the islands if the sea were to subside." [1]

The first of the difficulties above alluded to, namely, the non-extension of the shelves over certain parts of the glens, might be explained, said Mr. Darwin, by supposing in certain places a quick growth of green turf on a good soil, which prevented the rain from washing away any loose materials lying on the surface. But wherever the soil was barren, and where green sward took long to form, there may have been time for the removal of the gravel. In one case an intermediate shelf appears for a short distance (three quarters of a mile) on the face of the mountain called Tombhran, between the two upper shelves, and is seen nowhere else. It occurs where there was the longest space of open water, and where the waves may have acquired a more than ordinary power to heap up detritus.

The unequal number of the shelves in valleys communicating with each other, and in which the boundary rocks are similar in composition, and the general absence of any shelves at corresponding altitudes in glens on the opposite watershed, like that

[1] R. Chambers, *Ancient Sea Margins*, p. 114.

of the Spey, and in valleys where the waters flow eastward, are difficulties attending the marine theory which have never yet been got over. Mr. T. F. Jamieson, before cited, has, during a late visit to Lochaber, in 1861, observed many facts highly confirmatory of the hypothesis of glacier-lakes which, as I have already stated, was originally advanced by Mr. Agassiz. In the first place, he found much superficial scoring and polishing of rocks, and accumulation of boulders at those points where signs of glacial action ought to appear, if ice had once dammed up the waters of the glens in which the "roads" occur. Ben Nevis may have sent down its glaciers from the south, and Glen Arkaig from the north, for the mountains at the head of the last-mentioned glen are 3000 feet high, and may, together with other tributary glens, have helped to choke up the great Caledonian valley with ice, so as to block up for a time the mouths of the Spean, Roy, and Gluoy. The temporary conversion of these glens into glacier-lakes is the more conceivable, because the hills at their upper ends not being lofty nor of great extent, they may not have been filled with ice at a time when great glaciers were generated in other adjoining and much higher regions.

Secondly. The shelves, says Mr. Jamieson, are more precisely defined and unbroken than any of the raised beaches or acknowledged ancient coast-lines visible on the west of Scotland, as in Argyllshire, for example.

Thirdly. At the level of the lower shelf in Glen Roy, at points where torrents now cut channels through the shelf as they descend the hill-side, there are small delta-like extensions of the shelf, perfectly preserved, as if the materials, whether fine or coarse, had originally settled there in a placid lake, and had not been acted upon by tidal currents, mingling them with the sediment of other streams. These deltas are too entire to allow us to suppose that they have at any time since their origin been exposed to the waves of the sea.

Fourthly. The alluvium on the *cols* or watersheds, before alluded to, is such as would. have been formed if the waters of the rivers had been made to flow east, or out of the upper ends of the supposed glacier-lakes, instead of escaping at the lower ends, in a westerly direction, where the great blockages of ice are assumed to have occurred.

In addition to these arguments of Mr. Jamieson, I may mention that in Switzerland, at present, no testacea live in the cold waters of glacier-lakes; so that the entire absence of fossil shells, whether marine or freshwater, in the stratified materials of each shelf, would be accounted for if the theory above mentioned be embraced.

When I examined " the parallel roads " in 1825, in company with Dr. Buckland, neither this glacier theory nor Mr. Darwin's suggestion of ancient sea-margins had been proposed, and I have never since revisited Lochaber. But I retain in my memory a vivid recollection of the scenery and physical features of the district, and I now consider the glacier-lake theory as affording by far the most satisfactory solution of this difficult problem. The objection to it, which until lately appeared to be the most formidable, and which led Mr. Robert Chambers in his *Sea Margins,* to reject it entirely, was the difficulty of conceiving how the waters could be made to stand so high in Glen Roy as to allow the uppermost shelf to be formed. Grant a barrier of ice in the lower part of the glen of sufficient altitude to stop the waters from flowing westward, still, what prevented them from escaping over the *col* at the head of Glen Glaster? This *col* coincides exactly in level, as Mr. Milne Home first ascertained, with the second or middle shelf of Glen Roy. The difficulty here stated appears now to be removed by supposing that the higher lines or roads were formed before the lower ones, and when the quantity of ice was most in excess. We must imagine that at the time when the uppermost shelf of Glen Roy was forming in a shallow lake, the lower part of that glen was filled up with ice, and, according to Mr. Jamieson, a glacier from Loch Treig then protruded itself across Glen Spean and rested on the flank of the hill on the opposite side in such a manner as effectually to prevent any water from escaping over the Glen Glaster *col.* The proofs of such a glacier having actually existed at the point in question consist, he says, in numerous cross striæ observable in the bottom of Glen Spean, and in the presence of moraine matter in considerable abundance on the flanks of the hill extending to heights above the Glen Glaster *col.* When the ice shrank into less dimensions the second shelf would be formed, having its level determined by the *col* last mentioned, Glen Spean in the meantime being filled with a glacier. Finally, the ice blockage common to glens Roy, Spean, and Laggan, which consisted probably of a glacier from Ben Nevis, gave rise to the lowest and most extensive lake, the waters of which escaped over the pass of Muckul or the *col* at the head of Loch Laggan, which, as Mr. Jamieson has now ascertained, agrees precisely in level with the lowest of all the shelves, and where there are unequivocal signs of a river having flowed out for a considerable period.

Dr. Hooker has described some parallel terraces, very analogous in their aspect to those of Glen Roy, as existing in the higher valleys of the Himalaya, of which his pencil has given us several graphic illustrations. He believes these Indian shelves to have originated on the borders of glacier-lakes, the barriers of which were usually formed by the ice and moraines of lateral or tributary glaciers, which descended into and crossed the main valley, as we have supposed in the case of Glen Roy; but others he ascribes to the terminal moraine of the principal glacier itself, which had retreated during a series of milder seasons, so as to leave an interval between the ice and the terminal moraine. This interspace caused by the melting of ice becomes filled with water and forms a lake, the drainage of which usually takes place by percolation through the porous parts of the moraine, and not by a stream overflowing that barrier. Such a glacier-lake Dr. Hooker actually found in existence near the head of the Yangma valley in the Himalaya. It was moreover partially bounded by recently formed marginal terraces or parallel roads, implying changes of level in the barrier of ice and moraine matter.[1]

It has been sometimes objected to the hypothesis of glacier-lakes, as applied to the case of Glen Roy, that the shelves must have taken a very long period for their formation. Such a lapse of time, it is said, might be consistent with the theory of pauses or stationary periods in the rise of the land during an intermittent upward movement, but it is hardly compatible with the idea of so precarious and fluctuating a barrier as a mass of ice. But the reader will have seen that the permanency of level in such glacier-lakes has no necessary connection with minor changes in the height of the supposed dam of ice. If a glacier descending from higher mountains through a tributary glen enters the main valley in which there happens to be no glacier, the river is arrested in its course and a lake is formed. The dam may be constantly repaired and may vary in height several hundreds of feet without affecting the level of the lake, so long as the surplus waters escape over a *col* or parting ridge of rock. The height at which the waters remain stationary is determined solely by the elevation of the *col*, and not by the barrier of ice, provided the barrier is higher than the *col*.

[1] Hooker, *Himalayan Journal*, vol. i., p. 242; ii., pp. 119, 121, 166. I have also profited by the author's personal explanations.

But if we embrace the theory of glacier-lakes, we must be prepared to assume not only that the sea had nothing to do with the original formation of the " parallel roads," but that it has never, since the disappearance of the lakes, risen in any one of the glens up to the level of the lowest shelf, which is about 850 feet high; for in that case the remarkable persistency and integrity of the roads and deltas, before described, must have been impaired.

We have seen that 50 miles to the south of Lochaber, the glacier formations of Lanarkshire with marine shells of arctic character have been traced to the height of 524 feet. About 50 miles to the south-east in Perthshire are those stratified clays and sands, near Killiecrankie, which were once supposed to be of submarine origin, and which in that case would imply the former submergence of what is now dry land to the extent of 1550 feet, or several hundred feet beyond the highest of the parallel roads. Even granting that these laminated drifts may have had a different origin, as above suggested, there are still many facts connected with the distribution of erratics and the striation of rocks in Scotland which are not easily accounted for without supposing the country to have sunk, since the era of continental ice, to a greater depth than 525 feet, the highest point to which marine shells have yet been traced.

After what was said of the pressure and abrading power of a general crust of ice, like that now covering Greenland, it is almost superfluous to say that the parallel roads must have been of later date than such a state of things, for every trace of them must have been obliterated by the movement of such a mass of ice. It is no less clear that as no glacier-lakes can now exist in Greenland [Note 26], so there could have been none in Scotland, when the mountains were covered with one great crust of ice. It may, however, be contended that the parallel roads were produced when the general crust of ice first gave place to a period of separate glaciers, and that no period of deep submergence ever intervened in Lochaber after the time of the lakes. Even in that case, however, it is difficult not to suppose that the Glen Roy country participated in the downward movement which sank part of Lanarkshire 525 feet beneath the sea, subsequently to the first great glaciation of Scotland. Yet that amount of subsidence might have occurred, and even a more considerable one, without causing the sea to rise to the level of the lowest shelf, or to a height of 850 feet above the present sea-level.

This is a question on which I am not prepared at present to offer a decided opinion.

Whether the horizontality of the shelves or terrace-lines is really as perfect as has been generally assumed is a point which will require to be tested by a more accurate trigonometrical survey than has yet been made. The preservation of precisely the same level in the lowest line throughout the glens of Roy, Spean, and Laggan, for a distance of 20 miles east and west, and 10 or 12 miles north and south, would be very wonderful if ascertained with mathematical precision. Mr. Jamieson, after making in 1862 several measurements with a spirit-level, has been led to suspect a rise in the lowest shelf of one foot in a mile in a direction from west to east, or from the mouth of Glen Roy to a point 6 miles east of it in Glen Spean. To confirm such observations, and to determine whether a similar rate of rise continues eastward as far as the pass of Muckul, would be most important.

On the whole, I conclude that the Glen Roy terrace-lines and those of some neighbouring valleys, were formed on the borders of glacier-lakes, in times long subsequent to the principal glaciation of Scotland. They may perhaps have been nearly as late, especially the lowest of the shelves, as that portion of the Pleistocene period in which Man co-existed in Europe with the mammoth.

CHAPTER XIV

CHRONOLOGICAL RELATIONS OF THE GLACIAL PERIOD AND THE
EARLIEST SIGNS OF MAN'S APPEARANCE IN EUROPE—*continued*

Signs of extinct Glaciers in Wales—Great Submergence of Wales during
the Glacial Period proved by Marine Shells—Still greater Depression
inferred from stratified Drift—Scarcity of organic Remains in Glacial
Formations—Signs of extinct Glaciers in England—Ice Action of
Ireland—Maps illustrating successive Revolutions in Physical Geography during the Pleistocene Period—Southernmost Extent of
Erratics in England—Successive Periods of Junction and Separation
of England, Ireland, and the Continent—Time required for these
Changes—Probable Causes of the Upheaval and Subsidence of the
Earth's Crust—Antiquity of Man considered in relation to the Age
of the existing Fauna and Flora.

Extinct Glaciers in Wales

THE considerable amount of vertical movement in opposite
directions, which was suggested in the last chapter, as affording the most probable explanation of the position of some of
the stratified and fossiliferous drifts of Scotland, formed since
the commencement of the glacial period, will appear less startling
if it can be shown that independent observations lead us to
infer that a geographical revolution of still greater magnitude
accompanied the successive phases of glaciation through which
the Welsh mountains have passed.

That Wales was once an independent centre of the dispersion
of erratic blocks has long been acknowledged. Dr. Buckland
published in 1842 his reasons for believing that the Snowdonian
mountains in Caernarvonshire were formerly covered with
glaciers, which radiated from the central heights through the
seven principal valleys of that chain, where striæ and flutings
are seen on the polished rocks directed towards as many different
points of the compass. He also described the " moraines " of
the ancient glaciers, and the rounded masses of polished rock,
called in Switzerland " roches moutonnées." His views respecting the old extinct glaciers of North Wales were subsequently
confirmed by Mr. Darwin, who attributed the transport of many
of the larger erratic blocks to floating ice. Much of the Welsh
glacial drift had already been shown by Mr. Trimmer to have

had a submarine origin, and Mr. Darwin maintained that when the land rose again to nearly its present height, glaciers filled the valleys, and "swept them clean of all the rubbish left by the sea."[1]

Professor Ramsay, in a paper read to the Geological Society in 1851, and in a later work on the glaciation of North Wales, described three successive glacial periods, during the first of which the land was much higher than it now is, and the quantity of ice excessive; secondly, a period of submergence when the land was 2300 feet lower than at present, and when the higher mountain tops only stood out of the sea as a cluster of low islands, which nevertheless were covered with snow; and lastly, a third period when the marine boulder drift formed in the middle period was ploughed out of the larger valleys by a second set of glaciers, smaller than those of the first period. This last stage of glaciation may have coincided with that of the parallel roads of Glen Roy, spoken of in the last chapter. In Wales it was certainly preceded by submergence, and the rocks had been exposed to glacial polishing and friction before they sank.

Fortunately the evidence of the sojourn of the Welsh mountains beneath the waters of the sea is not deficient, as in Scotland, in that complete demonstration which the presence of marine shells affords. The late Mr. Trimmer discovered such shells on Moel Tryfan, in North Wales, in drift elevated more than 1300 feet above the level of the sea. It appears from his observations, and those of the late Edward Forbes, corroborated by others of Professor Ramsay and Mr. Prestwich, that about twelve species of shells, including *Fusus bamfius, F. antiquus, Venus striatula* (Forbes and Hanley), have been met with at heights of between 1000 and 1400 feet, in drift, reposing on a surface of rock which had been previously exposed to glacial friction and striation.[2] The shells, as a whole, are those of the glacial period, and not of the Norwich Crag. Two localities of these shells in Wales, in addition to that first pointed out by Mr. Trimmer, have since been observed by Professor Ramsay, who, however, is of opinion that the amount of submergence can by no means be limited to the extreme height to which the shells happen to have been traced; for drift of the same character as that of Moel Tryfan extends continuously to the height of 2300 feet. [Note 27.]

---

[1] *Philosophical Magazine*, ser. 3, vol. xxi., p. 180.
[2] Ramsay, *Quart. Jour. Geol. Soc.*, vol. viii., 1852, p. 372.

Rarity of Organic Remains in Glacial Formations

The general dearth of shells in such formations, below as well as above the level at which Mr. Trimmer first found them, deserves notice. Whether we can explain it or not, it is a negative character which seems to belong very generally to deposits formed in glacial seas. The porous nature of the strata, and the length of time during which they have been permeated by rain-water, may partly account, as we hinted in a former chapter, for the destruction of organic remains. But it is also possible that they were originally scarce, for we read of the waters of the sea being so freshened and chilled by the melting of ice-bergs in some Norwegian and Icelandic fjords, that the fish are driven away, and all the mollusca killed. The moraines of glaciers are always from the first devoid of shells, and if transported by ice-bergs to a distance, and deposited where the ice melts, may continue as barren of every indication of life as they were when they originated.

Nevertheless, it may be said, on the other hand, that herds of seals and walruses crowd the floating ice of Spitzbergen in latitude 80° north, of which Mr. Lamont has recently given us a lively picture,[1] and huge whales fatten on myriads of pteropods in polar regions. It had been suggested that the bottom of the sea, at the era of extreme submergence in Scotland and Wales, was so deep as to reach the zero of animal life, which, in part of the Mediterranean (the Ægean, for example), the late Edward Forbes fixed, after a long series of dredgings, at 300 fathoms. But the shells of the glacial drift of Scotland and Wales, when they do occur, are not always those of deep seas; and, moreover, our faith in the uninhabitable state of the ocean at great depths has been rudely shaken, by the recent discovery of Captain M'Clintock and Dr. Wallich, of starfish in water more than a thousand fathoms deep (7560 feet!), midway between Greenland and Iceland. That these radiata were really dredged up from the bottom, and that they had been living and feeding there, appeared from the fact that their stomachs were full of *Globigerina*, of which foraminiferous creatures, both living and dead, the oozy bed of the ocean at that vast depth was found to be exclusively composed. [Note 28.]

Whatever may be the cause, the fact is certain, that over large areas in Scotland, Ireland, and Wales, I might add through-

[1] *Seasons with the Sea-Horses*, 1861.

out the northern hemisphere on both sides of the Atlantic, the
stratified drift of the glacial period is very commonly devoid
of fossils, in spite of the occurrence here and there, at the height
of 500, 700, and even 1400 feet, of marine shells. These, when
met with, belong, with few exceptions, to known living species.
I am therefore unable to agree with Mr. Kjerulf that the amount
of former submergence can be measured by the extreme height
at which shells happen to have been found.

Glacial Formations in England

The mountains of Cumberland and Westmorland, and the
English lake district, afford equally unequivocal vestiges of ice-
action not only in the form of polished and grooved surfaces,
but also of those rounded bosses before mentioned as being so
abundant in the Alpine valleys of Switzerland, where glaciers
exist, or have existed. Mr. Hall has lately published a faithful

Fig. 38.

Dome-shaped rocks, or ' roches moutonnées,' in the valley of the Rothay,
near Ambleside, from a drawing by E. Hull, F.G.S.[1]

account of these phenomena, and has given a representation
of some of the English " roches moutonnées," which precisely
resemble hundreds of dome-shaped protuberances in North
Wales, Sweden, and North America.[2]

The marks of glaciation on the rocks, and the transportation
of erratics from Cumberland to the eastward, have been traced
by Professor Phillips over a large part of Yorkshire, extending
to a height of 1500 feet above the sea; and similar northern

[1] *Edinburgh New Philosophical Journal*, vol. xi., Pl. i., p. 31, 1860.
[2] Hull, *Edinburgh New Philosophical Journal*, July 1860.

drift has been observed in Lancashire, Cheshire, Derbyshire, Shropshire, Staffordshire, and Worcestershire. It is rare to find marine shells, except at heights of 200 or 300 feet; but a few instances of their occurrence have been noticed, especially of *Turritella communis* (a gregarious shell), far in the interior, at elevations of 500 feet, and even of 700 in Derbyshire, and some adjacent counties, as I learn from Mr. Binney and Mr. Prestwich.

Such instances are of no small theoretical interest, as enabling us to account for the scattering of large erratic blocks at equal or much greater elevations, over a large part of the northern and midland counties, such as could only have been conveyed to their present sites by floating ice. Of this nature, among others, is a remarkable angular block of syenitic greenstone, 4½ feet by 4 feet square, and 2 feet thick, which Mr. Darwin describes as lying on the summit of Ashley Heath, in Staffordshire, 803 feet above the sea, resting on New Red Sandstone.[1]

Signs of Ice-action and Submergence in Ireland during the Glacial Period

In Ireland we encounter the same difficulty as in Scotland in determining how much of the glaciation of the higher mountains should be referred to land glaciers, and how much to floating ice, during submergence. The signs of glacial action have been traced by Professor Jukes to elevations of 2500 feet in the Killarney district, and to great heights in other mountainous regions; but marine shells have rarely been met with higher than 600 feet above the sea, and that chiefly in gravel, clay, and sand in Wicklow and Wexford. They are so rare in the drift east of the Wicklow mountains, that an exception to the rule, lately observed at Ballymore Eustace, by Professor Jukes, is considered as a fact of no small geological interest. The wide extent of drift of the same character, spread over large areas in Ireland, shows that the whole island was, in some part of the glacial period, an archipelago, as represented in the maps, Figs. 39, 40.

Speaking of the Wexford drift, the late Professor E. Forbes states that Sir H. James found in it, together with many of the usual glacial shells, several species which are characteristic of the Crag; among others the reversed variety of *Fusus antiquus*, called *F. contrarius*, and the extinct species *Nucula*

---

[1] Ancient Glaciers of Caernarvonshire, *Philosophical Magazine*, series 3, xxi., p. 180.

Cobboldiæ, and *Turritella incrassata*. Perhaps a portion of this drift of the south of Ireland may belong to the close of the Pliocene period, and may be of a somewhat older date than the shells of the Clyde, alluded to in Chap. XIII. They may also correspond still more nearly in age with the fauna of the uppermost strata of the Norwich Crag, occurring at Chillesford. [Note 29.]

The scarcity of mammalian remains in the Irish drift favours the theory of its marine origin. In the superficial deposits of the whole island, I have only met with three recorded examples of the mammoth, one in the south near Dungarvan, where the bones of *Elephas primigenius*, two species of bear (*Ursus arctos* and *Ursus spelæus?*), the reindeer, horse, etc., were found in a cave;[1] another in the centre of the island near Belturbet, in the county of Cavan.

Perhaps the conversion into land of the bed of the glacial sea, and the immigration into the newly upheaved region of the elephant, rhinoceros, and hippopotamus, which co-existed with the fabricators of the St. Acheul flint hatchets, were events which preceded in time the elevation of the Irish drift, and the union of that island with England. Ireland may have continued for a longer time in the state of an archipelago, and was therefore for a much shorter time inhabited by the large extinct Pleistocene pachyderms.

In one of the reports of the Geological Survey of Ireland, published in 1859, Professor Jukes, in explanation of sheet 184 of the maps, alludes to beds of sand and gravel, and signs of the polishing and furrowing of the rocks in the counties of Kerry and Killarney, as high as 2500 feet above the sea, and supposes (perhaps with good reason) that the land was depressed even to that extent. He observes that above that elevation (2500 feet) the rocks are rough, and not smoothed, as if by ice. Some of the drift was traced as high as 1500 feet, the highest hills there exceeding 3400 feet. Mr. Jukes, however, is by no means inclined to insist on submergence to the extent of 2500 feet, as he is aware that ice, like that now prevailing in Greenland, might explain most, if not all, the appearances of glaciation in the highest regions.

Although the course taken by the Irish erratics in general is such that their transportation seems to have been due to floating ice or coast-ice, yet some granite blocks have travelled from south to north, as recorded by Sir R. Griffiths, namely,

[1] E. Brenan and Dr. Carte, Dublin, 1859.

those of the Ox Mountains in Sligo; a fact from which Mr.
Jamieson infers that those mountains formed at one time a
centre of dispersion. In the same part of Ireland, the general
direction in which the boulders have travelled is everywhere
from north-west to south-east, a course directly at right angles
to the prevailing trend of the present mountain ridges.

*Maps illustrating successive Revolutions in Physical
Geography during the Pleistocene Period*

The late Mr. Trimmer, before referred to, has endeavoured
to assist our speculations as to the successive revolutions in
physical geography, through which the British Islands have
passed since the commencement of the glacial period, by four
" sketch maps " as he termed them, in the first of which he
gave an ideal restoration of the original Continental period,
called by him the first elephantine period, or that of the forest
of Cromer, before described. He was not aware that the prevailing elephant of that era (*E. meridionalis*) was distinct from
the mammoth. At this era he conceived Ireland and England
to have been united with each other and with France, but much
of the area represented as land in the map, Fig. 41, was supposed to be under water. His second map, of the great submergence of the glacial period, was not essentially different
from our map, Fig. 39. His third map expressed a period of
partial re-elevation, when Ireland was reunited to Scotland
and the north of England; but England still separated from
France. This restoration appears to me to rest on insufficient
data, being constructed to suit the supposed area over which
the gigantic Irish deer, or *Megaceros*, migrated from east to
west, also to explain an assumed submergence of the district
called the Weald, in the south-east of England, which had
remained land during the grand glacial submergence.

The fourth map is a return to nearly the same continental
conditions as the first—Ireland, England, and the Continent
being united. This he called the second elephantine period;
and it would coincide very closely with that part of the Pleistocene era in which Man co-existed with the mammoth, and when,
according to Mr. Trimmer's hypothesis previously indicated
by Mr. Godwin-Austen, the Thames was a tributary of the
Rhine.[1]

[1] Joshua Trimmer, *Quart. Jour. Geol. Soc.*, vol. ix., 1853, Pl. 13, and
Godwin-Austen, *ibid.*, vol. vii., 1851, p. 134, and Pl. 7.

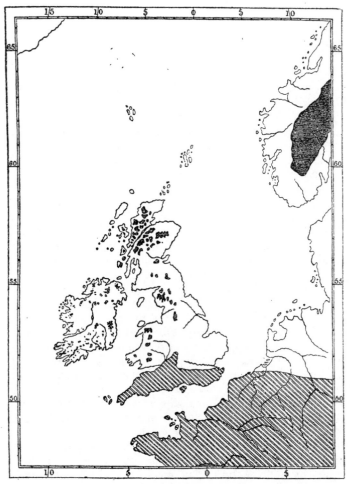

Fig. 39.

MAP OF THE BRITISH ISLES AND PART OF THE NORTH-WEST OF EUROPE,
SHOWING THE GREAT AMOUNT OF SUPPOSED SUBMERGENCE OF LAND
BENEATH THE SEA DURING PART OF THE GLACIAL PERIOD.

The submergence of Scotland is to the extent of 2000 feet, and of other
parts of the British Isles, 1300.

In the map, the dark shade expresses the land which alone remained
above water. The area shaded by diagonal lines is that which cannot be
shown to have been under water at the period of floating ice by the evidence of erratics, or by marine shells of northern species. How far the
several parts of the submerged area were simultaneously or successively
laid under water, in the course of the glacial period, cannot, in the present
state of our knowledge, be determined.

These geographical speculations were indulged in ten years after Edward Forbes had published his bold generalisations on the geological changes which accompanied the successive establishment of the Scandinavian, Germanic, and other living floras and faunas in the British Islands, and, like the theories of his predecessor, were the results of much reflection on a vast body of geological facts. It is by repeated efforts of this kind, made by geologists who are prepared for the partial failure of some of their first attempts, that we shall ultimately arrive at a knowledge of the long series of geographical revolutions which have followed each other since the beginning of the Pleistocene period.

The map, Fig. 39, will give some idea of the great extent of land which would be submerged, were we to infer, as many geologists have done, from the joint evidence of marine shells, erratics, glacial striæ and stratified drift at great heights, that Scotland was, during part of the glacial period, 2000 feet below its present level, and other parts of the British Isles, 1300 feet. A subsidence to this amount can be demonstrated in the case of North Wales by marine shells. In the lake district of Cumberland, in Yorkshire, and in Ireland, we must depend on proofs derived from glacial striæ and the transportation of erratics for so much of the supposed submergence as exceeds 600 feet. As to central England, or the country north of the Thames and Bristol Channel, marine shells of the glacial period sometimes reach as high as 600 and 700 feet, and erratics still higher, as we have seen above. But this region is of such moderate elevation above the sea, that it would be almost equally laid under water, were there a sinking of no more than 600 feet.

To make this last proposition clear, I have constructed, from numerous documents, many of them unpublished, the map, Fig. 40, which shows how that small amount of subsidence would reduce the whole of the British Isles to an archipelago of very small islands, with the exception of parts of Scotland, and the north of England and Wales, where four islands of considerable dimensions would still remain.

The map does not indicate a state of things supposed to have prevailed at any one moment of the past, because the district south of the Thames and the Bristol Channel seems to have remained land during the whole of the glacial period, at a time when the northern area was under water. The map simply represents the effects of a downward movement of a hundred fathoms, or 600 English feet, assumed to be uniform

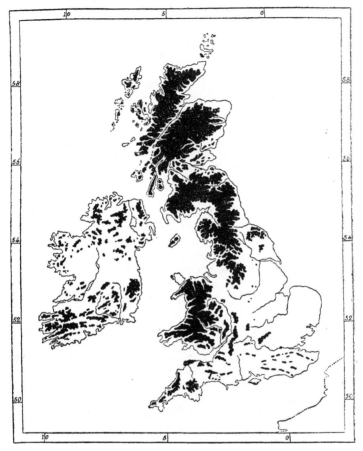

Fig. 40.

MAP SHOWING WHAT PARTS OF THE BRITISH ISLANDS WOULD REMAIN ABOVE
WATER AFTER A SUBSIDENCE OF THE AREA TO THE EXTENT OF 600 FEET.

The authorities to whom I am indebted for the information contained
in this map are—for

SCOTLAND—A. Geikie, Esq., F.G.S., and T. F. Jamieson, Esq., of Ellon,
Aberdeenshire.

ENGLAND—For the counties of Yorkshire, Lancashire, and Durham—
Col. Sir Henry James, R.E.
Dorsetshire, Hampshire, and Isle of Wight—H. W. Bristow, Esq.
Gloucestershire, Somersetshire, and part of Devon—R. Etheridge, Esq.
Kent and Sussex—Frederick Drew, Esq.
Isle of Man—W. Whitaker, Esq.

IRELAND—Reduced from a contour map constructed by Lieut. Larcom,
R.E., in 1837, for the Railway Commissioners.

over the whole of the British Isles. It shows the very different state of the physical geography of the area in question, when contrasted with the results of an opposite movement, or one of upheaval, to an equal amount, of which Sir Henry de la Beche had already given us a picture, in his excellent treatise called *Theoretical Researches*.[1] His map I have borrowed (Fig. 41), after making some important corrections in it.

If we are surprised when looking at the first map, Fig. 40, at the vast expanse of sea which so moderate a subsidence as 600 feet would cause, we shall probably be still more astonished to perceive, in Fig. 41, that a rise of the same number of feet would unite all the British Isles, including the Hebrides, Orkneys, and Shetlands, with one another and the Continent, and lay dry the sea now separating Great Britain from Sweden and Denmark.

It appears from soundings made during various Admiralty surveys, that the gained land thus brought above the level of the sea, instead of presenting a system of hills and valleys corresponding with those usually characterising the interior of most of our island, would form a nearly level terrace, or gently inclined plane, sloping outwards like those terraces of denudation and deposition which I have elsewhere described as occurring on the coasts of Sicily and the Morea.[2]

It seems that, during former and perhaps repeated oscillations of level undergone by the British Isles, the sea has had time to cut back the cliffs for miles in many places, while in others the detritus derived from wasting cliffs drifted along the shores, together with the sediment brought down by rivers and swept by currents into submarine valleys, has exerted a levelling power, filling up such depressions as may have pre-existed. Owing to this twofold action few marked inequalities of level have been left on the sea-bottom, the " silver-pits " off the mouth of the Humber offering a rare exception to the general rule, and even there the narrow depression is less than 300 feet in depth.

Beyond the 100 fathom line, the submarine slope surrounding the British coast is so much steeper that a second elevation of equal amount (or of 600 feet) would add but slightly to the area of gained land; in other words, the 100 and 200 fathom lines run very near each other.[3]

---

[1] Also repeated in De la Beche's *Geological Observer*.
[2] *Manual of Geology*, p. 74.
[3] De la Beche, *Geological Researches*, p. 191.

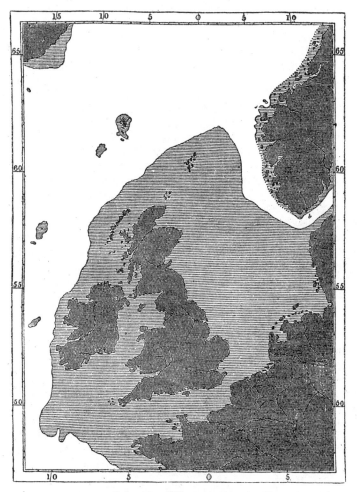

Fig. 41.

MAP OF PART OF THE NORTH-WEST OF EUROPE, INCLUDING THE BRITISH
ISLES, SHOWING THE EXTENT OF SEA WHICH WOULD BECOME LAND IF
THERE WERE A GENERAL RISE OF THE AREA TO THE EXTENT OF 600 FEET.

The darker shade expresses what is now land, the lighter shade the space
intervening between the present coast line and the 100 fathom line, which
would be converted by such a movement into land.

The original of this map will be found in Sir H. de la Beche's *Theoretical
Researches*, p. 190, 1834, but several important corrections have been introduced into it from recently published Admiralty Surveys, especially—

1st. A deep channel passing from the North Sea into the entrance of the
Baltic.

2nd. The more limited westerly extension of the West Coast of Ireland.

The naturalist would have been entitled to assume the former union, within the Pleistocene period, of all the British Isles with each other and with the Continent, as expressed in the map, Fig. 41, even if there had been no geological facts in favour of such a junction. For in no other way would he be able to account for the identity of the fauna and flora found throughout these lands. Had they been separated ever since the Miocene period, like Madeira, Porto Santo, and the Desertas, constituting the small Madeiran Archipelago, we might have expected to discover a difference in the species of land-shells, not only when Ireland was compared to England, but when different islands of the Hebrides were contrasted one with another, and each of them with England. It would not, however, be necessary, in order to effect the complete fusion of the animals and plants which we witness, to assume that all parts of the area formed continuous land at one and the same moment of time, but merely that the several portions were so joined within the Pleistocene era as to allow the animals and plants to migrate freely in succession from one district to another.

Southernmost Extent of Erratics in England

In reference to that portion of the south of England which is marked by diagonal lines in Fig. 39, the theory of its having been an area of dry land during the period of great submergence and floating ice does not depend merely on negative evidence, such as the absence of the northern drift or boulder clay on its surface; but we have also, in favour of the same conclusion, the remarkable fact of the presence of erratic blocks on the southern coast of Sussex, implying the existence there of an ancient coast-line at a period when the cold must have been at its height.

These blocks are to be seen in greatest number at Pagham and Selsea, 15 miles south of Chichester, in latitude 50° 40′ N.

They consist of fragments of granite, syenite and greenstone, as well as of Devonian and Silurian rocks, some of them of large size. I measured one of granite at Pagham, 27 feet in circumference. They are not of northern origin, but must have come from the coast of Normandy or Brittany, or from land which may once have existed to the south-west, in what is now the English Channel.

They were probably drifted into their present site by coast ice, and the yellow clay and gravel in which they are embedded are a littoral formation, as shown by the shells.   Beneath the gravel containing these large erratics, is a blue mud in which skeletons of *Elephas antiquus*, and other mammalia, have been observed.   Still lower occurs a sandy loam, from which Mr. R. G. Austen [1] has collected thirty-eight species of marine shells, all Recent, but forming an assemblage differing as a whole from that now inhabiting the English Channel.   The presence among them of *Lutraria rugosa* and *Pecten polymorphus*, not known to range farther north in the actual seas than the coast of Portugal, indicates a somewhat warmer temperature at the time when they flourished.   Subsequently, there must have been great cold when the Selsea erratics were drifted into their present position, and this cold doubtless coincided in time with a low temperature farther north. [Note 30.]   These trans-ported rocks of Sussex are somewhat older than a sea-beach with Recent marine shells which at Brighton is covered by Chalk rubble, called the " elephant-bed," which I cannot describe in this place, but I allude to it as one of many geological proofs of the former existence of a seashore in this region, and of ancient cliffs bounding the channel between France and England, all of older date than the close of the glacial period.   [Note 31.]

In order to form a connected view of the most simple series of changes in physical geography which can possibly account for the phenomena of the glacial period, and the period of the establishment of the present provinces of animals and plants, the following geographical states of the British and adjoining areas may be enumerated.

First, a continental period, towards the close of which the forest of Cromer flourished: when the land was at least 500 feet above its present level, perhaps much higher, and its extent probably greater than that given in the map, Fig. 41.

Secondly, a period of submergence, by which the land north of the Thames and Bristol Channel, and that of Ireland, was gradually reduced to such an archipelago as is pictured in map, Fig. 40; and finally to such a general prevalence of sea as is seen in map, Fig. 39.   This was the period of great submergence and of floating ice, when the Scandinavian flora, which occupied the lower grounds during the first continental period, may have obtained exclusive possession of the only lands not covered with perpetual snow.

Thirdly, a second continental period when the bed of the glacial sea, with its marine shells and erratic blocks, was laid

---

[1] *Quart. Jour. Geol. Soc.*, vol. xiii., 1857, p. 50.

dry, and when the quantity of land equalled that of the first period, and therefore probably exceeded that represented in the map, Fig. 41. During this period there were glaciers in the higher mountains of Scotland and Wales, and the Welsh glaciers, as we have seen, pushed before them and cleared out the marine drift with which some valleys had been filled during the period of submergence. The parallel roads of Glen Roy are referable to some part of the same era.

As a reason for presuming that the land which in map, Fig. 41, is only represented as 600 feet above its present level, was during part of this period much higher, Professor Ramsay has suggested that, as the previous depression far exceeded 100 fathoms (amounting in Wales to 1400 feet, as shown by marine shells, and to 2300, by stratified drift), it is not improbable that the upward movement was on a corresponding scale.

In passing from the period of chief submergence to this second continental condition of things, we may conceive a gradual change first from that of Map 39 to Map 40, then from the latter phase to that of Map 41, and finally to still greater accessions of land. During this last period the passage of the Germanic flora into the British area took place, and the Scandinavian plants, together with northern insects, birds, and quadrupeds, retreated into the higher grounds.

Judging from the evidence at present before us, the first appearance of Man, when, together with the mammoth and woolly rhinoceros, or with the *Elephas antiquus, Rhinoceros hemitœchus*, and *Hippopotamus major*, he ranged freely from all parts of the Continent into the British area, took place during this second continental period.

Fourthly, the next and last change comprised the breaking up of the land of the British area once more into numerous islands, ending in the present geographical condition of things. There were probably many oscillations of level during this last conversion of continuous land into islands, and such movements in opposite directions would account for the occurrence of marine shells at moderate heights above the level of the sea, notwithstanding a general lowering of the land. To the close of this era belong the marine deposits of the Clyde and the Carses of the Tay and Forth, before alluded to.

In a memoir by Professor E. Forbes, before cited, he observes, that the land of passage by which the plants and animals migrated into Ireland consisted of the upraised marine drift which had previously formed the bottom of the glacial sea.

Portions of this drift extend to the eastern shores of Wicklow and Wexford, others are found in the Isle of Man full of arctic shells, others on the British coast opposite Ireland. The fresh-water marl, containing numerous skeletons of the great deer, or *Megaceros*, overlie in the Isle of Man that marine glacial drift. Professor Forbes also remarks that the subsequent dis-junction of Ireland from England, or the formation of the Irish Channel, which is less than 400 feet in its greatest depth, pre-ceded the opening of the Straits of Dover, or the final separation of England from the Continent. This he inferred from the present distribution of species both in the animal and vegetable kingdoms. Thus, for example, there are twice as many reptiles in Belgium as in England, and the number inhabiting England is twice that found in Ireland. Yet the Irish species are all common to England, and all the English to Belgium. It is therefore assumed that the migration of species westward having been the work of time, there was not sufficient lapse of ages to complete the fusion of the continental and British reptilian fauna, before France was separated from England and England from Ireland.

For the same reason there are also a great number of birds of short flight, and small quadrupeds, inhabiting England which do not cross to Ireland, the Irish Channel seeming to have arrested them in their westward course.[1]

The depth of the Irish Channel in the narrower parts is only 360 feet, and the English Channel between Dover and Calais less than 200, and rarely anywhere more than 300 feet; so that vertical movements of slight amount compared to some of those previously considered, with the aid of denuding operations or the waste of sea cliffs, and the scouring out of the channel, might in time effect the insulation of the lands above alluded to.

Time required for successive Changes in Physical Geography in the Pleistocene Period

The time which it would require to bring about such changes of level, according to the average rate assumed in Chap. III., however vast, will not be found to exceed that which would best explain the successive fluctuations in terrestrial temperature, the glaciation of solid rocks, the transportation of erratics above and below the sea-level, the height of arctic shells above the

[1] E. Forbes, Fauna and Flora of British Isles, *Mem. Geol. Survey*, vol. i., 1846, p. 344.

sea, and last, not least, the migration of the existing species of animals and plants into their actual stations, and the extinction of some conspicuous forms which flourished during the Pleistocene ages. When we duly consider all these changes which have taken place since the beginning of the glacial epoch, or since the forest of Cromer and the *Elephas meridionalis* flourished, we shall find that the phenomena become more and more intelligible in proportion to the slowness of the rate of elevation and depression which we assume.

The submergence of Wales to the extent of 1400 feet, as proved by glacial shells, would require 56,000 years, at the rate of 2½ feet per century; but taking Professor Ramsay's estimate of 800 feet more, that depression being implied by the position of some of the stratified drift, we must demand an additional period of 32,000 years, amounting in all to 88,000; and the same time would be required for the re-elevation of the tract to its present height. But if the land rose in the second continental period as much as 600 feet above its present level, as in Fig. 41, this 600 feet, first of rising and then of sinking, would require 48,000 years more; the whole of the grand oscillation, comprising the submergence and re-emergence, having taken about 224,000 years for its completion; and this, even if there were no pause or stationary period, when the downward movement ceased, and before it was converted into an upward one.

I am aware that it may be objected that the average rate here proposed is a purely arbitrary and conjectural one, because, at the North Cape, it is supposed that there has been a rise of about 5 feet in a century, and at Spitsbergen, according to Mr. Lamont, a still faster upheaval during the last 400 years.[1] But, granting that in these and some exceptional cases (none of them as yet very well established) the rising or sinking has, for a time, been accelerated, I do not believe the average rate of motion to exceed that above proposed. Mr. Darwin, I find, considers that such a mean rate of upheaval would be as high as we could assume for the west coast of South America, where we have more evidence of sudden changes of level than anywhere else. He has not, however, attempted to estimate the probable rate of secular elevation in that or any other region.

Little progress has yet been made in divining the most probable causes of these great movements of the earth's crust; yet what little we know of the state of the interior leads us

[1] *Seasons with the Sea-Horses*, p. 202.

to expect that the gradual expansion or contraction of large portions of the solid crust may be the result of fluctuations in temperature, with which the existence of hundreds of active and thousands of extinct volcanoes is probably connected.

It is ascertained that solid rocks, such as granite and sandstone, expand and contract annually, even under such a moderate range of temperature as that of a Canadian winter and summer. If the heat should go on increasing through a thickness, say only of 10 miles of the earth's crust, the gradual upheaval of the incumbent mass may amount to many hundreds of feet; and the elevation may be carried still farther, by the complete fusion of part of the inferior rocks.

According to the experiments of Deville, the contraction of granite, in passing from a melted, or as some would say its plastic condition, to a solid state, must be more than 10 per cent.[1] So that we have at our command a source of depression on a grand scale, at every period when granitic rocks have originated in the interior of the earth's crust. All mineralogists are agreed that the passage of voluminous masses, from a liquid or pasty to a solid and crystalline state, must be an extremely slow process. It may often happen that, in the same series of superimposed rocks, some are expanding while still solid or while partially melting, while others are at the same time crystallising and contracting; so that the alterations of level at the surface may be the result of complicated and often of conflicting agencies. The more gradually we conceive such changes to take place, the more comprehensible they become in the eyes of the chemist and natural philosopher who speculates on the changes of the earth's interior; and the more fertile are they in the hands of the geologist in accounting for revolutions on the habitable surface.

We may presume, that after the movement has gone on for a long time in one determinate direction, whether of elevation or depression, the change to an opposite movement, implying the substitution of a heating for a refrigerating operation, or the reverse, would not take place suddenly; but would be marked by a period of inaction, or of slight movement, or such a state of quiescence, as prevails throughout large areas of dry land in the normal condition of the globe.

I see no reason for supposing that any part of the revolutions in physical geography, to which the maps above described have reference, indicate any catastrophes greater than those which

[1] *Bull. Soc. Geol. France*, 2nd series, vol. iv., p. 1312.

the present generation has witnessed. If Man was in existence when the Cromer forest was becoming submerged, he would have felt no more alarm than the Danish settlers on the east coast of Baffin's Bay, when they found the poles, which they had driven into the beach to secure their boats, had subsided below their original level.

Already, perhaps, the melting ice has thrown down till and boulders upon those poles, a counterpart of the boulder clay which overlies the forest-bed on the Norfolk cliffs.

We have seen that all the plants and shells, marine and freshwater, of the forest bed, and associated fluvio-marine strata of Norfolk, are specifically identical with those of the living European flora and fauna; so that if upon such a stratum a deposit of the present period, whether freshwater or marine, should be thrown down, it might lie conformably over it, and contain the same invertebrate fauna and flora. The strata so superimposed would, in ordinary geological language, be called contemporaneous, not only as belonging to the same epoch, but as appertaining strictly to the same subdivision of one and the same epoch; although they would in fact have been separated by an interval of several hundred thousand years.

If, in the lower of the two formations, some of the mammalia of the genera elephant and rhinoceros were found to be distinct in species from those of the same genera in the upper or " recent " stratum, it might appear as though there had been a sudden coming in of new forms, and a sudden dying out of old ones; for there would not have been time in the interval for any perceptible change in the invertebrate fauna, by which alone we usually measure the lapse of time in the older formations.

When we are contrasting the vertebrate contents of two sets of superimposed strata of the Cretaceous, Oolitic, or any other ancient formation in which the shells are identical in species, we ought never to lose sight of the possibility of their having been separated by such intervals or by two or three thousand centuries. That number of years may sometimes be of small moment in reference to the rate of fluctuation of species in the lower animals, but very important when the succession of forms in the highest classes of vertebrata is concerned.

If we reflect on the long series of events of the Pleistocene and Recent periods contemplated in this chapter, it will be remarked that the time assigned to the first appearance of Man, so far as our geological inquiries have yet gone, is extremely modern in relation to the age of the existing fauna and flora, or even to the time when most of the living species of animals and plants attained their actual geographical distribution. At the same time it will also be seen, that if the advent of Man in Europe occurred before the close of the second continental period, and antecedently to the separation of Ireland from England and of England from the Continent, the event would be sufficiently remote to cause the historical period to appear quite insignificant in duration, when compared to the antiquity of the human race.

CHAPTER XV

EXTINCT GLACIERS OF THE ALPS AND THEIR CHRONOLOGICAL
RELATION TO THE HUMAN PERIOD

Extinct Glaciers of Switzerland—Alpine Erratic Blocks on the Jura—Not
transported by floating Ice—Extinct Glaciers of the Italian Side of
the Alps—Theory of the Origin of Lake-Basins by the erosive Action
of Glaciers, considered—Successive Phases in the Development of
Glacial Action in the Alps—Probable Relation of these to the earliest
known Date of Man—Correspondence of the same with successive
Changes in the Glacial Condition of the Scandinavian and British
Mountains—Cold Period in Sicily and Syria.

Extinct Glaciers of Switzerland

WE have seen in the preceding chapters that the mountains
of Scandinavia, Scotland, and North Wales have served, during
the glacial period, as so many independent centres for the dispersion of erratic blocks, just as at present the ice-covered
continent of North Greenland is sending down ice in all directions to the coast, and filling Baffin's Bay with floating bergs,
many of them laden with fragments of rocks.

Another great European centre of ice-action during the Pleistocene period was the Alps of Switzerland, and I shall now
proceed to consider the chronological relations of the extinct
Alpine glaciers to those of more northern countries previously
treated of. [Note 32.]

The Alps lie far south of the limits of the northern drift
described in the foregoing pages, being situated between the
44th and 47th degrees of north latitude. On the flanks of
these mountains, and on the sub-Alpine ranges of hills or plains
adjoining them, those appearances which have been so often
alluded to, as distinguishing or accompanying the drift, between
the 50th and 70th parallels of north latitude, suddenly reappear
and assume, in a southern region, a truly arctic development.
Where the Alps are highest, the largest erratic blocks have been
sent forth; as, for example, from the regions of Mont Blanc
and Monte Rosa, into the adjoining parts of Switzerland and
Italy; while in districts where the great chain sinks in altitude,
as in Carinthia, Carniola, and elsewhere, no such rocky fragments, or a few only and of smaller bulk, have been detached
and transported to a distance.

In the year 1821, M. Venetz first announced his opinion that the Alpine glaciers must formerly have extended far beyond their present limits, and the proofs appealed to by him in confirmation of this doctrine were afterwards acknowledged by M. Charpentier, who strengthened them by new observations and arguments, and declared in 1836 his conviction that the glaciers of the Alps must once have reached as far as the Jura, and have carried thither their moraines across the great valley of Switzerland. M. Agassiz, after several excursions in the Alps with M. Charpentier, and after devoting himself some years to the study of glaciers, published in 1840 an admirable description of them and of the marks which attest the former action of great masses of ice over the entire surface of the Alps and the surrounding country.[1] He pointed out that the surface of every large glacier is strewed over with gravel and stones detached from the surrounding precipices by frost, rain, lightning, or avalanches. And he described more carefully than preceding writers the long lines of these stones, which settle on the sides of the glacier, and are called the lateral moraines; those found at the lower end of the ice being called terminal moraines. Such heaps of earth and boulders every glacier pushes before it when advancing, and leaves behind it when retreating. When the Alpine glacier reaches a lower and a warmer situation, about 3000 or 4000 feet above the sea, it melts so rapidly that, in spite of the downward movement of the mass, it can advance no farther. Its precise limits are variable from year to year, and still more so from century to century; one example being on record of a recession of half a mile in a single year. We also learn from M. Venetz, that whereas, between the eleventh and fifteenth centuries, all the Alpine glaciers were less advanced than now, they began in the seventeenth and eighteenth centuries to push forward, so as to cover roads formerly open, and to overwhelm forests of ancient growth.

These oscillations enable the geologist to note the marks which a glacier leaves behind it as it retrogrades; and among these the most prominent, as before stated, are the terminal moraines, or mounds of unstratified earth and stones, often divided by subsequent floods into hillocks, which cross the valley like ancient earthworks, or embankments made to dam up a river. Some of these transverse barriers were formerly pointed out by Saussure below the glacier of the Rhone, as

[1] Agassiz, *Études sur les Glaciers et Système Glaciaire*.

proving how far it had once transgressed its present boundaries. On these moraines we see many large angular fragments, which, having been carried along the surface of the ice, have not had their edges worn off by friction; but the greater number of the boulders, even those of large size, have been well rounded, not by the power of water, but by the mechanical force of the ice, which has pushed them against each other, or against the rocks flanking the valley. Others have fallen down the numerous fissures which intersect the glacier, where, being subject to the pressure of the whole mass of ice, they have been forced along, and either well rounded or ground down into sand, or even the finest mud, of which the moraine is largely constituted.

As the terminal moraines are the most prominent of all the monuments left by a receding glacier, so are they the most liable to obliteration; for violent floods or debacles are sometimes occasioned in the Alps by the sudden bursting of glacier-lakes, or those temporary sheets of water before alluded to which are caused by the damming up of a river by a glacier which has increased during a succession of cold seasons, and descending from a tributary into the main valley, has crossed it from side to side. On the failure of this icy barrier the accumulated waters, being let loose, sweep away and level many a transverse mound of gravel and loose boulders below, and spread their materials in confused and irregular beds over the river-plain.

Another mark of the former action of glaciers in situations where they exist no longer, is the polished, striated, and grooved surfaces of rocks before described. Stones which lie underneath the glacier and are pushed along by it sometimes adhere to the ice, and as the mass glides slowly along at the rate of a few inches, or at the utmost 2 or 3 feet per day, abrade, groove, and polish the rock, and the larger blocks are reciprocally grooved and polished by the rock on their lower sides. As the forces both of pressure and propulsion are enormous, the sand acting like emery polishes the surface; the pebbles, like coarse gravers, scratch and furrow it; and the large stones scoop out grooves in it. Lastly, projecting eminences of rock, called " roches moutonnées," are smoothed and worn into the shape of flattened domes where the glaciers have passed over them.

Although the surface of almost every kind of rock when exposed to the open air wastes away by decomposition, yet some retain for ages their polished and furrowed exterior: and if they are well protected by a covering of clay or turf, these marks of abrasion seem capable of enduring for ever. They have been traced in the Alps to great heights above the present glaciers, and to great horizontal distances beyond them.

Another effect of a glacier is to lodge a ring of stones round the summit of a conical peak which may happen to project through the ice. If the glacier is lowered greatly by melting, these circles of large angular fragments, which are called " perched blocks," are left in a singular situation near the top of a steep hill or pinnacle, the lower parts of which may be destitute of boulders.

Alpine erratic Blocks on the Jura

Now some or all the marks above enumerated,—the moraines, erratics, polished surfaces, domes, striæ, and perched rocks— are observed in the Alps at great heights above the present glaciers and far below their actual extremities; also in the great valley of Switzerland, 50 miles broad; and almost everywhere on the Jura, a chain which lies to the north of this valley. The average height of the Jura is about one-third that of the Alps, and it is now entirely destitute of glaciers; yet it presents almost everywhere moraines, and polished and grooved surfaces of rocks. The erratics, moreover, which cover it present a phenomenon which has astonished and perplexed the geologist for more than half a century. No conclusion can be more incontestable than that these angular blocks of granite, gneiss, and other crystalline formations, came from the Alps, and that they have been brought for a distance of 50 miles and upwards across one of the widest and deepest valleys of the world; so that they are now lodged on the hills and valleys of a chain composed of limestone and other formations, altogether distinct from those of the Alps. Their great size and angularity, after a journey of so many leagues, has justly excited wonder, for hundreds of them are as large as cottages; and one in particular, composed of gneiss, celebrated under the name of Pierre à Bot, rests on the side of a hill about 900 feet above the lake of Neufchatel, and is no less than 40 feet in diameter. But there are some far-transported masses of granite and gneiss which are still larger, and which have been found to contain 50,000 and 60,000 cubic feet of stone; and one limestone block at Devens, near Bex, which has travelled 30 miles, contains 161,000 cubic feet, its angles being sharp and unworn.

Von Buch, Escher, and Studer inferred, from an examination

of the mineral composition of the boulders, that those resting on the Jura, opposite the lakes of Geneva and Neufchatel, have come from the region of Mont Blanc and the Valais, as if they had followed the course of the Rhone to the lake of Geneva, and had then pursued their way uninterruptedly in a northerly direction.

M. Charpentier, who conceived the Alps in the period of greatest cold to have been higher by several thousand feet than they are now, had already suggested that the Alpine glaciers once reached continuously to the Jura, conveying thither the large erratics in question.[1] M. Agassiz, on the other hand, instead of introducing distinct and separate glaciers, imagined that the whole valley of Switzerland might have been filled with ice, and that one great sheet of it extended from the Alps to the Jura, the two chains being of the same height as now relatively to each other. To this idea it was objected that the difference of altitude, when distributed over a space of 50 miles, would give an inclination of two degrees only, or far less than that of any known glacier. In spite of this difficulty, the hypothesis has since received the support of Professor James Forbes in his very able work on the Alps published in 1843.

In 1841, I advanced jointly with Mr. Darwin[2] the theory that the erratics may have been transferred by floating ice to the Jura, at the time when the greater part of that chain and the whole of the Swiss valley to the south was under the sea. We pointed out that if at that period the Alps had attained only half their present altitude they would yet have constituted a chain as lofty as the Chilean Andes, which in a latitude corresponding to Switzerland now send down glaciers to the head of every sound, from which icebergs covered with blocks of granite are floated seaward. Opposite that part of Chile where the glaciers abound is situated the island of Chiloe 100 miles in length with a breadth of 30 miles, running parallel to the continent. The channel which separates it from the main land is of considerable depth and 25 miles broad. Parts of its surface, like the adjacent coast of Chile, are overspread with Recent marine shells, showing an upheaval of the land during a very modern period; and beneath these shells is a boulder deposit in which Mr. Darwin found large blocks of granite and syenite which had evidently come from the Andes.

A continuance in future of the elevatory movement now

[1] D'Archiac, *Histoire des Progrès*, etc., vol. ii., p. 249.
[2] See *Elements of Geology*, 2nd ed., 1841.

observed to be going on in this region of the Andes and of Chiloe might cause the former chain to rival the Alps in altitude and give to Chiloe a height equal to that of the Jura. The same rise might dry up the channel between Chiloe and the main land so that it would then represent the great valley of Switzerland.

Sir Roderick I. Murchison, after making several important geological surveys of the Alps, proposed in 1849 a theory agreeing essentially with that suggested by Mr. Darwin and myself, viz. that the erratics were transported to the Jura at a time when the great strath of Switzerland and many valleys receding far into the Alps were under water. He thought it impossible that the glacial detritus of the Rhone could ever have been carried to the Lake of Geneva and beyond it by a glacier, or that so vast a body of ice issuing from one narrow valley could have spread its erratics over the low country of the cantons of Vaud, Fribourg, Berne, and Soleure, as well as the slopes of the Jura, comprising a region of about 100 miles in breadth from south-west to north-east, as laid down in the map of Charpentier. He therefore imagined the granitic blocks to have been translated to the Jura by ice-floats when the intermediate country was submerged.[1] It may be remarked that this theory, provided the water be assumed to have been salt or brackish, demands quite as great an oscillation in the level of the land as that on which Charpentier had speculated, the only difference being that the one hypothesis requires us to begin with a subsidence of 2500 or 3000 feet, and the other with an elevation to the same amount. We should also remember that the crests or watersheds of the Alps and Jura are about 80 miles apart, and if once we suppose them to have been in movement during the glacial period it is very probable that the movements at such a distance may not have been strictly uniform. If so the Alps may have been relatively somewhat higher, which would have greatly facilitated the extension of Alpine glaciers to the flanks of the less elevated chain.

Five years before the publication of the memoir last mentioned, M. Guyot had brought forward a great body of new facts in support of the original doctrine of Charpentier, that the Alpine glaciers once reached as far as the Jura and that they had deposited thereon a portion of their moraines.[2] The scope of his observations and argument was laid with great

[1] *Quart. Jour. Geol. Soc.*, vol. vi., 1850, p. 65.
[2] *Bulletin de la Société des Sciences Naturelles de Neufchâtel*, 1845.

clearness before the British public in 1852 by Mr. Charles Maclaren, who had himself visited Switzerland for the sake of forming an independent opinion on a theoretical question of so much interest and on which so many eminent men of science had come to such opposite conclusions.[1]

M. Guyot had endeavoured to show that the Alpine erratics, instead of being scattered at random over the Jura and the great plain of Switzerland, are arranged in a certain determinate order strictly analogous to that which ought to prevail if they had once constituted the lateral, medial, and terminal moraines of great glaciers. The rocks chiefly relied on as evidence of this distribution consist of three varieties of granite, besides gneiss, chlorite-slate, euphotide, serpentine, and a peculiar kind of conglomerate, all of them foreign alike to the great strath between the Alps and Jura and to the structure of the Jura itself. In these two regions limestones, sandstones, and clays of the Secondary and Tertiary formations alone crop out at the surface, so that the travelled fragments of Alpine origin can easily be distinguished and in some cases the precise localities pointed out from whence they must have come.

The accompanying map or diagram slightly altered from one given by Mr. Maclaren will enable the reader more fully to appreciate the line of argument relied on by M. Guyot. The dotted area is that over which the Alpine fragments were spread by the supposed extinct glacier of the Rhone. The site of the present reduced glacier of that name is shown at A. From that point the boulders may first be traced to B, or Martigny, where the valley takes an abrupt turn at right angles to its former course. Here the blocks belonging to the right side of the river or derived from c d e have not crossed over to the left side at B, as they should have done had they been transported by floating ice, but continue to keep to the side to which they belonged, assuming that they once formed part of a right lateral moraine of a great extinct glacier. That glacier, after arriving at the lower end of the long narrow valley of the upper Rhone at F, filled the Lake of Geneva, F, I, with ice. From F, as from a great vomitory, it then radiated in all directions bearing along with it the moraines with which it was loaded and spreading them out on all sides over the great plain. But the principal icy mass moved straight onwards in a direct line towards the hill of Chasseron, G (precisely opposite F), where the Alpine erratics attain their maximum of height on the

Jura, that is to say 2015 English feet above the level of the Lake of Neufchâtel or 3450 feet above the sea. The granite blocks which have ascended to this eminence G came from the east shoulder of Mont Blanc *h*, having travelled in the direction B, F, G.

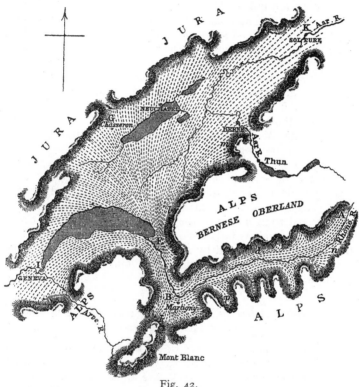

Fig. 42.

MAP SHOWING THE SUPPOSED COURSE OF THE ANCIENT AND NOW EXTINCT GLACIER OF THE RHONE, AND THE DISTRIBUTION OF THE ERRATIC BLOCKS AND DRIFT CONVEYED BY IT TO THE GREAT VALLEY OF SWITZERLAND AND THE JURA.

When these and the accompanying blocks resting on the south-eastern declivity of the Jura are traced from their culminating point G in opposite directions, whether westward towards Geneva or eastwards towards Soleure, they are found to decline in height from the middle of the arc G towards the two extremities I and K, both of which are at a lower level than G, by about 1500 feet. In other words the ice of the extinct glacier, having mounted up on the sloping flanks of the Jura in the line of greatest pressure to its highest elevation, began to decline laterally in the manner of a pliant or viscous mass with a gentle inclination till it reached two points distant from each other no less than 100 miles. [Note 33.]

In further confirmation of this theory M. Guyot observed that fragments derived from the right bank of the great valley of the Rhone *c d e* are found on the right side of the great Swiss basin or strath as at *l* and *m*, while those derived from the left bank *p h* occur on the left side of the basin or on the Jura between G and I; and those again derived from places farthest up on the left bank and nearest the source of the Rhone, as *n o*, occupy the middle of the great basin, constituting between *m* and K what M. Guyot calls the frontal or terminal moraine of the eastern prolongation of the old glacier.

A huge boulder of talcose granite, now at Steinhoff, 10 miles east from K, or Soleure, containing 61,000 French cubic feet, or equal in bulk to a mass measuring 40 feet in every direction, was ascertained by Charpentier from its composition to have been derived from *n*, one of the highest points on the left side of the Rhone valley far above Martigny. From this spot it must have gone all round by F, which is the only outlet to the deep valley, so as to have performed a journey of no less than 150 miles!

General Transportation of Erratics in Switzerland due to Glaciers and not to floating Ice

It is evident that the above described restriction of certain fragments of peculiar lithological character to that bank of the Rhone where the parent rocks are alone met with and the linear arrangement of the blocks in corresponding order on the opposite side of the great plain of Switzerland, are facts which harmonise singularly well with the theory of glaciers while they are wholly irreconcilable with that of floating ice. Against the latter hypothesis all the arguments which Charpentier originally brought forward in opposition to the first popular doctrine of a grand débacle or sudden flood rushing down from the Alps to the Jura might be revived. Had there ever been such a rush of muddy water, said he, the blocks carried down the basins of the principal Swiss rivers, such as the Rhone, Aar, Reuss, and Limmat, would all have been mingled confusedly together instead of having each remained in separate and distinct areas as they do and should do according to the glacial hypothesis.

M. Morlot presented me in 1857 with an unpublished map of Switzerland in which he had embodied the results of his own observations and those of MM. Guyot, Escher, and others, marking out by distinct colours the limits of the ice-transported detritus proper to each of the great river-basins. The arrangement of the drift and erratics thus depicted accords perfectly well with Charpentier's views and is quite irreconcilable with the supposition of the scattered blocks having been dispersed by floating ice when Switzerland was submerged.

As opposed to the latter hypothesis, I may also state that nowhere as yet have any marine shells or other fossils than those of a terrestrial character, such as the bones of the mammoth and a few other mammalia and some coniferous wood, been detected in those drifts, though they are often many hundreds of feet in thickness.

A glance at M. Morlot's map, above mentioned,[1] will show that the two largest areas, indicated by a single colour, are those over which the Rhone and the Rhine are supposed to have spread out in ancient times their enormous moraines. One of these only, that of the Rhone, has been exhibited in our diagram, Fig. 42. The distinct character of the drift in the two cases is such as it would be if two colossal glaciers should now come down from the higher Alps through the valleys traversed by those rivers, leaving their moraines in the low country. The space occupied by the glacial drift of the Rhine is equal in dimensions or rather exceeds that of the Rhone, and its course is not interfered with in the least degree by the Lake of Constance, 45 miles long, any more than is the dispersion of the erratics of the Rhone by the Lake of Geneva, about 50 miles in length. The angular and other blocks have in both instances travelled on precisely as if those lakes had no existence, or as if, which was no doubt the case, they had been filled with solid ice.

During my last visit to Switzerland in 1857, I made excursions, in company with several distinguished geologists, for the sake of testing the relative merits of the two rival theories above referred to, and I examined parts of the Jura above Neufchâtel in company with M. Desor, the country round Soleure with M. Langen, the southern side of the great strath near Lausanne with M. Morlot, the basin of the Aar around

[1] *See* map, *Quart. Jour. Geol. Soc.*, vol. xviii., 1862, p. 185, Pl. 18.

Berne with M. Escher von der Linth; and having satisfied myself that all the facts which I saw north of the Alps were in accordance with M. Guyot's views, I crossed to the Italian side of the great chain and became convinced that the same theory was equally applicable to the ancient moraines of the plains of the Po.

M. Escher pointed out to me at Trogen in Appenzel on the left bank of the Rhine fragments of a rock of a peculiar mineralogical character, commonly called the granite of Pontelyas, the natural position of which is well known near Trons, 100 miles from Trogen, on the left bank of the Rhine about 30 miles from the source of that river. All the blocks of this peculiar granite keep to the left bank, even where the valley turns almost at right angles to its former course near Mayenfeld below Chur, making a sharp bend resembling that of the valley of the Rhone at Martigny. The granite blocks, where they are traced to the low country, still keep to the left side of the Lake of Constance. That they should not have crossed over to the opposite river-bank below Chur is quite inexplicable if, rejecting the aid of land-ice, we appeal to floating ice as the transporting power.

In M. Morlot's map already cited we behold between the areas occupied by the glacial drift of the Rhine and Rhone three smaller yet not inconsiderable spaces distinguished by distinct colours, indicating the peculiar detritus brought down by the three great rivers, the Aar, Reuss, and Limmat. The ancient glacier of the first of these, the Aar, has traversed the lakes of Brienz and Thun and has borne angular, polished, and striated blocks of limestone and other rocks as far as Berne and somewhat below that city. The Reuss has also stamped the lithological character of its own mountainous region upon the lower part of its hydrographical basin by covering it with its peculiar Alpine drift. In like manner the old extinct glacier of the Limmat during its gradual retreat has left monuments of its course in the Lake of Zürich in the shape of terminal moraines, one of which has almost divided that great sheet of water into two lakes.

The ice-work done by the extinct glaciers, as contrasted with that performed by their dwarfed representatives of the present day, is in due proportion to the relative volume of the supposed glaciers, whether we measure them by the distances to which they have carried erratic blocks or the areas which they have strewed over with drift or the hard surfaces of rock and number of boulders which they have polished and striated. Instead of a length of 5, 10, or 20 miles and a thickness of 200, 300, or at the utmost 800 feet, those giants of the olden time must have been from 50 to 150 miles long and between 1000 and 3000 feet deep. In like manner the glaciation although identical in kind is on so small a scale in the existing Alpine glaciers as at first sight to disappoint a Swedish, Scotch, Welsh, or North American geologist. When I visited the terminal moraine of the glacier of the Rhone in 1859 and tried to estimate the number of angular or rounded pebbles and blocks which exhibited glacial polishing or scratches as compared to those bearing no such markings, I found that several thousand had to be reckoned before I arrived at the first which was so striated or polished as to differ from the stones of an ordinary torrent-bed. Even in the moraines of the glaciers of Zermatt, Viesch, and others, in which fragments of limestone and serpentine are abundant (rocks which most readily receive and most faithfully retain the signs of glaciation), I found, for one which displayed such indications, several hundreds entirely free from them. Of the most opposite character were the results obtained by me from a similar scrutiny of the boulders and pebbles of the terminal moraine of one of the old extinct glaciers, namely, that of the Rhone in the suburbs of Soleure. Thus at the point κ in the map, Fig. 42, I observed a mass of unstratified clay or mud, through which a variety of angular and rubbed stones were scattered and a marked proportion of the whole were polished and scratched and the clay rendered so compact, as if by the incumbent pressure of a great mass of ice, that it has been found necessary to blow it up with gunpowder in making railway cuttings through part of it. A limestone of the age of our Portland stone on which this old moraine rests, has its surface polished like a looking-glass, displaying beautiful sections of fossil shells of the genera *Nerinæa* and *Pteroceras*, while occasionally, besides finer striæ, there are deep rectilinear grooves, agreeing in direction with the course in which the extinct glacier would have moved according to the theory of M. Guyot, before explained.

Extinct Glaciers of the Italian Side of the Alps

To select another example from the opposite or southern side of the Alps. It will be seen in the elaborate map recently executed by Signor Gabriel de Mortillet of the ancient glaciers

of the Italian flank of the Alps that the old moraines descend
in narrow strips from the snow-covered ridges through the
principal valleys to the great basin of the Po, on reaching which
they expand and cover large circular or oval areas. Each of
these groups of detritus is observed (*see* map, Fig. 43) to contain

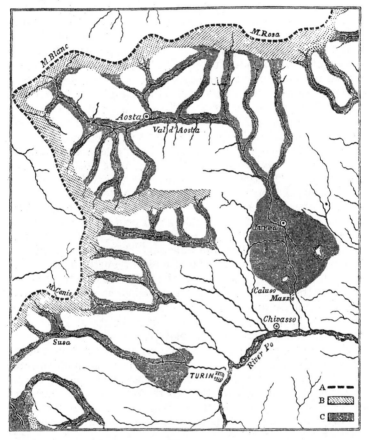

Fig. 43.

MAP OF THE MORAINES OF EXTINCT GLACIERS EXTENDING FROM THE ALPS
INTO THE PLAINS OF THE PO NEAR TURIN.

*From Map of the ancient Glaciers of the Italian side of the Alps by
Signor Gabriel de Mortillet.*

A Crest or watershed of the Alps.
B Snow-covered Alpine summits which fed the ancient glaciers.
C Moraines of ancient or extinct glaciers.

exclusively the wreck of such rocks as occur *in situ* on the Alpine heights of the hydrographical basins to which the moraines respectively belong.

I had an opportunity of verifying this fact, in company with Signor Gastaldi as my guide, by examining the erratics and boulder formation between Susa and Turin, on the banks of the Dora Riparia, which brings down the waters from Mont Cenis and from the Alps S.W. of it. I there observed striated fragments of dolomite and gypsum, which had come down from Mont Cenis and had travelled as far as Avigliana; also masses of serpentine brought from less remote points, some of them apparently exceeding in dimensions the largest erratics of Switzerland. I afterwards visited, in company with Signori Gastaldi and Michelotti, a still grander display of the work of a colossal glacier of the olden time, 20 miles N.E. of Turin, the moraine of which descended from the two highest of the Alps, Mont Blanc and Monte Rosa, and after passing through the valley of Aosta, issued from a narrow defile above Ivrea (*see* map, Fig. 43). From this vomitory the old glacier poured into the plains of the Po that wonderful accumulation of mud, gravel, boulders, and large erratics, which extend for 15 miles from above Ivrea to below Caluso and which, when seen in profile from Turin have the aspect of a chain of hills. In many countries, indeed, they might rank as an important range of hills, for where they join the mountains they are more than 1500 feet high, and retain more than half that height for a great part of their course, rising very abruptly from the plain, often with a slope of from 20° to 30°. This glacial drift reposes near the mountains on ancient metamorphic rocks and farther from them on marine Pliocene strata. Portions of the ridges of till and stratified matter have been cut up into mounds and hillocks by the action of the river, the Dora Baltea, and there are numerous lakes, so that the entire moraine much resembles, except in its greater height and width, the line of glacial drift of Perthshire and Forfarshire before described. Its complicated structure can only be explained by supposing that the ancient glacier advanced and retreated several times and left large lateral moraines, the more modern mounds within the limits of the older ones, and masses of till thrown down upon the rearranged and stratified materials of the first set of moraines. Such appearances accord well with the hypothesis of the successive phases of glacial action in Switzerland, to which I shall presently advert.

Contorted Strata of Glacial Drift south of Ivrea

At Mazzé near Caluso (*see* Fig. 43), the southern extremity of this great moraine has recently been cut through in making a tunnel for the railway which runs from Turin to Ivrea. In the fine section thus exposed Signor Gastaldi and I had an opportunity of observing the internal structure of the glacial formation. In close juxtaposition to a great mass of till with striated boulders, we saw stratified beds of alternating gravel, sand, and loam, which were so sharply bent that many of them had been twice pierced through in the same vertical cutting. Whether they had been thus folded by the mechanical power of an advancing glacier, which had pushed before it a heap of stratified matter, as the glacier of Zermatt has been sometimes known to shove forward blocks of stone through the walls of houses, or whether the melting of masses of ice, once inter-stratified with sand and gravel, had given rise to flexures in the manner before suggested; it is at least satisfactory to have detected this new proof of a close connection between ice-action and contorted stratification, such as has been described as so common in the Norfolk cliffs and which is also very often seen in Scotland and North America, where stratified gravel overlies till. I have little doubt that if the marine Pliocene strata which underlie a great part of the moraine below Ivrea were exposed to view in a vertical section, those fundamental strata would be found not to participate in the least degree in the plications of the sands and gravels of the overlying glacial drift.

To return to the marks of glaciation: in the moraine at Mazzé there are many large blocks of protogine and large and small ones of limestone and serpentine which have been brought down from Monte Rosa, through the gorge of Ivrea, after having travelled for a distance of 50 miles. Confining my attention to a part of the moraine where pieces of limestone and serpentine were very numerous, I found that no less than one-third of the whole number bore unequivocal signs of glacial action; a state of things which seems to bear some relation to the vast volume and pressure of the ice which once constituted the extinct glacier and to the distance which the stones had travelled. When I separated the pebbles of quartz, which were never striated, and those of granite, mica-schist, and diorite, which do not often exhibit glacial markings, and confined my attention to the serpentine alone I found no less than nineteen in twenty of the whole number polished and scratched; whereas in the terminal moraines of some modern glaciers, where the materials have travelled not more than 10 or 15, instead of 100 miles, scarce one in twenty even of the serpentine pebbles exhibit glacial polish and striation.

Theory of the Origin of Lake-basins by the erosive Action of Glaciers, considered

Geologists are all agreed that the last series of movements to which the Alps owe their present form and internal structure occurred after the deposition of the Miocene strata; and it has been usual to refer the origin of the numerous lake-basins of Alpine and sub-Alpine regions both in Switzerland and Northern Italy to the same movements; for it seemed not unnatural to suppose, that forces capable of modifying the configuration of the greatest European chain, by uplifting some of its component Tertiary strata (those of marine origin of the Miocene period) several thousand feet above their former level, after throwing them into vertical and contorted positions, must also have given rise to many superficial inequalities, in some of which large bodies of water would collect. M. Desor, in a memoir on the Swiss and Italian lakes, suggested that they may have escaped being obliterated by sedimentary deposition by having been filled with ice during the whole of the glacial period.

Subsequently to the retreat of the great glaciers we know that the lake-basins have been to a certain extent encroached upon and turned into land by river deltas; one of which, that of the Rhone at the head of the Lake of Geneva, is no less than 12 miles long and several miles broad, besides which there are many torrents on the borders of the same lake, forming smaller deltas.

M. Gabriel de Mortillet after a careful study of the glacial formations of the Alps agreed with his predecessors that the great lakes had existed before the glacial period, but came to the opinion in 1859 that they had all been first filled up with alluvial matter and then re-excavated by the action of ice, which during the epoch of intense cold had by its weight and force of propulsion scooped out the loose and incoherent alluvial strata, even where they had accumulated to a thickness of 2000 feet. Besides this erosion, the ice had carried the whole mass of mud and stones up the inclined planes, from the central depths to the lower outlets of the lakes and sometimes far beyond them. As some of these rock-basins are 500, others more than 2000 feet deep, having their bottoms in some cases 500, in others 1000 feet below the level of the sea, and having areas from 20 to 50 miles in length and from 4 to 12 in breadth, we may well be startled at the boldness of this hypothesis.

The following are the facts and train of reasoning which induced M. de Mortillet to embrace these views. At the lower ends of the great Italian lakes, such as Maggiore, Como, Garda, and others, there are vast moraines which are proved by their contents to have come from the upper Alpine valleys above the lakes. Such moraines often repose on an older stratified alluvium, made up of rounded and worn pebbles of precisely the same rocks as those forming the moraines, but not derived from them, being small in size, never angular, polished, or striated, and the whole having evidently come from a great distance. These older alluvial strata must, according to M. de Mortillet, be of pre-glacial date and could not have been carried past the sites of the lakes, unless each basin had previously been filled and levelled up with mud, sand, and gravel, so that the river channel was continuous from the upper to the lower extremity of each basin.

Professor Ramsay, after acquiring an intimate knowledge of the glacial phenomena of the British Isles, had taught many years before that small tarns and shallow rock-basins such as we see in many mountain regions owe their origin to glaciers which erode the softer rocks, leaving the harder ones standing out in relief and comparatively unabraded. Following up this idea after he had visited Switzerland and without any communication with M. de Mortillet or cognisance of his views, he suggested in 1859 that the lake-basins were not of pre-glacial date, but had been scooped out by ice during the glacial period, the excavation having for the most part been effected in Miocene sandstone, provincially called, on account of its softness, " molasse." By this theory he dispensed with the necessity of filling up pre-existing cavities with stratified alluvium, in the manner proposed by M. de Mortillet.

I will now explain to what extent I agree with, and on what points I feel compelled to differ from the two distinguished geologists above cited. First. It is no doubt true, as Professor Ramsay remarks, that heavy masses of ice, creeping for ages over a surface of dry land (whether this comprise hills, plateaus, and valleys, as in the case of Greenland, before described, or be confined to the bottoms of great valleys, as now in the higher

Alps), must often by their grinding action produce depressions, in consequence of the different degrees of resistance offered by rocks of unequal hardness. Thus, for example, where quartzose beds of mica-schist alternate with clay-slate, or where trap-dykes, often causing waterfalls in the courses of torrents, cut through sandstone or slate—these and innumerable other common associations of dissimilar stony compounds must give rise to a very unequal amount of erosion and consequently to lake-basins on a small scale. But the larger the size of any lake, the more certain it will be to contain within it rocks of every degree of hardness, toughness, and softness; and if we find a gradual deepening from the head towards the central parts and a shallowing again from the middle to the lower end, as in several of the great Swiss and Italian lakes, which are 30 or 40 miles in length, we require a power capable of acting with a considerable degree of uniformity on these masses of varying powers of resistance.

Secondly. Several of the great lakes are by no means in the line of direction which they ought to have taken had they been scooped out by the pressure and onward movement of the extinct glaciers. The Lake of Geneva, for instance, had it been the work of ice, would have been prolonged from the termination of the upper valley of the Rhone towards the Jura, in the direction from F to G of the map, Fig. 42, instead of running from F to I.

Thirdly. It has been ascertained experimentally, that in a glacier, as in a river, the rate of motion is accelerated or lessened, according to the greater or less slope of the ground; also, that the lower strata of ice, like those of water, move more slowly than those above them. In the Lago Maggiore, which is more than 2600 feet deep (797 metres), the ice, says Professor Ramsay, had to descend a slope of about 3° for the first 25 miles, and then to *ascend* for the last 12 miles (from the deepest part towards the outlet) at an angle of 5°. It is for those who are conversant with the dynamics of glacier motion to divine whether in such a case the discharge of ice would not be entirely effected by the superior and faster moving strata, and whether the lowest would not be motionless or nearly so, and would therefore exert very little, if any, friction on the bottom.

Fourthly. But the gravest objection to the hypothesis of glacial erosion on so stupendous a scale is afforded by the entire absence of lakes of the first magnitude in several areas where they ought to exist if the enormous glaciers which once occupied those spaces had possessed the deep excavating power ascribed to them. Thus in the area laid down on the map, Fig. 43, or that covered by the ancient moraine of the Dora Baltea, we see the monuments of a colossal glacier derived from Mont Blanc and Monte Rosa, which descended from points nearly 100 miles distant, and then emerging from the narrow gorge above Ivrea deployed upon the plains of the Po, advancing over a floor of marine Pliocene strata of no greater solidity than the Miocene sandstone and conglomerate in which the lake-basins of Geneva, Zürich, and some others are situated. Why did this glacier fail to scoop out a deep and wide basin rivalling in size the lakes of Maggiore or Como, instead of merely giving rise to a few ponds above Ivrea, which may have been due to ice action? There is one lake, it is true—that of Candia, near the southern extremity of the moraine—which is larger; but even this, as will be seen by the map, is quite of subordinate importance, and whether it is situated in a rock basin or is simply caused by a dam of moraine matter has not yet been fully made out.

There ought also to have been another great lake, according to the theory under consideration, in the space now occupied by the moraine of the Dora Riparia, between Susa and Turin (*see* map, Fig. 43). Signor Gastaldi has shown that all the ponds in that area consist exclusively of what M. de Mortillet has denominated *morainic lakes, i.e.* caused by barriers of glacier-mud and stones.

Fifthly. In proof of the great lakes having had no existence before the glacial period, Professor Ramsay observes that we do not find in the Alps any freshwater strata of an age intermediate between " the close of the Miocenic and the commencement of the glacial epoch." [1] But although such formations are scarce, they are by no means wholly wanting; and if it can be shown that any one of the principal lakes, that of Zürich for example, existed prior to the glacial era it will follow that in the Alps the erosive power of ice was not required to produce lake-basins on a large scale. The deposits alluded to on the borders of the Lake of Zürich are those of Utznach and Dürnten, situated each about 350 feet above the present level of the lake and containing valuable beds of lignite.

The first of them, that of Utznach, is a delta formed at the head of the ancient and once more extensive lake. The argillaceous and lignite-bearing strata, more than 100 feet in thickness, rest unconformably on highly inclined and sometimes

[1] *Quart. Jour. Geol. Soc.*, vol. xviii., 1862.

vertical Miocene molasse. These clays are covered conformably by stratified sand and gravel 60 feet thick, partly consolidated, in which the pebbles are of rocks belonging to the upper valleys of the Limmat and its tributaries, all of them small and not glacially striated and wholly without admixture of large angular stones. On the top of all repose very large erratic blocks, affording clear evidence that the colossal glacier which once filled the valley of the Limmat covered the old littoral deposit. The great age of the lignite is partly indicated by the bones of *Elephas antiquus* found in it.

I visited Utznach in company with M. Escher von der Linth in 1857, and during the same year examined the lignite of Dürnten, many miles farther down on the right bank of the lake, in company with Professor Heer and M. Marcou. The beds there are of the same age and within a few feet of the same height above the level of the lake. They might easily have been overlooked or confounded with the general glacial drift of the neighbourhood, had not the bed of lignite, which is from 5 to 12 feet thick, been worked for fuel, during which operation many organic remains came to light. Among these are the teeth of *Elephas antiquus*, determined by Dr. Falconer, and *Rhinoceros leptorhinus ?* (*R. megarhinus*, Christol), the wild bull and red deer (*Bos primigenius*, Boj., and *Cervus elaphus*, L.), the last two determined by Professor Rütimeyer. In the same beds I found many freshwater shells of the genera *Paludina*, *Limnæa*, etc., all of living species. The plants named by Professor Heer are also Recent and agree singularly with those of the Cromer buried forest, before described.

Among them are the Scotch and spruce firs, *Pinus sylvestris* and *Pinus abies*, and the buckbean, or *Menyanthes trifoliata*, etc., besides the common birch and other European plants.

Overlying this lignite are first, as at Utznach, stratified gravel not of glacial origin, about 30 feet thick; and secondly, highest of all, huge angular erratic blocks clearly indicating the presence of a great glacier posterior in date to all the organic remains above enumerated.

If any one of the existing Swiss lakes were now lowered by deepening its outlet, or by raising the higher portion of it relatively to the lower, we should see similar deltas of comparatively modern date exposed to view, some of them with embedded trunks of pines of the same species drifted down during freshets. Such deposits would be most frequent at the upper ends of the lakes, but a few would occur on either

bank not far from the shore where torrents once entered, agreeing in geographical position with the lignite formations of
Utznach and Dürnten.

There are other freshwater formations with lignite, besides
those on the Lake of Zürich, as those of Wetzikon near the
Pfaffikon Lake, of Kaltbrunnen, of Buchberg, and that of
Morschweil between St. Gall and Rorschach, but none probably older than the Dürnten beds. Like the buried forest of
Cromer they are all pre-glacial, yet they by no means represent
the older nor even the newer Pliocene period, but rather the
beginning of the Pleistocene. It is therefore true, as Professor
Ramsay remarks, that, as yet, no strata " of the age of the
English Crag " have been detected in any Alpine valley. In
other words, there are no freshwater formations yet known
corresponding in date to the Pliocene beds of the upper Val
d'Arno, above Florence—a fact from which we may infer
(though with diffidence, as the inference is based on negative
evidence), that, although the great Alpine valleys were eroded
in Pliocene times, the lake-basins were, nevertheless, of Pleistocene date—some of them formed before, others during, the
glacial epoch.

Sixthly. In what manner then did the great lake-basins
originate if they were not hollowed out by ice? My answer is,
they are all due to unequal movements of upheaval and subsidence. We have already seen that the buried forest of
Cromer, which by its organic contents seems clearly to be of
the same age as the lignite of Dürnten, was pre-glacial and
that it has undergone a great oscillation of level (about 500 feet
in both directions) since its origin, having first sunk to that
extent below the sea and then been raised up again to the sea-
level. In the countless Post-Miocene ages which preceded the
glacial period there was ample time for the slow erosion by
water of all the principal hydrographical basins of the Alps, and
the sites of all the great lakes coincide, as Professor Ramsay
truly says, with these great lines of drainage. The lake-cavities
do not lie in synclinal troughs, following the strike and foldings
of the strata, but often, as the same geologist remarks, cross
them at high angles; nor are they due to rents or gaping fissures,
although these, with other accidents connected with the disturbing movements of the Alps, may sometimes have determined
originally the direction of the valleys. The conformity of the
lake-basins to the principal watercourses is explicable if we
assume them to have resulted from inequalities in the upward

and downward movements of the whole country in Pleistocene times, after the valleys were eroded.

We know that in Sweden the rate of the rise of the land is far from uniform, being only a few inches in a century near Stockholm, while north of it and beyond Gefle it amounts to as many feet in the same number of years. Let us suppose with Charpentier that the Alps gained in height several thousand feet at the time when the intense cold of the glacial period was coming on. This gradual rise would be an era of aqueous erosion and of the deepening, widening, and lengthening of the valleys. It is very improbable that the elevation would be everywhere identical in quantity, but if it was never in excess in the outskirts as compared to the central region or crest of the chain, it would not give rise to lakes. When, however, the period of upheaval was followed by one of gradual subsidence, the movement not being everywhere strictly uniform, lake-basins would be formed wherever the rate of depression was in excess in the upper country. Let the region, for example, near the head waters of the great rivers sink at the rate of from 4 to 6 feet per century, while only half as much subsidence occurs towards the circumference of the mountains—the rate diminishing about an inch per mile in a distance, say of 40 miles—this might convert many of the largest and deepest valleys at their lower ends into lakes.

We have no certainty that such movements may not now be in progress in the Alps; for if they are as slow as we have assumed, they would be as insensible to the inhabitants as is the upheaval of Scandinavia or the subsidence of Greenland to the Swedes and Danes who dwell there. They only know of the progress of such geographical revolutions because a slight change of level becomes manifest on the margin of the sea. The lines of elevation or depression above supposed might leave no clear geological traces of their action on the high ridges and table-lands separating the valleys of the principal rivers; it is only when they cross such valleys that the disturbance caused in the course of thousands of years in the drainage becomes apparent. If there were no ice, the sinking of the land might not give rise to lakes. To accomplish this in the absence of ice, it is necessary that the rate of depression should be sufficiently fast to make it impossible for the depositing power of the river to keep pace with it, or in other words to fill up the incipient cavity as fast as it begins to form. Such levelling operations once complete, the running water, aided by sand and pebbles, will gradually cut a gorge through the newly raised rock so as to prevent it from forming a barrier. But if a great glacier fill the lower part of the valley all the conditions of the problem are altered. Instead of the mud, sand, and stones drifted down from the higher regions being left behind in the incipient basin, they all travel onwards in the shape of moraines on the top of the ice, passing over and beyond the new depression, so that when at the end of fifty or a thousand centuries the glacier melts, a large and deep basin representing the difference in the movement of two adjoining mountain areas—namely, the central and the circumferential—is for the first time rendered visible.

By adopting this hypothesis, we concede that there is an intimate connection between the glacial period and a predominance of lakes, in producing which the action of ice is threefold; first, by its direct power in scooping out shallow basins where the rocks are of unequal hardness; an operation which can by no means be confined to the land, for it must extend to below the level of high water a thousand feet and more in such fjords as have been described as filled with ice in Greenland.

Secondly. The ice will act indirectly by preventing cavities caused by inequalities of subsidence or elevation from becoming the receptacles first of water and then of sediment, by which the cavities would be levelled up and the lakes obliterated.

Thirdly. The ice is also an indirect cause of lakes, by heaping up mounds of moraine matter and thus giving rise to ponds and even to sheets of water several miles in diameter.

The comparative scarcity, therefore, of lakes of Pleistocene date in tropical countries, and very generally south of the fortieth and fiftieth parallels of latitude, may be accounted for by the absence of glacial action in such regions.

Post-glacial Lake-dwelling in the North of Italy

We learn from M. de Mortillet that in the peat which has filled up one of the " morainic lakes " formed by the ancient glacier of the Ticino, M. Moro has discovered at Mercurago the piles of a lake-dwelling like those of Switzerland, together with various utensils and a canoe hollowed out of the trunk of a tree. From this fact we learn that south of the Alps as well as north of them a primitive people having similar habits flourished after the retreat of the great glaciers.

Successive Phases of Glacial Action in the Alps, and their
Relation to the Human Period [Note 34]

According to the geological observations of M. Morlot, the following successive phases in the development of ice-action in the Alps are plainly recognisable:—

First. There was a period when the ice was in its greatest excess, when the glacier of the Rhone not only reached the Jura, but climbed to the height of 2015 feet above the Lake of Neufchâtel, and 3450 feet above the sea, at which time the Alpine ice actually entered the French territory at some points, penetrating by certain gorges, as through the defile of the Fort de l'Ecluse, among others.

Second. To this succeeded a prolonged retreat of the great glaciers, when they evacuated not only the Jura and the low country between that chain and the Alps, but retired some way back into the Alpine valleys. M. Morlot supposes their diminution in volume to have accompanied a general subsidence of the country to the extent of at least 1000 feet. The geological formations of the second period consist of stratified masses of sand and gravel, called the " ancient alluvium " by MM. Necker and Favre, corresponding to the " older or lower diluvium " of some writers. Their origin is evidently due to the action of rivers, swollen by the melting of ice, by which the materials of parts of the old moraines were rearranged and stratified and left usually at considerable heights above the level of the present valley plains.

Third. The glaciers again advanced and became of gigantic dimensions, though they fell far short of those of the first period. That of the Rhone, for example, did not again reach the Jura, though it filled the Lake of Geneva and formed enormous moraines on its borders and in many parts of the valley between the Alps and Jura.

Fourth. A second retreat of the glaciers took place when they gradually shrank nearly into their present limits, accompanied by another accumulation of stratified gravels which form in many places a series of terraces above the level of the alluvial plains of the existing rivers.

In the gorge of the Dranse, near Thonon, M. Morlot discovered no less than three of these glacial formations in direct superposition, namely, at the bottom of the section, a mass of compact till or boulder-clay (No. 1) 12 feet thick, including striated boulders of Alpine limestone, and covered by regularly stratified ancient alluvium (No. 2) 150 feet thick, made up of rounded pebbles in horizontal beds. This mass is in its turn overlaid by a second formation (No. 3) of unstratified boulder clay, with erratic blocks and striated pebbles, which constituted the left lateral moraine of the great glacier of the Rhone when it advanced for the second time to the Lake of Geneva. At a short distance from the above section terraces (No. 4) composed of stratified alluvium are seen at the heights of 20, 50, 100, and 150 feet above the Lake of Geneva, which by their position can be shown to be posterior in date to the upper boulder-clay and therefore belong to the fourth period, or that of the last retreat of the great glaciers. In the deposits of this fourth period the remains of the mammoth have been discovered, as at Morges, for example, on the Lake of Geneva. The conical delta of the Tinière, mentioned in Chap. II. as containing at different depths monuments of the Roman as well as of the antecedent bronze and stone ages, is the work of alluvial deposition going on when the terrace of 50 feet was in progress. This modern delta is supposed by M. Morlot to have required 10,000 years for its accumulation. At the height of 150 feet above the lake, following up the course of the same torrent, we come to a more ancient delta, about ten times as large, which is therefore supposed to be the monument of about ten times as many centuries, or 100,000 years, all referable to the fourth period mentioned in the preceding page, or that which followed the last retreat of the great glaciers.[1]

If the lower flattened cone of Tinière be referred in great part to the age of the oldest lake-dwellings, the higher one might perhaps correspond with the Pleistocene period of St. Acheul, or the era when Man and the *Elephas primigenius* flourished together; but no human remains or works of art have as yet been found in deposits of this age or in any alluvium containing the bones of extinct mammalia in Switzerland.

Upon the whole, it is impossible not to be struck with an apparent correspondence in the succession of events of the glacial period of Switzerland and that of the British Isles before described. The time of the first Alpine glaciers of colossal dimensions, when that chain perhaps was several thousand feet higher than now, may have agreed with the first continental period when Scotland was invested with a universal crust of ice.

---

[1] Morlot, Terrain quaternaire du Bassin de Léman, *Bulletin de la Société Vaudoise des Sciences Naturelles*, No. 44.

The retreat of the first Alpine glaciers, caused partly by a lowering of that chain, may have been synchronous with the period of great submergence and floating ice in England. The second advance of the glaciers may have coincided in date with the re-elevation of the Alps, as well as of the Scotch and Welsh mountains; and lastly, the final retreat of the Swiss and Italian glaciers may have taken place when Man and the extinct mammalia were colonising the north-west of Europe and beginning to inhabit areas which had formed the bed of the glacial sea during the era of chief submergence.

But it must be confessed that in the present state of our knowledge these attempts to compare the chronological relations of the periods of upheaval and subsidence of areas so widely separated as are the mountains of Scandinavia, the British Isles, and the Alps, or the times of the advance and retreat of glaciers in those several regions and the greater or less intensity of cold, must be looked upon as very conjectural.

We may presume with more confidence that when the Alps were highest and the Alpine glaciers most developed, filling all the great lakes of northern Italy and loading the plains of Piedmont and Lombardy with ice, the waters of the Mediterranean were chilled and of a lower average temperature than now. Such a period of refrigeration is required by the conchologist to account for the prevalence of northern shells in the Sicilian seas about the close of the Pliocene or commencement of the Pleistocene period. For such shells as *Cyprina islandica, Panopœa norvegica* (= *P. bivonœ,* Philippi), *Leda pygmœa,* Münst, and some others, enumerated among the fossils of the latest Tertiary formations of Sicily by Philippi and Edward Forbes, point unequivocally to a former more severe climate. Dr. Hooker also in his late journey to Syria (in the autumn of 1860) found the moraines of extinct glaciers, on which the whole of the ancient cedars of Lebanon grow, to descend 4000 feet below the summit of that chain. The temperature of Syria is now so much milder that there is no longer perpetual snow even on the summit of Lebanon, the height of which was ascertained to be 10,200 feet above the Mediterranean.[1]

Such monuments of a cold climate in latitudes so far south as Syria and the north of Sicily, between 33° and 38° N., may be confidently referred to an early part of the glacial period, or to times long anterior to those of Man and the extinct mammalia of Abbeville and Amiens.

[1] Hooker, *Natural History Review,* No. 5, January 1862, p. 11.

CHAPTER XVI

HUMAN REMAINS IN THE LOESS, AND THEIR PROBABLE AGE

Nature, Origin, and Age of the Loess of the Rhine and Danube—Impalpable Mud produced by the grinding Action of Glaciers—Dispersion
of this Mud at the Period of the Retreat of the great Alpine Glaciers
—Continuity of the Loess from Switzerland to the Low Countries—
Characteristic organic Remains not Lacustrine—Alpine Gravel in the
Valley of the Rhine covered by Loess—Geographical Distribution of
the Loess and its Height above the Sea—Fossil Mammalia—Loess
of the Danube—Oscillations in the Level of the Alps and lower Country
required to explain the Formation and Denudation of the Loess—
More rapid Movement of the inland Country—The same Depression
and Upheaval might account for the Advance and Retreat of the
Alpine Glaciers—Himalayan Mud of the Plains of the Ganges compared to European Loess—Human Remains in Loess near Maestricht,
and their probable Antiquity.

Nature and Origin of the Loess

INTIMATELY connected with the subjects treated of in the last
chapter, is the nature, origin, and age of certain loamy deposits,
commonly called loess, which form a marked feature in the superficial deposits of the basins of the Rhine, Danube, and some
other large rivers draining the Alps, and which extend down
the Rhine into the Low Countries, and were once perhaps continuous with others of like composition in the north of France.
[Note 35.]

It has been reported of late years that human remains have
been detected at several points in the loess of the Meuse around
and below Maestricht. I have visited the localities referred to;
but, before giving an account of them, it will be desirable to
explain what is meant by the loess, a step the more necessary as a
French geologist for whose knowledge and judgment I have
great respect, tells me he has come to the conclusion that " the
loess " is " a myth," having no real existence in a geological
sense or as holding a definite place in the chronological series.

No doubt it is true that in every country, and at all geological
periods, rivers have been depositing fine loam on their inundated
plains in the manner explained above in Chap. III., where the
Nile mud was spoken of. This mud of the plains of Egypt,
according to Professor Bischoff's chemical analysis agrees closely

in composition with the loess of the Rhine.[1] I have also shown
when speaking of the fossil man of Natchez, how identical
in mineral character and in the genera of its terrestrial and
amphibious shells is the ancient fluviatile loam of the Mississippi
with the loess of the Rhine. But granting that loam presenting
the same aspect has originated at different times and in distinct
hydrographical basins, it is nevertheless true that during the
glacial period the Alps were a great centre of dispersion, not only
of erratics, as we have seen in the last chapter, and of gravel
which was carried farther than the erratics, but also of very fine
mud which was transported to still greater distances and in
greater volume down the principal river-courses between the
mountains and the sea.

Mud produced by Glaciers

They who have visited Switzerland are aware that every
torrent which issues from an icy cavern at the extremity of a
glacier is densely charged with an impalpable powder, produced
by the grinding action to which the subjacent floor of rock
and the stones and sand frozen into the ice are exposed in the
manner before described. We may therefore readily conceive
that a much greater volume of fine sediment was swept along
by rivers swollen by melting ice at the time of the retreat of the
gigantic glaciers of the olden time. The fact that a large proportion of this mud, instead of being carried to the ocean where
it might have formed a delta on the coast or have been dispersed far and wide by the tides and currents, has accumulated
in inland valleys, will be found to be an additional proof of the
former occurrence of those grand oscillations in the level of the
Alps and parts of the adjoining continent which were required
to explain the alternate advance and retreat of the glaciers, and
the superposition of more than one boulder clay and stratified
alluvium.

The position of the loess between Basle and Bonn is such as
to imply that the great valley of the Rhine had already acquired
its present shape, and in some places, perhaps more than its
actual depth and width, previously to the time when it was
gradually filled up to a great extent with fine loam. The greater
part of this loam has been since removed, so that a fringe only
of the deposit is now left on the flanks of the boundary hills, or

[1] *Chemical and Physical Geology*, vol. i., p. 132.

occasionally some outliers in the middle of the great plain of the Rhine where it expands in width.

These outliers are sometimes on such a scale as to admit of minor hills and valleys, having been shaped out of them by the action of rain and small streamlets, as near Freiburg in the Breisgau and other districts.

Fossil Shells of the Loess

The loess is generally devoid of fossils, although in many places they are abundant, consisting of land-shells, all of living species, and comprising no small part of the entire molluscous fauna now inhabiting the same region. The three shells most frequently met with are those represented in the annexed figures. The slug, called *Succinea*, is not strictly aquatic, but lives in damp places, and may be seen in full activity far from rivers, in meadows where the grass is wet with rain or dew; but

Fig. 44.   Fig. 45.   Fig. 46.

Succinea oblonga.   *Pupa muscorum*.   *Helix hispida*, Lin.; *H. plebeia*, Drap.

shells of the genera *Limnæa, Planorbis, Paludina, Cyclas*, and others, requiring to be constantly in the water, are extremely exceptional in the loess, occurring only at the bottom of the deposit where it begins to alternate with ancient river-gravel on which it usually reposes.

This underlying gravel consists in the valley of the Rhine for the most part of pebbles and boulders of Alpine origin, showing that there was a time when the rivers had power to convey coarse materials for hundreds of miles northwards from Switzerland towards the sea; whereas at a later period an entire change was brought about in the physical geography of the same district, so that the same river deposited nothing but fine mud, which accumulated to a thickness of 800 feet or more above the original alluvial plain.

But although most of the fundamental gravel was derived from the Alps, there has been observed in the neighbourhood of the principal mountain chains bordering the great valley, such as the Black Forest, Vosges, and Odenwald, an admixture of detritus characteristic of those several chains. We cannot doubt therefore that as some of these mountains, especially the Vosges, had during the glacial period their own glaciers, a part of the fine mud of their moraines must have been mingled with loess of Alpine origin; although the principal mass of the latter must have come from Switzerland, and can in fact be traced continuously from Basle to Belgium.

Geographical Distribution of the Loess

It was stated in the last chapter that at the time of the greatest extension of the Swiss glaciers the Lake of Constance and all the other great lakes were filled with ice, so that gravel and mud could pass freely from the upper Alpine valley of the Rhine to the lower region between Basle and the sea, the great lake intercepting no part of the moraines whether fine or coarse. On the other hand the Aar with its great tributaries the Limmat and the Reuss does not join the Rhine till after it issues from the Lake of Constance; and by their channels a large part of the Alpine gravel and mud could always have passed without obstruction into the lower country, even after the ice of the great lake had melted.

It will give the reader some idea of the manner in which the Rhenish loess occurs, if he is told that some of the earlier scientific observers imagined it to have been formed in a vast lake which occupied the valley of the Rhine from Basle to Mayence, sending up arms or branches into what are now the valleys of the Main, Neckar, and other large rivers. They placed the barrier of this imaginary lake in the narrow and picturesque gorge of the Rhine between Bingen and Coblenz: and when it was objected that the lateral valley of the Lahn, communicating with that gorge, had also been filled with loess, they were compelled to transfer the great dam farther down and to place it below Bonn. Strictly speaking it must be placed much farther north, or in the 51st parallel of latitude, where the limits of the loess have been traced out by MM. Omalius D'Halloy, Dumont, and others, running east and west by Cologne, Juliers, Louvain, Oudenarde, and Courtrai in Belgium to Cassel, near Dunkirk in France. This boundary line may not indicate the original seaward extent of the formation, as it may have stretched still farther north and its present abrupt termination may only show how far it was cut back at some former period by the denuding action of the sea.

Even if the imbedded fossil shells of the loess had been

lacustrine, instead of being, as we have seen, terrestrial and amphibious, the vast height and width of the required barrier would have been fatal to the theory of a lake: for the loess is met with in great force at an elevation of no less than 1600 feet above the sea, covering the Kaiserstuhl, a volcanic mountain which stands in the middle of the great valley of the Rhine, near Freiburg in Breisgau. The extent to which the valley has there been the receptacle of fine mud afterwards removed is most remarkable.

The loess of Belgium was called " Hesbayan mud " in the geological map of the late M. Dumont, who, I am told, recognised it as being in great part composed of Alpine mud. M. d'Archiac, when speaking of the loess, observes that it envelopes Hainault, Brabant, and Limburg like a mantle everywhere uniform and homogeneous in character, filling up the lower depressions of the Ardennes and passing thence into the north of France, though not crossing into England. In France, he adds, it is found on high plateaus 600 feet above some of the rivers, such as the Marne; but as we go southwards and eastwards of the basin of the Seine, it diminishes in quantity, and finally thins out in those directions.[1] It may even be a question whether the " *limon des plateaux*," or upland loam of the Somme valley, before alluded to,[2] may not be a part of the same formation. As to the higher and lower level gravels of that valley, which, like that of the Seine, contain no foreign rocks, we have seen that they are each of them covered by deposits of loess or inundation-mud belonging respectively to the periods of the gravels, whereas the upland loam is of much older date, more widely spread, and occupying positions often independent of the present lines of drainage. To restore in imagination the geographical outline of Picardy, to which rivers charged with so much homogeneous loam and running at such heights may once have belonged is now impossible.[3]

In the valley of the Rhine, as I before observed, the body of the loess, instead of having been formed at successively lower and lower levels as in the case of the basin of the Somme, was deposited in a wide and deep pre-existing basin, or strath, bounded by lofty mountain chains such as the Black Forest, Vosges, and Odenwald. In some places the loam accumulated to such a depth as first to fill the valley and then to spread over the adjoining table-lands, as in the case of the Lower

---

[1] D'Archiac, *Histoire des Progrès*, vol. ii., pp. 169, 170.
[2] No. 4, Fig. 7, p. 84.   [3] *See above*, p. 105.

Eifel, where it encircled some of the modern volcanic cones of loose pumice and ashes. In these instances it does not appear to me that the volcanoes were in eruption during the time of the deposition of the loess, as some geologists have supposed. The interstratification of loam and volcanic ejectamenta was probably occasioned by the fluviatile mud having gradually enveloped the cones of loose scoriæ after they were completely formed. I am the more inclined to embrace this view after having seen the junction of granite and loess on the steep slopes of some of the mountains bounding the great plain of the Rhine on its right bank in the Bergstrasse. Thus between Darmstadt and Heidelberg perpendicular sections are seen of loess 200 feet thick, at various heights above the river, some of them at elevations of 800 feet and upwards. In one of these may be seen, resting on the hill side of Melibocus in the Odenwald, the usual yellow loam free from pebbles at its contact with a steep slope of granite, but divided into horizontal layers for a short distance from the line of junction. In these layers, which abut against the granite, a mixture of mica and of unrounded grains of quartz and felspar occur, evidently derived from the disintegration of the crystalline rock, which must have decomposed in the atmosphere before the mud had reached this height. Entire shells of *Helix*, *Pupa*, and *Succinea*, of the usual living species, are embedded in the granitic mixture. We may therefore be sure that the valley bounded by steep hills of granite existed before the tranquil accumulation of this vast body of loess.

During the re-excavation of the basin of the Rhine successive deposits of loess of newer origin were formed at various heights; and it is often difficult to distinguish their relative ages, especially as fossils are often entirely wanting, and the mineral composition of the formation is so uniform.

The loess in Belgium is variable in thickness, usually ranging from 10 to 30 feet. It caps some of the highest hills or tableland around Brussels at the height of 300 feet above the sea. In such places it usually rests on gravel and rarely contains shells, but when they occur they are of Recent species. I found the *Succinea oblonga*, before mentioned, and *Helix hispida* in the Belgian loess at Neerepen, between Tongres and Hasselt, where M. Bosquet had previously obtained remains of an elephant referred to *E. primigenius*. This pachyderm and *Rhinoceros tichorhinus* are cited as characterising the loess in various parts of the valley of the Rhine. Several perfect skeletons of the marmot have been disinterred from the loess of Aix-la-Chapelle. But much remains to be done in determining the species of mammalia of this formation and the relative altitudes above the valley-plain at which they occur.

If we ascend the basin of the Neckar, we find that it is filled with loess of great thickness, far above its junction with the Rhine. At Canstadt near Stuttgart, loess resembling that of the Rhine contains many fossil bones, especially those of *Elephas primigenius*, together with some of *Rhinoceros tichorhinus*, the species having been lately determined by Dr. Falconer. At this place the loess is covered by a thick bed of travertine, used as a building stone, the product of a mineral spring. In the travertine are many fossil plants, all Recent except two, an oak and poplar, the leaves of which Professor Heer has not been able to identify with any known species.

Below the loess of Canstadt, in which bones of the mammoth are so abundant, is a bed of gravel evidently an old river channel now many feet above the level of the Neckar, the valley having there been excavated to some depth below its ancient channel so as to lie in the underlying red sandstone of Keuper. Although the loess, when traced from the valley of the Rhine into that of the Neckar, or into any other of its tributaries, often undergoes some slight alteration in its character, yet there is so much identity of composition as to suggest the idea that the mud of the main river passed far up the tributary valleys, just as that of the Mississippi during floods flows far up the Ohio, carrying its mud with it into the basin of that river. But the uniformity of colour and mineral composition does not extend indefinitely into the higher parts of every basin. In that of the Neckar, for example, near Tübingen, I found the fluviatile loam or brick-earth, enclosing the usual *Helices* and *Succineæ*, together with the bones of the mammoth, very distinct in colour and composition from ordinary Rhenish loess, and such as no one could confound with Alpine mud. It is mottled with red and green, like the New Red Sandstone or Keuper, from which it has clearly been derived.

Such examples, however, merely show that where a basin is so limited in size that the detritus is derived chiefly or exclusively from one formation, the prevailing rock will impart its colour and composition in a very decided manner to the loam; whereas, in the basin of a great river which has many tributaries, the loam will consist of a mixture of almost every variety of rock, and will therefore exhibit an average result nearly the same in all countries. Thus, the loam which fills to a great depth the wide valley of the Saone, which is bounded on the west side by an escarpment of Inferior Oolite, and by the chain of the Jura on the east, is very like the loess found in the continuation of the same great basin after the junction of the Rhone, by which a large supply of Alpine mud has been added and intermixed.

In the higher parts of the basin of the Danube, loess of the same character as that of the Rhine, and which I believe to be chiefly of Alpine origin, attains a far greater elevation above the sea than any deposits of Rhenish loess; but the loam which, according to M. Stur, fills valleys on the north slope of the Carpathians almost up to the watershed between Galicia and Hungary, may be derived from a distinct source.

Oscillations of Level required to explain the Accumulation and Denudation of the Loess

A theory, therefore, which attempts to account for the position of the loess cannot be satisfactory unless it be equally applicable to the basins of the Rhine and Danube. So far as relates to the source of so much homogeneous loam, there are many large tributaries of the Danube which, during the glacial period, may have carried an ample supply of moraine-mud from the Alps to that river; and in regard to grand oscillations in the level of the land, it is obvious that the same movements both downward and upward of the great mountain-chain would be attended with analogous effects, whether the great rivers flowed northwards or eastwards. In each case fine loam would be accumulated during subsidence and removed during the up-heaval of the land. Changes, therefore, of level analogous to those on which we have been led to speculate when endeavour-ing to solve the various problems presented by the glacial phenomena, are equally available to account for the nature and geological distribution of the loess. But we must suppose that the amount of depression and re-elevation in the central region was considerably in excess of that experienced in the lower countries, or those nearer the sea, and that the rate of sub-sidence in the latter was never so considerable as to cause submergence, or the admission of the sea into the interior of the continent by the valleys of the principal rivers.

We have already assumed that the Alps were loftier than

now, when they were the source of those gigantic glaciers which reached the flanks of the Jura. At that time gravel was borne to the greatest distances from the central mountains through the main valleys, which had a somewhat steeper slope than now, and the quantity of river-ice must at that time have aided in the transportation of pebbles and boulders. To this state of things gradually succeeded another of an opposite character, when the fall of the rivers from the mountains to the sea became less and less, while the Alps were slowly sinking, and the first retreat of the great glaciers was taking place. Suppose the depression to have been at the rate of 5 feet in a century in the mountains and only as many inches in the same time nearer the coast, still, in such areas as the eye could survey at once, comprising a small part only of Switzerland or of the basin of the Rhine, the movement might appear to be uniform and the pre-existing valleys and heights might seem to remain relatively to each other as before.

Such inequality in the rate of rising or sinking, when we contemplate large continental spaces, is quite consistent with what we know of the course of nature in our own times as well as at remote geological epochs. Thus in Sweden, as before stated, the rise of land now in progress is nearly uniform as we proceed from north to south for moderate distances; but it greatly diminishes southwards if we compare areas hundreds of miles apart; so that instead of the land rising about 5 feet in a hundred years as at the North Cape, it becomes less than the same number of inches at Stockholm, and farther south the land is stationary, or, if not, seems rather to be descending than ascending.[1]

To cite an example of high geological antiquity, M. Hébert has demonstrated that, during the Oolitic and Cretaceous periods, similar inequalities in the vertical movements of the earth's crust took place in Switzerland and France. By his own observations and those of M. Lory he has proved that the area of the Alps was rising and emerging from beneath the ocean towards the close of the Oolitic epoch, and was above water at the commencement of the Cretaceous era; while, on the other hand, the area of the Jura, about 100 miles to the north, was slowly sinking at the close of the Oolitic period, and had become submerged at the commencement of the Cretaceous. Yet these oscillations of level were accomplished without any perceptible derangement in the strata, which remained all the

[1] *Principles of Geology*, chap. xxx., 9th ed., p. 519 *et seq.*

while horizontal, so that the Lower Cretaceous or Neocomian beds were deposited conformably on the Oolitic.[1]

Taking for granted then that the depression was more rapid in the more elevated region, the great rivers would lose century after century some portion of their velocity or carrying power, and would leave behind them on their alluvial plains more and more of the moraine-mud with which they were charged, till at length, in the course of thousands or some tens of thousands of years, a large part of the main valleys would begin to resemble the plains of Egypt where nothing but mud is deposited during the flood season. The thickness of loam containing shells of land and amphibious mollusca might in this way accumulate to any extent, so that the waters might overflow some of the heights originally bounding the valley and deposits of " platform mud," as it has been termed in France, might be extensively formed. At length, whenever a re-elevation of the Alps at the time of the second extension of the glaciers took place, there would be renewed denudation and removal of such loess; and if, as some geologists believe, there has been more than one oscillation of level in the Alps since the commencement of the glacial period, the changes would be proportionally more complicated and terraces of gravel covered with loess might be formed at different heights and at different periods.

Himalayan Mud of the Ganges compared to European Loess

Some of the revolutions in physical geography above suggested for the continent of Europe during the Pleistocene epoch, may have had their counterparts in India in the Recent period. The vast plains of Bengal are overspread with Himalayan mud, which as we ascend the Ganges extends inland for 1200 miles from the sea, continuing very homogeneous on the whole, though becoming more sandy as it nears the hills. They who sail down the river during a season of inundation see nothing but a sheet of water in every direction, except here and there where the tops of trees emerge above its level. To what depth the mud extends is not known, but it resembles the loess in being generally devoid of stratification, and of shells, though containing occasionally land shells in abundance, as well as calcareous concretions, called kunkur, which may be compared to the nodules of carbonate of lime sometimes observed to form layers in the Rhenish loess. I am told by

[1] *Bulletin de la Société Géologique de France,* 2 series, vol. xvi., 1859, p. 596.

Colonel Strachey and Dr. Hooker that above Calcutta, in the Hooghly, when the flood subsides, the Gangetic mud may be seen in river cliffs 80 feet high, in which they were unable to detect organic remains, a remark which I found to hold equally in regard to the Recent mud of the Mississippi.

Dr. Wallich, while confirming these observations, informs me that at certain points in Bengal, farther inland, he met with land-shells in the banks of the great river. Borings have been made at Calcutta, beginning not many feet above the sea-level, to the depth of 300 and 400 feet; and wherever organic remains were found in the strata pierced through they were of a fluviatile or terrestrial character, implying that during a long and gradual subsidence of the country the sediment thrown down by the Ganges and Brahmaputra had accumulated at a sufficient rate to prevent the sea from invading that region.

At the bottom of the borings, after passing through much fine loam, beds of pebbles, sand, and boulders were reached, such as might belong to an ancient river channel; and the bones of a crocodile and the shell of a freshwater tortoise were met with at the depth of 400 feet from the surface. No pebbles are now brought down within a great distance of this point, so that the country must once have had a totally different character and may have had its valleys, hills, and rivers, before all was reduced to one common level by the accumulation upon it of fine Himalayan mud. If the latter were removed during a gradual re-elevation of the country, many old hydrographical basins might reappear, and portions of the loam might alone remain in terraces on the flanks of hills, or on platforms, attesting the vast extent in ancient times of the muddy envelope. A similar succession of events has, in all likelihood, occurred in Europe during the deposition and denudation of the loess of the Pleistocene period, which, as we have seen in a former chapter, was long enough to allow of the gradual development of almost any amount of such physical changes.

Human Remains in the Loess near Strasburg

M. Ami Boué, well known by his numerous works on geology and a well-practised observer in every branch of the science, disinterred in the year 1823 with his own hands many bones of a human skeleton from ancient undisturbed loess at Lahr, nearly opposite Strasburg, on the right side of the great valley of the Rhine. No skull was detected, but the tibia, fibula, and several other bones were obtained in a good state of preservation and shown at the time to Cuvier, who pronounced them to be human.

Human Remains in Loess near Maestricht

The banks of the Meuse at Maestricht, like those of the Rhine at Bonn and Cologne, are slightly elevated above the level of the alluvial plain. On the right bank of the Meuse, opposite Maestricht, the difference of level is so marked that a bridge with many arches has been constructed to keep up, during the flood season, a communication between the higher parts of the alluvial plain and the hills or bluffs which bound it. This plain is composed of modern loess, undistinguishable in mineral character from that of higher antiquity, before alluded to, and entirely without signs of successive deposition and devoid of terrestrial or fluviatile shells. It is extensively worked for brick-earth to the depth of about 8 feet. The bluffs before alluded to often consist of a terrace of gravel, from 30 to 40 feet in thickness, covered by an older loess, which is continuous as we ascend the valley to Liège. In the suburbs of that city patches of loess are seen at the height of 200 feet above the level of the Meuse. The table-land in that region, composed of Carboniferous and Devonian rocks, is about 450 feet high, and is not overspread with loess.

A terrace of gravel covered with loess has been mentioned as existing on the right bank of the Meuse at Maestricht. Answering to it another is also seen on the left bank below that city, and a promontory of it projecting into the alluvial plain of the Meuse and approaching to within a hundred yards of the river, was cut through during the excavation of a canal running from Maestricht to Hocht, between the years 1815 and 1823. This section occurs at the village of Smeermass, and is about 60 feet deep, the lower 40 feet consisting of stratified gravel and the upper of 20 feet of loess. The number of molars, tusks, and bones (probably parts of entire skeletons) of elephants obtained during these diggings, was extraordinary. Not a few of them are still preserved in the museums of Maestricht and Leyden, together with some horns of deer, bones of the ox-tribe and other mammalia, and a human lower jaw, with teeth. According to Professor Crahay, who published an account of it at the time, this jaw, which is now preserved at Leyden, was found at the depth of 19 feet from the surface, where the loess joins the underlying gravel, in a stratum of sandy loam resting on gravel and overlaid by some pebbly and sandy beds. The stratum is said to have been intact and undisturbed, but the human jaw was isolated, the nearest tusk of an elephant being six yards removed from it in horizontal distance.

Most of the other mammalian bones were found, like these human remains, in or near the gravel, but some of the tusks and teeth of elephants were met with much nearer the surface. I visited the site of these fossils in 1860 in company with M. van Binkhorst, and we found the description of the ground, published by the late Professor Crahay of Louvain, to be very correct.[1] The projecting portion of the terrace, which was cut through in making the canal, is called the hill of Caberg, which is flat-topped, 60 feet high, and has a steep slope on both sides towards the alluvial plain. M. van Binkhorst (who is the author of some valuable works on the palæontology of the Maestricht Chalk) has recently visited Leyden, and ascertained that the human fossil above mentioned is still entire in the museum of the University. Although we had no opportunity of verifying the authenticity of Professor Crahay's statements, we could see no reason for suspecting the human jaw to belong to a different geological period from that of the extinct elephant. If this were granted, it might have no claims to a higher antiquity than the human remains which Dr. Schmerling disentombed from the Belgian caverns; but the fact of their occurring in a Pleistocene alluvial deposit in the open plains, would be one of the first examples of such a phenomenon. The top of the hill of Caberg is not so high above the Meuse as is the terrace of St. Acheul with its flint implements above the Somme, but at St. Acheul no human bones have yet been detected.

In the museum at Maestricht are preserved a human frontal and a pelvic bone, stained of a dark peaty colour; the frontal very remarkable for its lowness and the prominence of the superciliary ridges, which resemble those of the Borreby skull, Fig. 5. These remains may be the same as those alluded to by Professor Crahay in his memoir, where he says that in a black deposit in the suburbs of Hocht were found leaves, nuts, and freshwater shells in a very perfect state, and a human skull of a dark colour. They were of an age long posterior to that of the loess containing the bones of elephants and in which the human jaw now at Leyden is said to have been embedded.

---

[1] M. van Binkhorst has shown me the original MS. read to the Maestricht Athenæum in 1823. The memoir was published in 1836 in the *Bulletin de l'Académie Royale de Belgique*, vol. iii., p. 43.

CHAPTER XVII

POST-GLACIAL DISLOCATIONS AND FOLDINGS OF CRETACEOUS AND
DRIFT STRATA IN THE ISLAND OF MÖEN, IN DENMARK

Geological Structure of the Island of Möen—Great Disturbances of the
Chalk posterior in Date to the Glacial Drift, with Recent Shells—M.
Puggaard's Sections of the Cliffs of Möen—Flexures and Faults
common to the Chalk and Glacial Drift—Different Direction of the
Lines of successive Movement, Fracture, and Flexure—Undisturbed
Condition of the Rocks in the adjoining Danish Islands—Unequal
Movements of Upheaval in Finmark—Earthquake of New Zealand
in 1855—Predominance in all Ages of uniform Continental Movements
over those by which the Rocks are locally convulsed.

In the preceding chapters I have endeavoured to show that
the study of the successive phases of the glacial period in
Europe, and the enduring marks which they have left on many
of the solid rocks and on the character of the superficial drift
are of great assistance in enabling us to appreciate the vast
lapse of ages which are comprised in the Pleistocene epoch.
They enlarge at the same time our conception of the antiquity,
not only of the living species of animals and plants but of their
present geographical distribution, and throw light on the chronological relations of these species to the earliest date yet ascertained for the existence of the human race.   That date, it will
be seen, is very remote if compared to the times of history and
tradition, yet very modern if contrasted with the length of
time during which all the living testacea, and even many of the
mammalia, have inhabited the globe.

In order to render my account of the phenomena of the
glacial epoch more complete, I shall describe in this chapter
some other changes in physical geography and in the internal
structure of the earth's crust, which have happened in the
Pleistocene period, because they differ in kind from any previously alluded to, and are of a class which were thought by the
earlier geologists to belong exclusively to epochs anterior to
the origin of the existing fauna and flora.   Of this nature are
those faults and violent local dislocations of the rocks, and those
sharp bendings and foldings of the strata, which we so often
behold in mountain chains, and sometimes in low countries also,
especially where the rock-formations are of ancient date.

Post-glacial Dislocations and Foldings of Cretaceous and drift Strata in the Island of Möen, Denmark

A striking illustration of such convulsions of Pleistocene date may be seen in the Danish island of Möen, which is situated about 50 miles south of Copenhagen. The island is about 60 miles in circumference, and consists of white Chalk, several hundred feet thick, overlaid by boulder clay and sand, or glacial drift which is made up of several subdivisions, some unstratified and others stratified, the whole having a mean thickness of 60 feet, but sometimes attaining nearly twice that thickness. In one of the oldest members of the formation fossil marine shells of existing species have been found.

Throughout the greater part of Möen the strata of the drift are undisturbed and horizontal, as are those of the subjacent Chalk; but on the north-eastern coast they have been throughout a certain area bent, folded, and shifted, together with the beds of the underlying Cretaceous formation. Within this area they have been even more deranged than is the English Chalk-with-flints along the central axis of the Isle of Wight in Hampshire, or of Purbeck in Dorsetshire. The whole displacement of the Chalk is evidently posterior in date to the origin of the drift, since the beds of the latter are horizontal where the fundamental Chalk is horizontal, and inclined, curved, or vertical where the Chalk displays signs of similar derangement. Although I had come to these conclusions respecting the structure of Möen in 1835, after devoting several days in company with Dr. Forchhammer to its examination,[1] I should have hesitated to cite the spot as exemplifying convulsions on so grand a scale, of such extremely modern date, had not the island been since thoroughly investigated by a most able and reliable authority, the Danish geologist, Professor Puggaard, who has published a series of detailed sections of the cliffs.

These cliffs extend through the north-eastern coast of the island, called Möens Klint,[2] where the Chalk precipices are bold and picturesque, being 300 and 400 feet high, with tall beech-trees growing on their summits, and covered here and there at their base with huge taluses of fallen drift, verdant with wild shrubs and grass, by which the monotony of a continuous range of white Chalk cliffs is prevented.

[1] Leyll, *Geological Transactions*, 2nd series, vol. ii., p. 243.
[2] Puggaard, *Geologie d. Insel Möen*, Bern, 1851; and *Bulletin de la Société Géologique de France*, 1851.

In the low part of the island, at A, Fig. 47, or the southern extremity of the line of section above alluded to, the drift is horizontal, but when we reach B, a change, both in the height of the cliffs and in the inclination of the strata, begins to be perceptible, and the Chalk No. 1 soon makes its appearance from beneath the overlying members of the drift Nos. 2, 3, 4, and 5.

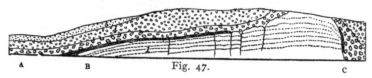

A   B   Fig. 47.   C

Southern extremity of Möens Klint (Puggaard).

A Horizontal drift.
B Chalk and overlying drift beginning to rise.
C First flexure and fault. Height of cliff at this point, 180 feet.

This Chalk, with its layers of flints, is so like that of England as to require no description. The incumbent drift consists of the following subdivisions, beginning with the lowest:

No. 2. Stratified loam and sand, 5 feet thick, containing at one spot near the base of the cliff, at S, Fig. 48, *Cardium edule, Tellina solidula,* and *Turritella,* with fragments of other shells. Between No. 2 and the Chalk No. 1, there usually intervenes a breccia of broken flints.

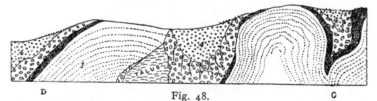

D   Fig. 48.   G

Section of Möens Klint (Puggaard), continued from Fig. 47.

S Fossil shells of recent species in the drift at this point.
G Greatest height near G, 280 feet.

No. 3. Unstratified blue clay or till, with small pebbles and fragments of Scandinavian rocks occasionally scattered through it, 20 feet thick.

No. 4. A second unstratified mass of yellow and more sandy clay 40 feet thick, with pebbles and angular polished and striated blocks of granite and other Scandinavian rocks, transported from a distance.

No. 5. Stratified sands and gravel, with occasionally large erratic blocks; the whole mass varying from 40 to 100 feet in thickness, but this only in a few spots.

The angularity of many of the blocks in Nos. 3 and 4, the glaciated surfaces of others, and the transportation from a distance attested by their crystalline nature, prove them to belong to the northern drift or glacial period.

It will be seen that the four subdivisions 2, 3, 4, and 5 begin to rise at B, Fig. 47, and that at C, where the cliff is 180 feet high, there is a sharp flexure shared equally by the Chalk and the incumbent drift. Between D and G, Fig. 48, we observe a great fracture in the rocks with synclinal and anticlinal folds,

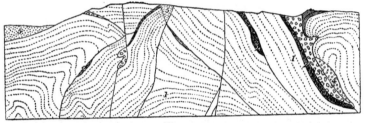

Fig. 49.

Post-glacial disturbances of vertical, folded, and shifted strata of Chalk and drift, in the Dronningestol, Möen, height 400 feet (Puggaard).

1 Chalk, with flints.
2 Marine stratified loam, lowest member of glacial formation.
3 Blue clay or till, with erratic blocks unstratified.
4 Yellow sandy till, with pebbles and glaciated boulders.
5 Stratified sand and gravel with erratics.

exhibited in cliffs nearly 300 feet high, the drift beds participating in all the bendings of the Chalk; that is to say, the three lower members of the drift, including No. 2, which, at the point s in this diagram, contains the shells of Recent species before alluded to.

Near the northern end of the Möens Klint, at a place called " Taler," more than 300 feet high, are seen similar folds, so sharp that there is an appearance of four distinct alternations of the glacial and Cretaceous formations in vertical or highly inclined beds; the Chalk at one point bending over so that the position of all the beds is reversed.

But the most wonderful shiftings and faultings of the beds are observable in the Dronningestol part of the same cliff, 400 feet in perpendicular height, where, as shown in Fig. 49, the drift is thoroughly entangled and mixed up with the dislocated Chalk.

If we follow the lines of fault, we may see, says M. Puggaard, along the planes of contact of the shifted beds, the marks of polishing and rubbing which the Chalk flints have undergone, as have many stones in the gravel of the drift, and some of these have also been forced into the soft Chalk. The manner in which the top of some of the arches of bent Chalk have been cut off in this and several adjoining sections, attests the great denudation which accompanied the disturbances, portions of the bent strata having been removed, probably while they were emerging from beneath the sea.

M. Puggaard has deduced the following conclusions from his study of these cliffs.

First. The white Chalk, when it was still in horizontal stratification, but after it had suffered considerable denudation, subsided gradually, so that the lower beds of drift No. 2, with their littoral shells, were superimposed on the Chalk in a shallow sea.

Second. The overlying unstratified boulder clays 3 and 4 were thrown down in deeper water by the aid of floating ice coming from the north.

Third. Irregular subsidences then began, and occasionally partial failures of support, causing the bending and sometimes the engulfment of overlying masses both of the Chalk and drift, and causing the various dislocations above described and depicted. The downward movement continued till it exceeded 400 feet, for upon the surface even of No. 5, in some parts of the island, lie huge erratics 20 feet or more in diameter, which imply that they were carried by ice in a sea of sufficient depth to float large icebergs. But these big erratics, says Puggaard, never enter into the fissures as they would have done had they been of date anterior to the convulsions.

Fourth. After this subsidence, the re-elevation and partial denudation of the Cretaceous and glacial beds took place during a general upward movement, like that now experienced in parts of Sweden and Norway.

In regard to the lines of movement in Möen, M. Puggaard believes, after an elaborate comparison of the cliffs with the interior of the island, that they took at least three distinct directions at as many successive eras, all of post-glacial date; the first line running from E.S.E. to W.N.W., with lines of fracture at right angles to them; the second running from

S.S.E. to N.N.W., also with fractures in a transverse direction; and lastly, a sinking in a N. and S. direction, with other subsidences of contemporaneous date running at right angles or E. and W.

When we approach the north-west end of Möens Klint, or the range of coast above described, the strata begin to be less bent and broken, and after travelling for a short distance beyond we find the Chalk and overlying drift in the same horizontal position as at the southern end of the Möens Klint. What makes these convulsions the more striking is the fact that in the other adjoining Danish islands, as well as in a large part of Möen itself, both the Secondary and Tertiary formations are quite undisturbed.

It is impossible to behold such effects of reiterated local movements, all of post-Tertiary date, without reflecting that, but for the accidental presence of the stratified drift, all of which might easily have been missing, where there has been so much denudation, even if it had once existed, we might have referred the verticality and flexures and faults of the rocks to an ancient period, such as the era between the Chalk with flints and the Maestricht Chalk, or to the time of the latter formation, or to the Eocene, or Miocene, or Pliocene eras, even the last of them long prior to the commencement of the glacial epoch. Hence we may be permitted to suspect that in some other regions, where we have no such means at our command for testing the exact date of certain movements, the time of their occurrence may be far more modern than we usually suppose. In this way some apparent anomalies in the position of erratic blocks, seen occasionally at great heights above the parent rocks from which they have been detached, might be explained, as well as the irregular direction of certain glacial furrows like those described by Professor Keilhau and Mr. Hörbye on the mountains of the Dovrefjeld in latitude 62° N., where the striation and friction is said to be independent of the present shape and slope of the mountains.[1] Although even in such cases it remains to be proved whether a general crust of continental ice, like that of Greenland described by Rink (*see above*, Chap. XIII.), would not account for the deviation of the furrows and striæ from the normal directions which they ought to have followed had they been due to separate glaciers filling the existing valleys.

It appears that in general the upward movements in Scan-

---

[1] *Observations sur les Phénomènes d'Érosion en Norwège*, 1857.

dinavia, which have raised sea-beaches containing marine shells of Recent species to the height of several hundred feet, have been tolerably uniform over very wide spaces; yet a remarkable exception to this rule was observed by M. Bravais at Altenfjord in Finmark, between latitude 70° and 71° N. An ancient water-level, indicated by a sandy deposit forming a terrace and by marks of the erosion of the waves, can be followed for 30 miles from south to north along the borders of a fjord rising gradually from a height of 85 feet to an elevation of 220 feet above the sea, or at the rate of about 4 feet in a mile.[1]

To pass to another and very remote part of the world, we have witnessed so late as January 1855 in the northern island of New Zealand a sudden and permanent rise of land on the northern shores of Cook's Straits, which at one point, called Muko-muka, was so unequal as to amount to 9 feet vertically, while it declined gradually from this maximum of upheaval in a distance of about 23 miles north-west of the greatest rise, to a point where no change of level was perceptible. Mr. Edward Roberts of the Royal Engineers, employed by the British Government at the time of the shock in executing public works on the coast, ascertained that the extreme upheaval of certain ancient rocks followed a line of fault running at least 90 miles from south to north into the interior; and what is of great geological interest, immediately to the east of this fault the country, consisting of Tertiary strata, remained unmoved or stationary; a fact well established by the position of a line of Nullipores marking the sea-level before the earthquake, both on the surface of the Tertiary and Palæozoic rocks.[2]

The repetition of such unequal movements, especially if they recurred at intervals along the same lines of fracture, would in the course of ages cause the strata to dip at a high angle in one direction, while towards the opposite point of the compass they would terminate abruptly in a steep escarpment.

But it is probable that the multiplication of such movements in the post-Tertiary period has rarely been so great as to produce results like those above described in Möen, for the principal movements in any given period seem to be of a more uniform kind, by which the topography of limited districts and the position of the strata are not visibly altered except in their

---

[1] *Proceedings of the Geological Society*, 1845, vol. iv., p. 94.
[2] *Bulletin de la Société Géologique de France*, vol. xiii., 1856, p. 660, where I have described the facts communicated to me by Messrs. Roberts and Walter Mantell.

height relatively to the sea. Were it otherwise we should not find conformable strata of all ages, including the primary fossiliferous of shallow-water origin, which must have remained horizontal throughout vast areas during downward movements of several thousand feet going on at the period of their accumulation. Still less should we find the same primary strata, such as the Carboniferous, Devonian, or Silurian, still remaining horizontal over thousands of square leagues, as in parts of North America and Russia, having escaped dislocation and flexure throughout the entire series of epochs which separate Palæozoic from Recent times. Not that they have been motionless, for they have undergone so much denudation, and of such a kind, as can only be explained by supposing the strata to have been subjected to great oscillations of level, and exposed in some cases repeatedly to the destroying and planing action of the waves of the sea.

It seems probable that the successive convulsions in Möen were contemporary with those upward and downward movements of the glacial period which were described in the thirteenth and some of the following chapters, and that they ended before the upper beds of No. 5, Fig. 49, with its large erratic blocks, were deposited, as some of those beds occurring in the disturbed parts of Möen appear to have escaped the convulsions to which Nos. 2, 3, and 4 were subjected. If this be so, the whole derangement, although Pleistocene, may have been anterior to the human epoch, or rather to the earliest date to which the existence of man has as yet been traced back.

CHAPTER XVIII

THE GLACIAL PERIOD IN NORTH AMERICA

Post-glacial Strata containing Remains of *Mastodon giganteus* in North America—Scarcity of Marine Shells in Glacial Drift of Canada and the United States—Greater southern Extension of Ice-action in North America than in Europe—Trains of Erratic Blocks of vast Size in Berkshire, Massachusetts—Description of their Linear Arrangement and Points of Departure—Their Transportation referred to Floating and Coast Ice—General Remarks on the Causes of former Changes of Climate at successive geological Epochs—Supposed Effects of the Diversion of the Gulf Stream in a Northerly instead of North-Easterly Direction—Development of extreme Cold on the opposite Sides of the Atlantic in the Glacial Period not strictly simultaneous—Effect of Marine Currents on Climate—Pleistocene Submergence of the Sahara.

On the North American continent, between the arctic circle and the 42nd parallel of latitude, we meet with signs of ice-action on a scale as grand as, if not grander than, in Europe; and there also the excess of cold appears to have been first felt at the close of the Tertiary, and to have continued throughout a large portion of the Pleistocene period. [Note 36.]

The general absence of organic remains in the North American glacial formation makes it as difficult as in Europe to determine what mammalia lived on the continent at the time of the most intense refrigeration, or when extensive areas were becoming strewed over with glacial drift and erratic blocks, but it is certain that a large proboscidean now extinct, the *Mastodon giganteus*, Cuv., together with many other quadrupeds, some of them now living and others extinct, played a conspicuous part in the post-glacial era. By its frequency as a fossil species, this pachyderm represents the European *Elephas primigenius*, although the latter also occurs fossil in the United States and Canada, and abounds, as I learn from Sir John Richardson, in latitudes farther north than those to which the mastodon has been traced.

In the state of New York, the mastodon is not unfrequently met with in bogs and lacustrine deposits formed in hollows in the drift, and therefore, in a geological position, much resembling that of Recent peat and shell-marl in the British Isles, Denmark, or the valley of the Somme, as before described.

Sometimes entire skeletons have been discovered within a few feet of the surface, in peaty earth at the bottom of small ponds, which the agriculturists had drained. The shells in these cases belong to freshwater genera, such as *Limnæa, Physa, Planorbis, Cyclas*, and others, differing from European species, but the same as those now proper to ponds and lakes in the same parts of America.

I have elsewhere given an account of several of these localities which I visited in 1842,[1] and can state that they certainly have a more modern aspect than almost all the European deposits in which remains of the mammoth occur, although a few instances are cited of *Elephas primigenius* having been dug out of peat in Great Britain. Thus I was shown a mammoth's tooth in the museum at Torquay in Devonshire which is believed to have been dredged up from a deposit of vegetable matter now partially submerged beneath the sea. A more elevated part of the same peaty formation constitutes the bottom of the valley in which Tor Abbey stands. This individual elephant must certainly have been of more modern date than his fellows found fossil in the gravel of the Brixham cave, before described, for it flourished when the physical geography of Devonshire, unlike that of the cave period, was almost identical with that now established.

I cannot help suspecting that many tusks and teeth of the mammoth, said to have been found in peat, may be as spurious as are the horns of the rhinoceros cited more than once in the *Memoirs of the Wernerian Society*, as having been obtained from shell-marl in Forfarshire and other Scotch counties; yet, between the period when the mammoth was most abundant and that when it died out, there must have elapsed a long interval of ages when it was growing more and more scarce; and we may expect to find occasional stragglers buried in deposits long subsequent in date to others, until at last we may succeed in tracing a passage from the Pleistocene to the Recent fauna, by geological monuments, which will fill up the gap before alluded to as separating the era of the flint tools of Amiens and Abbeville from that of the peat of the valley of the Somme.

How far the lacustrine strata of North America above mentioned may help to lessen this hiatus, and whether some individuals of the *Mastodon giganteus* may have come down to the confines of the historical period, is a question not so easily answered as might at first sight be supposed. A geologist

---

[1] *Travels in North America*, vol. i., p. 55, London, 1845; and *Manual of Geology*, ch. xii., 5th ed., p. 144.

might naturally imagine that the fluviatile formation of Goat Island, seen at the falls of Niagara, and at several points below the falls,[1] was very modern, seeing that the fossil shells contained in it are all of species now inhabiting the waters of the Niagara, and seeing also that the deposit is more modern than the glacial drift of the same locality. In fact, the old river bed, in which bones of the mastodon occur, holds the same position relatively to the boulder formation as the strata of shell-marl and bog-earth with bones of mastodon, so frequent in the State of New York, bear to the glacial drift, and all may be of contemporaneous date. But in the case of the valley of the Niagara we happen to have a measure of time which is wanting in the other localities, namely, the test afforded by the recession of the falls, an operation still in progress, by which the deep ravine of the Niagara, 7 miles long, between Queenstown and Goat Island has been hollowed out. This ravine is not only post-glacial, but also posterior in date to the fluviatile or mastodon-bearing beds. The individual therefore found fossil near Goat Island flourished before the gradual excavation of the deep and long chasm, and we must reckon its antiquity, not by thousands, but by tens of thousands of years, if I have correctly estimated the minimum of time which was required for the erosion of that great ravine.[2]

The stories widely circulated of bones of the mastodon having been observed with their surfaces pierced as if by arrow-heads or bearing the marks of wounds inflicted by some stone implement, must in future be more carefully inquired into, for we can scarcely doubt that the mastodon in North America lived down to a period when the mammoth co-existed with Man in Europe. But I need say no more on this subject, having already explained my views in regard to the evidence of the antiquity of Man in North America when treating of the human bone discovered at Natchez on the Mississippi.

In Canada and the United States we experience the same difficulty as in Europe when we attempt to distinguish between glacial formations of submarine and those of supra-marine origin. In the New World, as in Scotland and England, marine shells of this era have rarely been traced higher than 500 feet above the sea, and 700 feet seems to be the maximum to which at present they are known to ascend. In the same countries,

---

[1] *Travels in North America,* by the author, vol. i., ch. ii., and vol. ii., ch. xix.
[2] *Principles of Geology,* 9th ed., p. 2; and *Travels in North America,* vol. i., p. 32, 1845.

erratic blocks have travelled from N. to S., following the same direction as the glacial furrows and striæ imprinted almost everywhere on the solid rocks underlying the drift. Their direction rarely deviates more than fifteen degrees E. or W. of the meridian, so that we can scarcely doubt, in spite of the general dearth of marine shells, that icebergs floating in the sea and often running aground on its rocky bottom were the instruments by which most of the blocks were conveyed to southern latitudes.

There are, nevertheless, in the United States, as in Europe, several groups of mountains which have acted as independent centres for the dispersion of erratics, as, for example, the White Mountains, latitude 44° N., the highest of which, Mount Washington, rises to about 6300 feet above the sea; and according to Professor Hitchcock some of the loftiest of the hills of Massachusetts once sent down their glaciers into the surrounding lower country.

Great southern Extension of Trains of Erratic Blocks in Berkshire, Massachusetts, U.S., latitude 42° N.

Having treated so fully in this volume of the events of the glacial period, I am unwilling to conclude without laying before the reader the evidence displayed in North America of ice-action in latitudes farther south by about ten degrees than any seen on an equal scale in Europe. This extension southwards of glacial phenomena in regions where there are no snow-covered mountains like the Alps to explain the exception, nor any hills of more than moderate elevation, constitutes a feature of the western as compared to the eastern side of the Atlantic, and must be taken into account when we speculate on the causes of the refrigeration of the northern hemisphere during the Pleistocene period.

In 1852, accompanied by Mr. James Hall, state geologist of New York, author of many able and well-known works on geology and palæontology, I examined the glacial drift and erratics of the county of Berkshire, Massachusetts, and those of the adjoining parts of the state of New York, a district about 130 miles inland from the Atlantic coast and situated due west of Boston in latitude 42° 25′ north. This latitude corresponds in Europe to that of the north of Portugal. Here numerous detached fragments of rock are seen, having a linear arrangement or being continuous in long parallel trains, running nearly in straight lines over hill and dale for distances of 5, 10, and 20 miles, and sometimes greater distances. Seven of the more conspicuous of these trains, from 1 to 7 inclusive, Fig. 50, are laid down in the accompanying map or ground plan.[1] It will be remarked that they run in a N.W. and S.E. direction, or almost transversely to the ranges of hills A, B, and C, which run N.N.E. and S.S.W. The crests of these chains are about 800 feet in height above the intervening valleys. The blocks of the northernmost train, No. 7, are of limestone derived from the calcareous chain B; those of the two trains next to the south, Nos. 6 and 5, are composed exclusively in the first part of their course of a green chloritic rock of great toughness, but after they have passed the ridge B, a mixture of calcareous blocks is observed. After traversing the valley for a distance of 6 miles these two trains pass through depressions or gaps in the range C, as they had previously done in crossing the range B, showing that the dispersion of the erratics bears some relation to the acutal inequalities of the surface, although the course of the same blocks is perfectly independent of the more leading features of the geography of the country, or those by which the present lines of drainage are determined. The greater number of the green chloritic fragments in trains 5 and 6 have evidently come from the ridge A, and a large proportion of the whole from its highest summit d, where the crest of the ridge has been worn into those dome-shaped masses called " roches moutonnées," already alluded to, and where several fragments having this shape, some of them 30 feet long, are seen *in situ*, others only slightly removed from their original position, as if they had been just ready to set out on their travels. Although smooth and rounded on their tops they are angular on their lower parts, where their outline has been derived from the natural joints of the rock. Had these blocks been conveyed from d by glaciers, they would have radiated in all directions from a centre, whereas not one even of the smaller ones is found to the westward of A, though a very slight force would have made them roll down to the base of that ridge, which is very steep on its western declivity. It is clear, therefore, that the propelling power, whatever it may have been, acted exclusively in a south-easterly direction. Professor Hall, and I observed one of the green blocks—24 feet long, poised

---

[1] This ground plan, and a farther account of the Berkshire erratics, was given in an abstract of a lecture delivered by me to the Royal Institution of Great Britain, April 27, 1855, and published in their *Proceedings*.

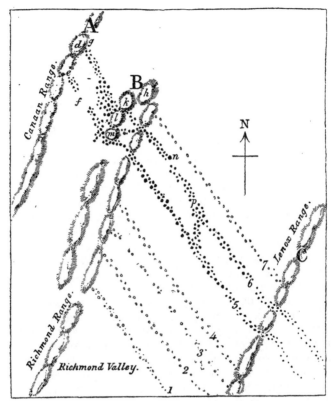

Fig. 50.

MAP SHOWING THE RELATIVE POSITION AND DIRECTION OF SEVEN TRAINS OF
ERRATIC BLOCKS IN BERKSHIRE, MASSACHUSETTS, AND IN PART OF THE
STATE OF NEW YORK.

Distance in a straight line, between the mountain ranges A *and* C, *about
eight miles.*

A Canaan range, in the State of New York. The crest consists of green
    chloritic rock.

B Richmond range, the western division of which consists in Merriman's
    Mount of the same green rock as A, but in a more schistose form,
    while the eastern division is composed of slaty limestone.

C The Lenox range, consisting in part of mica-schist, and in some
    districts of crystalline limestone.

d Knob in the range A, from which most of the train No. 6 is supposed
    to have been derived.

e Supposed starting point of the train No. 5 in the range A.

f Hiatus of 175 yards, or space without blocks.

g Sherman's House.

h Perry's Peak.

k Flat Rock.

l Merriman's Mount.

upon another about 19 feet in length.   The largest of all on the west flank of *m*, or Dupey's Mount, called the Alderman, is above 90 feet in diameter, and nearly 300 feet in circumference. We counted at some points between forty and fifty blocks visible at once, the smallest of them larger than a camel.

The annexed drawing represents one of the best known of train No. 6, being that marked *n* on the map.   According to our measurement it is 52 feet long by 40 in width, its height above the drift in which it is partially buried being 15 feet. At the distance of several yards occurs a smaller block, 3 or 4 feet in height, 20 feet long, and 14 broad, composed of the same compact chloritic rock, and evidently a detached fragment from the bigger mass, to the lower and angular part of which it would fit on exactly.   This erratic *n* has a regularly rounded top, worn and smoothed like the " roches moutonnées " before mentioned, but no part of the attrition can have occurred since it left its parent rock, the angles of the lower portion being quite sharp and unblunted.

From railway cuttings through the drift of the neighbourhood and other artificial excavations, we may infer that the position of the block *n*, if seen in a vertical section, would be as represented in Fig. 52.   The deposit *c* in that section consists of sand, mud, gravel, and stones, for the most part un-

---

m Dupey's Mount.

n Largest block of train, No. 6.   *See* Figs. 51 and 52.

p Point of divergence of part of the train No. 6, where a branch is sent off to No. 5.

No. 1 The most southerly train examined by Messrs. Hall and Lyell, between Stockbridge and Richmond, composed of blocks of black slate, blue limestone and some of the green Canaan rock, with here and there a boulder of white quartz.

No. 2 Train composed chiefly of large limestone masses, some of them divided into two or more fragments by natural joints.

No. 3 Train composed of blocks of limestone and the green Canaan rock; passes south of the Richmond Station on the Albany and Boston railway; is less defined than Nos. 1 and 2.

No. 4 Train chiefly of limestone blocks, some of them thirty feet in diameter, running to the north-west of the Richmond Station, and passing south of the Methodist Meeting-house, where it is intersected by a railway cutting.

No. 5 South train of Dr. Reid, composed entirely of large blocks of the green chloritic Canaan rock; passes north of the Old Richmond Meeting-house, and is three-quarters of a mile north of the preceding train (No. 4).

No. 6 The great or principal train (north train of Dr. Reid), composed of very large blocks of the Canaan rock, diverges at *p*, and unites by a branch with train No. 5.

No. 7 A well-defined train of limestone blocks, with a few of the Canaan rock, traced from the Richmond to the slope of the Lenox range.

stratified, resembling the till or boulder clay of Europe. It varies in thickness from 10 to 50 feet, being of greater depth in the valleys. The uppermost portion is occasionally, though rarely, stratified. Some few of the imbedded stones have flattened, polished, striated, and furrowed sides. They consist

Fig. 51.

Erratic dome-shaped block of compact chloritic rock (*n* map, Fig. 50), near the Richmond Meeting-house, Berkshire, Massachusetts, lat. 42° 25 N'. Length, 52 feet; width, 40 feet; height above the soil, 15 feet.

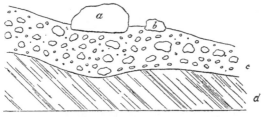

Fig. 52.

Section showing position of the block, Fig. 51.

a The large block.   Fig. 51 and *n*, map, p. 282.
b Fragment detached from the same.
c Unstratified drift with boulders.
d Silurian limestone in inclined stratification.

invariably, like the seven trains above mentioned, of kinds of rock confined to the region lying to the N.W., none of them having come from any other quarter. Whenever the surface of the underlying rock has been exposed by the removal of the superficial detritus, a polished and furrowed surface is seen, like that underneath a glacier, the direction of the furrows being from N.W. to S.E., or corresponding to the course of the large erratics.

As all the blocks, instead of being dispersed from a centre, have been carried in one direction and across the ridges A, B, C and the intervening valleys, the hypothesis of glaciers is out of the question. I conceive, therefore, that the erratics were conveyed to the places they now occupy by coast ice, when the country was submerged beneath the waters of a sea cooled by icebergs coming annually from arctic regions.

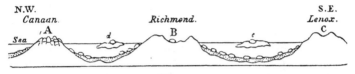

Fig. 53

d, e Masses of floating ice carrying fragments of rock.

Suppose the highest peaks of the ridges A, B, C in the annexed diagram to be alone above water, forming islands, and *d e* to be masses of floating ice, which drifted across the Canaan and Richmond valleys at a time when they were marine channels, separating islands or rather chains of islands, having a N.N.E. and S.S.W. direction. A fragment of ice such as *d*, freighted with a block from A, might run aground and add to the heap of erratics at the N.W. base of the island (now ridge) B, or, passing through a sound between B and the next island of the same group, might float on till it reached the channel between B and C. Year after year two such exposed cliffs in the Canaan range as *d* and *e* of the map, Fig. 50, undermined by the waves, might serve as the points of departure of blocks, composing the trains Nos. 5 and 6. It may be objected that oceanic currents could not always have had the same direction; this may be true, but during a short season of the year when the ice was breaking up the prevailing current may have always run S.E.

If it be asked why the blocks of each train are not more scattered, especially when far from their source, it may be observed that after passing through sounds separating islands, they issued again from a new and narrow starting-point; moreover, we must not exaggerate the regularity of the trains, as their width is sometimes twice as great in one place in as another; and No. 6 sends off a branch at *p*, which joins No. 5. There are also stragglers, or large blocks here and there in the spaces between the two trains. As to the distance to which any given block would be carried, that must have depended on a variety of circumstances; such as the strength of the current, the direction of the wind, the weight of the block or the quantity and draught of the ice attached to it. The smaller fragments would, on the whole, have the best chance of going farthest; because, in the first place, they were more numerous, and then, being lighter, they required less ice to float them, and would not ground so readily on shoals, or if stranded, would be more easily started again on their travels. Many of the blocks, which at first sight seem to consist of single masses, are found when examined to be made up of two, three, or more pieces divided by natural joints. In case of a second removal by ice, one or more portions would become detached and be drifted to different points further on. Whenever this happened, the original size would be lessened, and the angularity of the block previously worn by the breakers would be restored, and this tendency to split may explain why some of the far-transported fragments remain very angular.

These various considerations may also account for the fact that the average size of the blocks of all the seven trains laid down on the plan, Fig. 50, lessens sensibly in proportion as we recede from the principal points of departure of particular kinds of erratics, yet not with any regularity, a huge block now and then recurring when the rest of the train consists of smaller ones.

All geologists acquainted with the district now under consideration are agreed that the mountain ranges A, B, and C, as well as the adjoining valleys, had assumed their actual form and position before the drift and erratics accumulated on and in them and before the surface of the fixed rocks was polished and furrowed. I have the less hesitation in ascribing the transporting power to coast-ice, because I saw in 1852 an angular block of sandstone, 8 feet in diameter, which had been brought down several miles by ice only three years before to the mouth of the Petitcodiac estuary, in Nova Scotia, where it joins the Bay of Fundy; and I ascertained that on the shores of the same bay, at the South Joggins, in the year 1850, much larger blocks had been removed by coast-ice, and after they had floated half a mile, had been dropped in salt water by the side of a pier built for loading vessels with coal, so that it was necessary at low tide to blast these huge ice-borne rocks with gunpowder in order that the vessels might be able to draw up alongside the pier. These recent exemplifications of the vast carrying powers of ice occurred in latitude 46° N. (corresponding to that of Bordeaux), in a bay never invaded by icebergs.

I may here remark that a sheet of ice of moderate thickness, if it extend over a wide area, may suffice to buoy up the largest erratics which fall upon it. The size of these will depend, not on the intensity of the cold but on the manner in which the rock is jointed, and the consequent dimensions of the blocks into which it splits when falling from an undermined cliff.

When I first endeavoured in the *Principles of Geology* in 1830,[1] to explain the causes, both of the warmer and colder climates which have at former periods prevailed on the globe, I referred to successive variations in the height and position of the land and its extent relatively to the sea in polar and equatorial latitudes—also to fluctuations in the course of oceanic currents and other geographical conditions, by the united influence of which I still believe the principal revolutions in the meteorological state of the atmosphere at different geological periods have been brought about. The Gulf Stream was particularly alluded to by me as moderating the winter climate of northern Europe and as depending for its direction on temporary and accidental peculiarities in the shape of the land, especially that of the narrow Straits of Bahama, which a slight modification in the earth's crust would entirely alter.

Mr. Hopkins, in a valuable essay on the causes of former changes of climate,[2] has attempted to calculate how much the annual temperature of Europe would be lowered if this Gulf Stream were turned in some other and new direction, and estimates the amount at about six or seven degrees of Fahrenheit. He also supposes that if at the same time a considerable part of northern and central Europe were submerged, so that a cold current from the arctic seas should sweep over it, an additional refrigeration of three or four degrees would be produced. He has speculated in the same essay on the effects which would be experienced in the eastern hemisphere if the same mighty current of warm water, instead of crossing the Atlantic, were made to run northwards from the Gulf of Mexico through the region now occupied by the valley of the Mississippi, and so onwards to the arctic regions.

After reflecting on what has been said in the thirteenth

---

[1] 1st ed., ch. vii.; 9th ed. *ibid.*
[2] Hopkins, *Quart. Jour. Geol. Soc.*, vol. viii., 1852, p. 56.

chapter of the submergence and re-elevation of the British Isles and the adjoining parts of Europe, and the rising and sinking of the Alps and the basins of some of the great rivers flowing from that chain, since the commencement of the glacial period, a geologist will not be disposed to object to the theory above adverted to, on the score of its demanding too much conversion of land into sea, or almost any amount of geographical change in Pleistocene times. But a difficulty of another kind presents itself. We have seen that, during the glacial period, the cold in Europe extended much farther south than it does at present, and in this chapter we have demonstrated that in North America the cold also extended no less than 10° of latitude still farther southwards than in Europe; so that if a great body of heated water, instead of flowing north-eastward, were made to pass through what is now the centre of the American continent towards the Arctic Circle, it could not fail to mitigate the severity of the winter's cold in precisely those latitudes where the cold was greatest and where it has left monuments of ice-action surpassing in extent any exhibited on the European side of the ocean.

In the actual state of the globe, the isothermal lines, or lines of equal winter temperature when traced westward from Europe to North America bend 10° south, there being a marked excess of winter cold in corresponding latitudes west of the Atlantic. During the glacial period, viewing it as a whole, we behold signs of a precisely similar deflection of these same isothermal lines when followed from east to west; so that if, in the hope of accounting for the former severity of glacial action in Europe, we suppose the absence of the Gulf Stream and imagine a current of equivalent magnitude to have flowed due north from the Gulf of Mexico, we introduce, as we have just hinted, a source of heat into precisely that part of the continent where the extreme conditions of refrigeration are most manifest. Viewed in this light, the hypothesis in question would render the glacial phenomena described in the present chapter more perplexing and anomalous than ever. But here another question arises, whether the eras at which the maximum of cold was attained on the opposite sides of the Atlantic were really contemporaneous? We have now discovered not only that the glacial period was of vast duration, but that it passed through various phases and oscillations of temperature; so that, although the chief polishing and furrowing of the rocks and transportation of erratics in Europe and North America may have taken place contemporaneously, according to the ordinary language of geology, or when the same testacea and the same Pleistocene assemblage of mammalia flourished, yet the extreme development of cold on the opposite sides of the ocean may not have been strictly simultaneous, but on the contrary the one may have preceded or followed the other by a thousand or more than a thousand centuries.

It is probable that the greatest refrigeration of Norway, Sweden, Scotland, Wales, the Vosges, and the Alps coincided very nearly in time; but when the Scandinavian and Scotch mountains were encrusted with a general covering of ice, similar to that now enveloping Greenland, this last country may not have been in nearly so glacial a condition as now, just as we find that the old icy crust and great glaciers, which have left their mark on the mountains of Norway and Sweden, have now disappeared, precisely at a time when the accumulation of ice in Greenland is so excessive. In other words, we see that in the present state of the northern hemisphere, at the distance of about 1500 miles, two meridional zones enjoying very different conditions of temperature may co-exist, and we are, therefore, at liberty to imagine some former alternations of colder and milder climates on the opposite sides of the ocean throughout the Pleistocene era of a compensating kind, the cold on the one side balancing the milder temperature on the other. By assuming such a succession of events we can more easily explain why there has not been a greater extermination of species, both terrestrial and aquatic, in polar and temperate regions during the glacial epoch, and why so many species are common to pre-glacial and post-glacial times.

The numerous plants which are common to the temperate zones N. and S. of the equator have been referred by Mr. Darwin and Dr. Hooker to migrations which took place along mountain chains running from N. to S. during some of the colder phases of the glacial epoch.[1] Such an hypothesis enables us to dispense with the doctrine that the same species ever originated independently in two distinct and distant areas; and it becomes more feasible if we admit the doctrine of the co-existence of meridional belts of warmer and colder climate, instead of the simultaneous prevalence of extreme cold both in the eastern and western hemisphere. It also seems necessary, as colder currents of water always flow to lower latitudes, while warmer

---

[1] Darwin, *Origin of Species*, ch. xi., p. 365; Hooker, *Flora of Australia*, Introduction, p. xviii., 1859.

ones are running towards polar regions, that some such compensation should take place, and that an increase of cold in one region must to a certain extent be balanced by a mitigation of temperature elsewhere.

Sir John F. Herschel, in his recent work on *Physical Geography*, when speaking of the open sea which is caused in part of the polar regions by the escape of ice through Behring's Straits, and the flow of warmer water northwards through the same channel, observes that these straits, by which the continents of Asia and North America are now parted, "are only thirty miles broad where narrowest and only twenty-five fathoms in their greatest depth." But " this narrow channel," he adds, " is yet important in the economy of nature, inasmuch as it allows a portion of the circulating water from a warmer region to find its way into the polar basin, aiding thereby not only to mitigate the extreme rigour of the polar cold, but to prevent in all probability a continual accretion of ice, which else might rise to a mountainous height." [1]

Behring's Straits, here alluded to, happen to agree singularly in width and depth with the Straits of Dover, the difference in depth not being more than 3 or 4 feet; so that at the rate of upheaval, which is now going on in many parts of Scandinavia, of $2\frac{1}{2}$ feet in a century, such straits might be closed in 3000 years, and a vast accumulation of ice to the northward commence forthwith.

But, on the other hand, although such an accumulation might spread its refrigerating influence for many miles southwards beyond the new barrier, the warm current which now penetrates through the straits, and which at other times is chilled by floating ice issuing from them, would when totally excluded from all communication with the icy sea have its temperature raised and its course altered, so that the climate of some other area must immediately begin to improve.

There is still another probable cause of a vast change in the temperature of central Europe in comparatively modern times, to which no allusion has yet been made; namely, the conversion of the great desert of the Sahara from sea into land since the commencement of the Pleistocene period. When that vast region was still submerged, no sirocco blowing for days in succession carried its hot blasts from a wide expanse of burning sand across the Mediterranean. The south winds were comparatively cool, allowing the snows of the Alps to augment to

[1] Herschel's *Physical Geography*, p. 41, 1861.

an extent which the colossal dimensions of the moraines of extinct glaciers can alone enable us to estimate.

The scope and limits of this volume forbid my pursuing these speculations and reasonings farther; but I trust I have said enough to show that the monuments of the glacial period, when more thoroughly investigated, will do much towards expanding our views as to the antiquity of the fauna and flora now contemporary with Man, and will therefore enable us the better to determine the time at which Man began in the northern hemisphere to form part of the existing fauna.   [Note 37.]

CHAPTER XIX

Recapitulation of Results arrived at in the earlier Chapters—Ages of Stone and Bronze—Danish Peat and Kitchen-Middens—Swiss Lake-Dwellings—Local Changes in Vegetation and in the wild and domesticated Animals and in Physical Geography coeval with the Age of Bronze and the later Stone Period—Estimates of the positive Date of some Deposits of the later Stone Period—Ancient Division of the Age of Stone of St. Acheul and Aurignac—Migrations of Man in that Period from the Continent to England in Post-Glacial Times—Slow Rate of Progress in barbarous Ages—Doctrine of the superior Intelligence and Endowments of the original Stock of Mankind considered—Opinions of the Greeks and Romans, and their Coincidence with those of the modern Progressionist.

THE ages of stone and bronze, so called by archæologists, were spoken of in the earlier chapters of this work. That of bronze has been traced back to times anterior to the Roman occupation of Helvetia, Gaul, and other countries north of the Alps. When weapons of that mixed metal were in use, a somewhat uniform civilisation seems to have prevailed over a wide extent of central and northern Europe, and the long duration of such a state of things in Denmark and Switzerland is shown by the gradual ·improvement which took place in the useful and ornamental arts. Such progress is attested by the increasing variety of the forms, and the more perfect finish and tasteful decoration of the tools and utensils obtained from the more modern deposits of the bronze age, those from the upper layers of peat, for example, as compared to those found in the lower ones. The great number also of the Swiss lake-dwellings of the bronze age (about seventy villages having been already discovered), and the large population which some of them were capable of containing, afford indication of a considerable lapse of time, as does the thickness of the stratum of mud in which in some of the lakes the works of art are entombed. The unequal antiquity, also, of the settlements is occasionally attested by the different degrees of decay which the wooden stakes or piles have undergone, some of them projecting more above the mud than others, while all the piles of the antecedent age of stone have rotted away quite down to the level of the mud, such part of them only as was originally driven into the bed of the lake having escaped decomposition.[1]

Among the monuments of the stone period, which immediately preceded that of bronze, the polished hatchets called celts are abundant, and were in very general use in Europe before metallic tools were introduced. We learn, from the Danish peat and shell-mounds, and from the older Swiss lake-settlements, that the first inhabitants were hunters who fed almost entirely on game, but their food in after ages consisted more and more of tamed animals and still later a more complete change to a pastoral state took place, accompanied as population increased by the cultivation of some cereals.

Both the shells and quadrupeds belonging to the later stone period and to the age of bronze consist exclusively of species now living in Europe, the fauna being the same as that which flourished in Gaul at the time when it was conquered by Julius Cæsar, even the *Bos primigenius*, the only animal of which the wild type is lost, being still represented, according to Cuvier, Bell, and Rütimeyer, by one of the domesticated races of cattle now in Europe.

These monuments, therefore, whether of stone or bronze, belong to what I have termed geologically the Recent period, the definition of which some may think rather too dependent on negative evidence, or on the non-discovery hitherto of extinct mammalia, such as the mammoth, which may one day turn up in a fossil state in some of the oldest peaty deposits, as indeed it is already said to have done at some spots, though I have failed as yet to obtain authentic evidence of the fact.[2] No doubt some such exceptional cases may be met with in the course of future investigations, for we are still imperfectly acquainted with the entire fauna of the age of stone in Denmark as we may infer from an opinion expressed by Steenstrup, that some of the instruments exhumed by antiquaries from the Danish peat are made of the bones and horns of the elk and reindeer. Yet no skeleton or uncut bone of either of those species has hitherto been observed in the same peat.

Nevertheless, the examination made by naturalists of the various Danish and Swiss deposits of the Recent period has been so searching, that the finding in them of a stray elephant

---

[1] Troyon, *Habitations lacustres*. Lausanne, 1860.

[2] A molar of *E. primigenius*, in a very fresh state, in the museum at Torquay, believed to have been washed up by the waves of the sea out of the submerged mass of vegetable matter at the extremity of the valley in which Tor Abbey stands, is the best case I have seen. *See above*, p. 278.

or rhinoceros, should it ever occur, would prove little more than that some few individuals lingered on, when the species was on the verge of extinction, and such rare exceptions would not render the classification above proposed inappropriate.

At the time when many wild quadrupeds and birds were growing scarce and some of them becoming locally extirpated in Denmark, great changes were taking place in the vegetation. The pine, or Scotch fir, buried in the oldest peat, gave place at length to the oak, and the oak, after flourishing for ages, yielded in its turn to the beech, the periods when these three forest trees predominated in succession tallying pretty nearly with the ages of stone, bronze, and iron in Denmark. In the same country also, during the stone period, various fluctuations, as we have seen, occurred in physical geography. Thus, on the ocean side of certain islands, the old refuse-heaps, or "kitchen-middens," were destroyed by the waves, the cliffs having wasted away, while on the side of the Baltic, where the sea was making no encroachment or where the land was sometimes gaining on the sea, such mounds remained uninjured. It was also shown that the oyster, which supplied food to the primitive people, attained its full size in parts of the Baltic where it cannot now exist owing to a want of saltness in the water, and that certain marine univalves and bivalves, such as the common periwinkle, mussel, and cockle, of which the castaway shells are found in the mounds, attained in the olden time their full dimensions, like the oysters, whereas the same species, though they still live on the coast of the inland sea adjoining the mounds, are dwarfed and never half their natural size, the water being rendered too fresh for them by the influx of so many rivers.

Some archæologists and geologists of merit have endeavoured to arrive at positive dates, or an exact estimate of the minimum of time assignable to the later age of stone. These computations have been sometimes founded on changes in the level of the land, or on the increase of peat, as in the Danish bogs, or on the conversion of water into land by alluvial deposits, since certain lake-settlements in Switzerland were abandoned. Alterations also in the geographical distribution or preponderance of certain living species of animals and plants have been taken into account in corroboration, as have the signs of progress in human civilisation, as serving to mark the lapse of time during the stone and bronze epochs.

M. Morlot has estimated with care the probable antiquity of three superimposed vegetable soils cut open at different depths in the delta of the Tinière, each containing human bones or works or art, belonging successively to the Roman, bronze, and later stone periods. According to his estimate, an antiquity of 7000 years at least must be assigned to the oldest of these remains, though believed to be long posterior in date to the time when the mammoth and other extinct mammalia flourished together with Man in Europe. Such computations of past time must be regarded as tentative in the present state of our knowledge and much collateral evidence will be required to confirm them; yet the results appear to me already to afford a rough approximation to the truth.

Between the newer or Recent division of the stone period and the older division, which has been called the Pleistocene, there was evidently a vast interval of time—a gap in the history of the past, into which many monuments of intermediate date will one day have to be intercalated. Of this kind are those caves in the south of France, in which M. Lartet has lately found bones of the reindeer, associated with works of art somewhat more advanced in style than those of St. Acheul or of Aurignac. In the valley of the Somme we have seen that peat exists of great thickness, containing in its upper layers Roman and Celtic memorials, the whole of which has been of slow growth, in basins or depressions conforming to the present contour and drainage levels of the country, and long posterior in date to older gravels, containing bones of the mammoth and a large number of flint implements of a very rude and antique type. Some of those gravels were accumulated in the channels of rivers which flowed at higher levels by 100 feet than the present streams, and before the valley had attained its present depth and form. No intermixture has been observed in those ancient river beds of any of the polished weapons, called "celts," or other relics of the more modern times, or of the second or Recent stone period, nor any interstratified peat; and the climate of those Pleistocene ages, when Man was a denizen of the north-west of France and of southern and central England, appears to have been much more severe in winter than it is now in the same region, though far less cold than in the glacial period which immediately preceded.

We may presume that the time demanded for the gradual dying out or extirpation of a large number of wild beasts which figure in the Pleistocene strata and are missing in the Recent fauna was of protracted duration, for we know how tedious a task it is in our own times, even with the aid of fire-arms, to exterminate a noxious quadruped, a wolf, for example, in any region comprising within it an extensive forest or a mountain chain. In many villages in the north of Bengal, the tiger still occasionally carries off its human victims, and the abandonment of late years by the natives of a part of the Sunderbunds or lower delta of the Ganges, which they once peopled, is attributed chiefly to the ravages of the tiger. It is probable that causes more general and powerful than the agency of Man, alterations in climate, variations in the range of many species of animals, vertebrate and invertebrate, and of plants, geographical changes in the height, depth, and extent of land and sea, some or all of these combined, have given rise in a vast series of years to the annihilation, not only of many large mammalia, but to the disappearance of the *Cyrena fluminalis*, once common in the rivers of Europe, and to the different range or relative abundance of other shells which we find in the European drifts.

That the growing power of Man may have lent its aid as the destroying cause of many Pleistocene species, must, however, be granted; yet, before the introduction of fire-arms, or even the use of improved weapons of stone, it seems more wonderful that the aborigines were able to hold their own against the cave-lion, hyæna, and wild bull, and to cope with such enemies, than that they failed to bring about their speedy extinction.

It is already clear that Man was contemporary in Europe with two species of elephant, now extinct, *E. primigenius* and *E. antiquus*, two also of rhinoceros, *R. tichorhinus* and *R. hemitœhcus* (Falc.), at least one species of hippopotamus, the cave-bear, cave-lion, and cave-hyæna, various bovine, equine, and cervine animals now extinct, and many smaller Carnivora, Rodentia, and Insectivora. While these were slowly passing away, the musk ox, reindeer, and other arctic species which have survived to our times were retreating northwards from the valleys of the Thames and Seine to their present more arctic haunts.

The human skeletons of the Belgian caverns of times coeval with the mammoth and other extinct mammalia do not betray any signs of a marked departure in their structure, whether of skull or limb, from the modern standard of certain living races of the human family. As to the remarkable Neanderthal skeleton (Chap. V.), it is at present too isolated and exceptional, and its age too uncertain, to warrant us in relying on its abnormal and ape-like characters, as bearing on the question whether the farther back we trace Man into the past, the more we shall find him approach in bodily conformation to those species of the anthropoid quadrumana which are most akin to him in structure.

In the descriptions already given of the geographical changes which the British Isles have undergone since the commencement of the glacial period (as illustrated by several maps, Figs. 39-41), it has been shown that there must have been a free communication by land between the Continent and these islands, and between the several islands themselves, within the Pleistocene epoch, in order to account for the Germanic fauna and flora having migrated into every part of the area, as well as for the Scandinavian plants and animals to have retreated into the higher mountains. During some part of the Pleistocene ages, the large pachyderms and accompanying beasts of prey, now extinct, wandered from the Continent to England; and it is highly probable that France was united with some part of the British Isles as late as the period of the gravels of St. Acheul and the era of those engulfed rivers which, in the basin of the Meuse near Liège, swept into many a rent and cavern the bones of Man and of the mammoth and cave-bear. There have been vast geographical revolutions in the times alluded to, and oscillations of land, during which the English Channel, which can be shown by the Pagham erratics and the old Brighton beach (p. 222), to be of very ancient origin, may have been more than once laid dry and again submerged. During some one of these phases, Man may have crossed over, whether by land or in canoes, or even on the ice of a frozen sea (as Mr. Prestwich has hinted), for the winters of the period of the higher-level gravels of the valley of the Somme were intensely cold.

The primitive people, who co-existed with the elephant and rhinoceros in the valley of the Ouse at Bedford, and who made use of flint tools of the Amiens type, certainly inhabited part of England which had already emerged from the waters of the glacial sea and the fabricators of the flint tools of Hoxne, in Suffolk, were also, as we have seen, post-glacial. We may likewise presume that the people of Pleistocene date, who have left their memorials in the valley of the Thames, were of corresponding antiquity, posterior to the boulder clay but anterior to the time when the rivers of that region had settled into their present channels.

The vast distance of time which separated the origin of the higher and lower gravels of the valley of the Somme, both of them rich in flint implements of similar shape (although those of oval form predominate in the newer gravels), leads to the conclusion that the state of the arts in those early times remained stationary for almost indefinite periods. There may, however, have been different degrees of civilisation and in the art of fabricating flint tools, of which we cannot easily detect the signs in the first age of stone, and some contemporary tribes may have been considerably in advance of others. Those hunters, for example, who feasted on the rhinoceros and buried their dead with funeral rites at Aurignac may have been less barbarous than the savages of St. Acheul, as some of their weapons and utensils have been thought to imply. To a European who looks down from a great eminence on the products of the humble arts of the aborigines of all times and countries, the stone knives and arrows of the Red Indian of North America, the hatchets of the native Australian, the tools found in the ancient Swiss lake-dwellings or those of the Danish kitchen-middens and of St. Acheul, seem nearly all alike in rudeness and very uniform in general character. The slowness of the progress of the arts of savage life is manifested by the fact that the earlier instruments of bronze were modelled on the exact plan of the stone tools of the preceding age, although such shapes would never have been chosen had metals been known from the first. The reluctance or incapacity of savage tribes to adopt new inventions has been shown in the East by their continuing to this day to use the same stone implements as their ancestors, after that mighty empires, where the use of metals in the arts was well known, had flourished for three thousand years in their neighbourhood.

We see in our own times that the rate of progress in the arts and sciences proceeds in a geometrical ratio as knowledge increases, and so when we carry back our retrospect into the past, we must be prepared to find the signs of retardation augmenting in a like geometrical ratio; so that the progress of a thousand years at a remote period may correspond to that of a century in modern times, and in ages still more remote Man would more and more resemble the brutes in that attribute which causes one generation exactly to imitate in all its ways the generation which preceded it.

The extent to which even a considerably advanced state of civilisation may become fixed and stereotyped for ages, is the wonder of Europeans who travel in the East. One of my friends declared to me, that whenever the natives expressed to him a wish " that he might live a thousand years," the idea struck him as by no means extravagant, seeing that if he were doomed to sojourn for ever among them, he could only hope to exchange in ten centuries as many ideas, and to witness as much progress as he could do at home in half a century.

It has sometimes happened that one nation has been conquered by another less civilised though more warlike, or that during social and political revolutions, people have retrograded in knowledge. In such cases, the traditions of earlier ages, or of some higher and more educated caste which has been destroyed, may gave rise to the notion of degeneracy from a primæval state of superior intelligence, or of science supernaturally communicated. But had the original stock of mankind been really endowed with such superior intellectual powers and with inspired knowledge and had possessed the same improvable nature as their posterity, the point of advancement which they would have reached ere this would have been immeasurably higher. We cannot ascertain at present the limits, whether of the beginning or the end, of the first stone period when Man co-existed with the extinct mammalia, but that it was of great duration we cannot doubt. During those ages there would have been time for progress of which we can scarcely form a conception, and very different would have been the character of the works of art which we should now be endeavouring to interpret,—those relics which we are now disinterring from the old gravel-pits of St. Acheul, or from the Liège caves. In them, or in the upraised bed of the Mediterranean, on the south coast of Sardinia, instead of the rudest pottery or flint tools so irregular in form as to cause the unpractised eye to doubt whether they afford unmistakable evidence of design, we should now be finding sculptured forms surpassing in beauty the masterpieces of Phidias or Praxiteles; lines of buried railways or electric telegraphs from which the best engineers of our day might gain invaluable hints; astronomical instruments and microscopes of more advanced construction than any known in Europe, and other indications of perfection in the arts and sciences such as the nineteenth century has not yet witnessed. Still farther would the triumphs of inventive genius be found to have been carried, when the later deposits, now assigned to the ages of bronze and iron, were formed. Vainly should we be straining our imaginations to guess the possible uses and meaning of such relics—machines, perhaps, for navigating the air or exploring the depths of the ocean, or for calculating arithmetical problems beyond the wants or even the conception of living mathematicians.

The opinion entertained generally by the classical writers of Greece and Rome, that Man in the first stage of his existence was but just removed from the brutes, is faithfully expressed by Horace in his celebrated lines, which begin—

Quum prorepserunt primis animalia terris.—*Sat.*, lib. i., 3, 99.

The picture of transmutation given in these verses, however severe and contemptuous the strictures lavishly bestowed on it by Christian commentators, accords singularly with the train of thought which the modern doctrine of progressive development has encouraged.

" When animals," he says, " first crept forth from the newly formed earth, a dumb and filthy herd, they fought for acorns and lurking-places with their nails and fists, then with clubs, and at last with arms, which, taught by experience, they had forged. They then invented names for things and words to express their thoughts, after which they began to desist from war, to fortify cities and enact laws." They who in later times have embraced a similar theory, have been led to it by no deference to the opinions of their pagan predecessors, but rather in spite of very strong prepossessions in favour of an opposite hypothesis, namely, that of the superiority of their original progenitors, of whom they believe themselves to be the corrupt and degenerate descendants.

So far as they are guided by palæontology, they arrive at this result by an independent course of reasoning; but they have been conducted partly to the same goal as the ancients by ethnological considerations common to both, or by reflecting in what darkness the infancy of every nation is enveloped and that true history and chronology are the creation, as it were, of yesterday.

CHAPTER XX

THEORIES OF PROGRESSION AND TRANSMUTATION

Antiquity and Persistency in Character of the existing Races of Mankind—Theory of their Unity of Origin considered—Bearing of the Diversity of Races on the Doctrine of Transmutation—Difficulty of defining the Terms " Species " and " Race "—Lamarck's Introduction of the Element of Time into the Definition of a Species—His Theory of Variation and Progression—Objections to his Theory, how far answered—Arguments of modern writers in favour of Progression in the Animal and Vegetable World—The old Landmarks supposed to indicate the first Appearance of Man, and of different Classes of Animals, found to be erroneous—Yet the Theory of an advancing Series of organic Beings not inconsistent with Facts—Earliest known Fossil Mammalia of low Grade—No Vertebrata as yet discovered in the oldest fossiliferous Rocks—Objections to the Theory of Progression considered—Causes of the Popularity of the Doctrine of Progression as compared with that of Transmutation.

WHEN speaking in a former work of the distinct races of mankind,[1] I remarked that, " if all the leading varieties of the human family sprang originally from a single pair " (a doctrine, to which then, as now, I could see no valid objection), " a much greater lapse of time was required for the slow and gradual formation of such races as the Caucasian, Mongolian, and Negro, than was embraced in any of the popular systems of chronology."

In confirmation of the high antiquity of two of these, I referred to pictures on the walls of ancient temples in Egypt, in which, a thousand years or more before the Christian era, " the Negro and Caucasian physiognomies were portrayed as faithfully, and in as strong contrast, as if the likenesses of these races had been taken yesterday." In relation to the same subject, I dwelt on the slight modification which the Negro has undergone, after having been transported from the tropics and settled for more than two centuries in the temperate climate of Virginia. I therefore concluded that, " if the various races were all descended from a single pair, we must allow for a vast series of antecedent ages, in the course of which the long-continued influence of external circumstances gave rise to peculiarities increased in many successive generations and at length fixed by hereditary transmission."

[1] *Principles of Geology*, 7th ed., p. 637, 1847; *see also* 9th ed., p. 660.

So long as physiologists continued to believe that Man had not existed on the earth above six thousand years, they might with good reason withhold their assent from the doctrine of a unity of origin of so many distinct races; but the difficulty becomes less and less, exactly in proportion as we enlarge our ideas of the lapse of time during which different communities may have spread slowly, and become isolated, each exposed for ages to a peculiar set of conditions, whether of temperature, or food, or danger, or ways of living. The law of the geometrical rate of the increase of population which causes it always to press hard on the means of subsistence, would ensure the migration in various directions of offshoots from the society first formed abandoning the area where they had multiplied. But when they had gradually penetrated to remote regions by land or water—drifted sometimes by storms and currents in canoes to an unknown shore—barriers of mountains, deserts, or seas, which oppose no obstacle to mutual intercourse between civilised nations, would ensure the complete isolation for tens or thousands of centuries of tribes in a primitive state of barbarism.

Some modern ethnologists, in accordance with the philosophers of antiquity, have assumed that men at first fed on the fruits of the earth, before even a stone implement or the simplest form of canoe had been invented. They may, it is said, have begun their career in some fertile island in the tropics, where the warmth of the air was such that no clothing was needed and where there were no wild beasts to endanger their safety. But as soon as their numbers increased they would be forced to migrate into regions less secure and blest with a less genial climate. Contests would soon arise for the possession of the most fertile lands, where game or pasture abounded and their energies and inventive powers would be called forth, so that at length they would make progress in the arts.

But as ethnologists have failed, as yet, to trace back the history of any one race to the area where it originated, some zoologists of eminence have declared their belief that the different races, whether they be three, five, twenty, or a much greater number (for on this point there is an endless diversity of opinion),[1] have all been primordial creations, having from the first been stamped with the characteristic features, mental and bodily, by which they are now distinguished, except where intermarriage has given rise to mixed or hybrid races. Were

[1] See *Transactions of Ethnological Society*, vol. i., 1861.

we to admit, say they, a unity of origin of such strongly marked varieties as the Negro and European, differing as they do in colour and bodily constitution, each fitted for distinct climates and exhibiting some marked peculiarities in their osteological, and even, in some details of cranial and cerebral conformation, as well as in their average intellectual endowments—if, in spite of the fact that all these attributes have been faithfully handed down unaltered for hundreds of generations, we are to believe that, in the course of time, they have all diverged from one common stock, how shall we resist the arguments of the transmutationist, who contends that all closely allied species of animals and plants have in like manner sprung from a common parentage, albeit that for the last three or four thousand years they may have been persistent in character? Where are we to stop, unless we make our stand at once on the independent creation of those distinct human races, the history of which is better known to us than that of any of the inferior animals?

So long as Geology had not lifted up a part of the veil which formerly concealed from the naturalist the history of the changes which the animate creation had undergone in times immediately antecedent to the Recent period, it was easy to treat these questions as too transcendental, or as lying too far beyond the domain of positive science to require serious discussion. But it is no longer possible to restrain curiosity from attempting to pry into the relations which connect the present state of the animal and vegetable worlds, as well as of the various races of mankind, with the state of the fauna and flora which immediately preceded.

In the very outset of the inquiry, we áre met with the difficulty of defining what we mean by the terms " species " and " race; " and the surprise of the unlearned is usually great, when they discover how wide is the difference of opinion now prevailing as to the significance of words in such familiar use. But, in truth, we can come to no agreement as to such definitions, unless we have previously made up our minds on some of the most momentous of all the enigmas with which the human intellect ever attempted to grapple.

It is now thirty years since I gave an analysis in the first edition of my *Principles of Geology* (vol. ii., 1832) of the views which had been put forth by Lamarck, in the beginning of the century, on this subject. In that interval the progress made in zoology and botany, both in augmenting the number of known animals and plants, and in studying their physiology and geographical distribution and above all in examining and describing fossil species, is so vast that the additions made to our knowledge probably exceed all that was previously known; and what Lamarck then foretold has come to pass; the more new forms have been multiplied, the less are we able to decide what we mean by a variety, and what by a species. In fact, zoologists and botanists are not only more at a loss than ever how to define a species, but even to determine whether it has any real existence in nature, or is a mere abstraction of the human intellect, some contending that it is constant within certain narrow and impassable limits of variability, others that it is capable of indefinite and endless modification.

Before I attempt to explain a great step, which has recently been made by Mr. Darwin and his fellow-labourers in this field of inquiry, I think it useful to recapitulate in this place some of the leading features of Lamarck's system, without attempting to adjust the claims of some of his contemporaries (Geoffroy St. Hilaire in particular) to share in the credit of some of his original speculations.

From the time of Linnæus to the commencement of the present century, it seemed a sufficient definition of the term species to say that " a species consisted of individuals all resembling each other, and reproducing their like by generation." But Lamarck after having first studied botany with success, had then turned his attention to conchology, and soon became aware that in the newer (or Tertiary) strata of the earth's crust there were a multitude of fossil species of shells, some of them identical with living ones, others simply varieties of the living, and which as such were entitled to be designated, according to the ordinary rules of classification, by the same names. He also observed that other shells were so nearly allied to living forms that it was difficult not to suspect that they had been connected by a common bond of descent. He therefore proposed that the element of time should enter into the definition of a species, and that it should run thus: " A species consists of individuals all resembling each other, and reproducing their like by genera-tion, *so long as the surrounding conditions do not undergo changes sufficient to cause their habits, characters, and forms to vary.*" He came at last to the conclusion that none of the animals and plants now existing were primordial creations, but were all derived from pre-existing forms, which, after they may have gone on for indefinite ages reproducing their like, had at length, by the influence of alterations in climate and in the animate world been made to vary gradually, and adapt themselves to new circumstances, some of them deviating in the course of ages so far from their original type as to have claims to be regarded as new species.

In support of these views, he referred to wild and cultivated plants and to wild and domesticated animals, pointing out how their colour, form, structure, physiological attributes and even instincts were gradually modified by exposure to new soils and climates, new enemies, modes of subsistence, and kinds of food.

Nor did he omit to notice that the newly acquired peculiarities may be inherited by the offspring for an indefinite series of generations, whether they be brought about naturally—as when a species, on the extreme verge of its geographical range, comes into competition with new antagonists and is subjected to new physical conditions; or artificially—as when by the act of the breeder or horticulturist peculiar varieties of form or disposition are selected.

But Lamarck taught not only that species had been constantly undergoing changes from one geological period to another, but that there also had been a progressive advance of the organic world from the earliest to the latest times, from beings of the simplest to those of more and more complex structure, and from the lowest instincts up to the highest, and finally from brute intelligence to the reasoning powers of Man. The improvement in the grade of being had been slow and continuous, and the human race itself was at length evolved out of the most highly organised and endowed of the inferior mammalia.

In order to explain how, after an indefinite lapse of ages, so many of the lowest grades of animal or plant still abounded, he imagined that the germs or rudiments of living things, which he called monads, were continually coming into the world and that there were different kinds of these monads for each primary division of the animal and vegetable kingdoms. This last hypothesis does not seem essentially different from the old doctrine of equivocal or spontaneous generation; it is wholly unsupported by any modern experiments or observation, and therefore affords us no aid whatever in speculating on the commencement of vital phenomena on the earth.

Some of the laws which govern the appearance of new varieties were clearly pointed out by Lamarck. He remarked, for example, that as the muscles of the arm become strengthened by exercise or enfeebled by disuse, some organs may in this way, in the course of time, become entirely obsolete, and others previously weak become strong and play a new or more leading part in the organisation of a species. And so with instincts, where animals experience new dangers they become more cautious and cunning, and transmit these acquired faculties to their posterity. But not satisfied with such legitimate speculations, the French philosopher conceived that by repeated acts of volition animals might acquire new organs and attributes, and that in plants, which could not exert a will of their own, certain subtle fluids or organising forces might operate so as to work out analogous effects.

After commenting on these purely imaginary causes, I pointed out in 1832, as the two great flaws in Lamarck's attempt to explain the origin of species, first, that he had failed to adduce a single instance of the initiation of a new organ in any species of animal or plant; and secondly, that variation, whether taking place in the course of nature or assisted artificially by the breeder and horticulturist, had never yet gone so far as to produce two races sufficiently remote from each other in physiological constitution as to be sterile when intermarried, or, if fertile, only capable of producing sterile hybrids, etc.[1]

To this objection Lamarck would, no doubt, have answered that there had not been time for bringing about so great an amount of variation; for when Cuvier and some other of his contemporaries appealed to the embalmed animals and plants taken from Egyptian tombs, some of them 3000 years old, which had not experienced in that long period the slightest modification in their specific characters, he replied that the climate and soil of the valley of the Nile had not varied in the interval, and that there was therefore no reason for expecting that we should be able to detect any change in the fauna and flora. " But if," he went on to say, " the physical geography, temperature, and other conditions of life had been altered in Egypt as much as we know from geology has happened in other regions, some of the same animals and plants would have deviated so far from their pristine types as to be thought entitled to take rank as new and distinct species."

Although I cited this answer of Lamarck in my account of his theory,[2] I did not at the time fully appreciate the deep conviction which it displays of the slow manner in which

---

[1] *Principles of Geology*, 1st ed., vol. ii., ch. ii.   [2] *Ibid.*, p. 587.

geological changes have taken place and the insignificance of thirty or forty centuries in the history of a species, and that, too, at a period when very narrow views were entertained of the extent of past time by most of the ablest geologists, and when great revolutions of the earth's crust, and its inhabitants, were generally attributed to sudden and violent catastrophes.

While in 1832 I argued against Lamarck's doctrine of the gradual transmutation of one species into another, I agreed with him in believing that the system of changes now in progress in the organic world would afford, when fully understood, a complete key to the interpretation of all the vicissitudes of the living creation in past ages. I contended against the doctrine, then very popular, of the sudden destruction of vast multitudes of species and the abrupt ushering into the world of new batches of plants and animals.

I endeavoured to sketch out (and it was, I believe, the first systematic attempt to accomplish such a task) the laws which govern the extinction of species, with a view of showing that the slow but ceaseless variations now in progress in physical geography, together with the migration of plants and animals into new regions, must in the course of ages give rise to the occasional loss of some of them and eventually cause an entire fauna and flora to die out; also that we must infer from geological data that the places thus left vacant from time to time are filled up without delay by new forms adapted to new conditions, sometimes by immigration from adjoining provinces, sometimes by new creations. Among the many causes of extinction enumerated by me were the power of hostile species, diminution of food, mutations in climate, the conversion of land into sea and of sea into land, etc. I firmly opposed Brocchi's hypothesis of a decline in the vital energy of each species; [1] maintaining that there was every reason to believe that the reproductive powers of the last surviving representatives of a species were as vigorous as those of their predecessors, and that they were as capable, under favourable circumstances, of re-peopling the earth with their kind. The manner in which some species are now becoming scarce and dying out, one after the other, appeared to me to favour the doctrine of the fixity of the specific character, showing a want of pliancy and capability of varying, which ensured their annihilation whenever changes adverse to their well-being occurred; time not being allowed for such a transformation as might be conceived capable of

[1] *Principles of Geology*, 1st ed., vol. ii., ch. viii.; and 9th ed., p. 668.

adapting them to the new circumstances, and of converting them into what naturalists would call new species.[1]

But while rejecting transmutation, I was equally opposed to the popular theory that the creative power had diminished in energy, or that it had been in abeyance ever since Man had entered upon the scene. That a renovating force which had been in full operation for millions of years should cease to act while the causes of extinction were still in full activity, or even intensified by the accession of Man's destroying power, seemed to me in the highest degree improbable. The only point on which I doubted was whether the force might not be intermittent instead of being, as Lamarck supposed, in ceaseless operation. Might not the births of new species, like the deaths of old ones, be sudden? Might they not still escape our observation? If the coming in of one new species, and the loss of one other which had endured for ages, should take place annually, still, assuming that there are a million of animals and plants living on the globe, it would require, I observed, a million of years to bring about a complete revolution in the fauna and flora. In that case, I imagined that, although the first appearance of a new form might be as abrupt as the disappearance of an old one, yet naturalists might never yet have witnessed the first entrance on the stage of a large and conspicuous animal or plant, and as to the smaller kinds, many of them may be conceived to have stolen in unseen, and to have spread gradually over a wide area, like species migrating into new provinces.[2]

It may now be useful to offer some remarks on the very different reception which the twin branches of Lamarck's development theory, namely, progression and transmutation, have met with, and to inquire into the causes of the popularity of the one and the great unpopularity of the other. We usually test the value of a scientific hypothesis by the number and variety of the phenomena of which it offers a fair or plausible explanation. If transmutation, when thus tested, has decidedly the advantage over progression and yet is comparatively in disfavour, we may reasonably suspect that its reception is retarded, not so much by its own inherent demerits, as by some apprehended consequences which it is supposed to involve and which run counter to our preconceived opinions.

---

[1] Laws of Extinction, *Principles of Geology*, 1st ed., 1832 vol., ii., chaps. v. to xi. inclusive; and 9th ed., chaps. xxxvii. to xlii. inclusive. 1853.

[2] *Principles of Geology*, 1st ed., 1832, vol. ii., ch. xi.; and 9th ed., p. 706.

Theory of Progression

In treating of this question, I shall begin with the doctrine of progression, a concise statement of which, so far as it relates to the animal kingdom, was thus given twelve years ago by Professor Sedgwick, in the preface to his *Discourse on the Studies of the University of Cambridge.*

" There are traces," he says, " among the old deposits of the earth of an organic progression among the successive forms of life. They are to be seen in the absence of mammalia in the older, and their very rare appearance in the newer Secondary groups; in the diffusion of warm-blooded quadrupeds (frequently of unknown genera) in the older Tertiary system, and in their great abundance (and frequently of known genera) in the upper portions of the same series; and lastly, in the recent appearance of Man on the surface of the earth."

" This historical development," continues the same author, " of the forms and functions of organic life during successive epochs, seems to mark a gradual evolution of creative power, manifested by a gradual ascent towards a higher type of being." " But the elevation of the fauna of successive periods was not made by transmutation, but by creative additions; and it is by watching these additions that we get some insight into Nature's true historical progress, and learn that there was a time when Cephalopoda were the highest types of animal life, the primates of this world; that Fishes next took the lead, then Reptiles; and that during the secondary period they were anatomically raised far above any forms of the reptile class now living in the world. Mammals were added next, until Nature became what she now is, by the addition of Man." [1]

Although in the half century which has elapsed between the time of Lamarck and the publication of the above summary, new discoveries have caused geologists to assign a higher antiquity both to Man and the oldest fossil mammalia, fish, and reptiles than formerly, yet the generalisation, as laid down by the Woodwardian Professor, as to progression, still holds good in all essential particulars.

The progressive theory was propounded in the following terms by the late Hugh Miller in his *Footprints of the Creator.*

" It is of itself an extraordinary fact without reference to

---

[1] Professor Sedgwick's *Discourse on the Studies of the University of Cambridge,* Preface to 5th ed., pp. xliv., cliv., ccxvi., 1850.

other considerations, that the order adopted by Cuvier in his *Animal Kingdom*, as that in which the four great classes of vertebrate animals, when marshalled according to their rank and standing, naturally range, should be also that in which they occur in order of time. The brain, which bears an average proportion to the spinal cord of not more than two to one, comes first—it is the brain of the fish; that which bears to the spinal cord an average proportion of two and a half to one succeeded it—it is the brain of the reptile; then came the brain averaging as three to one—it is that of the bird. Next in succession came the brain that averages as four to one—it is that of the mammal; and last of all there appeared a brain that averages as twenty-three to one—reasoning, calculating Man had come upon the scene." [1]

M. Agassiz, in his *Essay on Classification*, has devoted a chapter to the " Parallelism between the Geological Succession of Animals and Plants and' their present relative Standing; " in which he has expressed a decided opinion that within the limits of the orders of each great class there is a coincidence between their relative rank in organisation and the order of succession of their representatives in time.[2]

Professor Owen, in his Palæontology, has advanced similar views, and has remarked, in regard to the vertebrata that there is much positive as well as negative evidence in support of the doctrine of an advance in the scale of being, from ancient to more modern geological periods. We observe, for example, in the Triassic, Oolitic, and Cretaceous strata, not only an absence of placental mammalia, but the presence of innumerable reptiles, some of large size, terrestrial and aquatic, herbivorous and predaceous, fitted to perform the functions now discharged by the mammalia.

The late Professor Bronn, of Heidelberg, after passing in review more than 24,000 fossil animals and plants, which he had classified and referred each to their geological position in his *Index Palæontologicus*, came to the conclusion that, in the course of time, there had been introduced into the earth more and more highly organised types of animal and vegetable life; the modern species being, on the whole, more specialised, *i.e.* having separate organs, or parts of the body, to perform different functions, which, in the earlier periods and in beings

---

[1] *Footprints of the Creator*, Edinburgh, 1849, p. 283.
[2] *Contributions to Natural History of United States*, Part I.—Essay on Classification, p. 108.

of simpler structure, were discharged in common by a single part or organ.

Professor Adolphe Brongniart, in an essay published in 1849 on the botanical classification and geological distribution of the genera of fossil plants,[1] arrives at similar results as to the progress of the vegetable world from the earliest periods to the present. He does not pretend to trace an exact historical series from the sea-weed to the fern, or from the fern again to the conifers and cycads, and lastly from those families to the palms and oaks, but he, nevertheless, points out that the cryptogamic forms, especially the acrogens, predominate among the fossils of the primary formations, the Carboniferous especially, while the gymnosperms or coniferous and cycadeous plants abound in all the strata, from the Trias to the Wealden inclusive; and lastly, the more highly developed angiosperms, both mono-cotyledonous and dicotyledonous, do not become abundant until the Tertiary period. It is a remarkable fact, as he justly observes, that the angiospermous exogens, which comprise four-fifths of living plants—a division to which all our native European trees, except the Coniferæ, belong, and which embrace all the *Compositæ*, *Leguminosæ*, *Umbelliferæ*, *Cruciferæ*, Heaths, and so many other families—are wholly unrepresented by any fossils hitherto discovered in the Primary and Secondary formations from the Silurian to the Oolitic inclusive. It is not till we arrive at the Cretaceous period that they begin to appear, sparingly at first, and only playing a conspicuous part, together with the palms and other endogens, in the Tertiary epoch.

When commenting on the eagerness with which the doctrine of progression was embraced from the close of the last century to the time when I first attempted, in 1830, to give some account of the prevailing theories in geology, I observed that far too much reliance was commonly placed on the received dates of the first appearances of certain orders or classes of animals or plants, such dates being determined by the age of the stratum in which we then happened to have discovered the earliest memorials of such types. At that time (1830), it was taken for granted that Man had not co-existed with the mammoth and other extinct mammalia, yet now that we have traced back the signs of his existence to the Pleistocene era, and may anticipate the finding of his remains on some future day in the Pliocene period, the theory of progression is not shaken; for we cannot

---

[1] Tableau des Genres de Végétaux fossiles, etc., *Dictionnaire Universel d'Histoire Naturelle*, Paris, 1849.

expect to meet with human bones in the Miocene formations, where all the species and nearly all the genera of mammalia belong to types widely differing from those now living; and had some other rational being, representing Man, then flourished, some signs of his existence could hardly have escaped unnoticed, in the shape of implements of stone or metal, more frequent and more durable than the osseous remains of any of the mammalia.

In the beginning of this century it was one of the canons of the popular geological creed that the first warm-blooded quadrupeds which had inhabited this planet were those derived from the Eocene gypsum of Montmartre in the suburbs of Paris, almost all of which Cuvier had shown to belong to extinct genera. This dogma continued in force for more than a quarter of a century, in spite of the discovery in 1818 of a marsupial quadruped in the Stonesfield strata, a member of the Lower Oolite, near Oxford. Some disputed the authority of Cuvier himself as to the mammalian character of the fossil; others, the accuracy of those who had assigned to it so ancient a place in the chronological series of rocks. In 1832 I pointed out that the occurrence of this single fossil in the Oolite was "fatal to the theory of successive development" as then propounded.[1] Since that period great additions have been made to our knowledge of the existence of land quadrupeds in the olden times. We have ascertained that, in Eocene strata older than the gypsum of Paris, no less than four distinct sets of placental mammalia have flourished; namely, first, those of the Headon series in the Isle of Wight, from which fourteen species have been procured; secondly, those of the antecedent Bagshot and Bracklesham beds, which have yielded, together with the contemporaneous "calcaire grossier" of Paris, twenty species; thirdly, the still older beds of Kyson, near Ipswich, and those of Herne Bay, at the mouth of the Thames, in which seven species have been found; and fourthly, the Woolwich and Reading beds, which have supplied ten species.[2]

We can scarcely doubt that we should already have traced back the evidence of this class of fossils much farther had not our inquiries been arrested, first by the vast gap between the Tertiary and Secondary formations, and then by the marine nature of the Cretaceous rocks.

The mammalia next in antiquity, of which we have any

---

[1] *Principles of Geology*, 2nd ed., i., 173.
[2] Lyell's supplement to 5th ed. of *Elements*, 1857.

cognisance, are those of the Upper Oolite of Purbeck, discovered between the years 1854 and 1857, and comprising no
less than fourteen species, referable to eight or nine genera;
one of them, *Plagiaulax*, considered by Dr. Falconer to have
been a herbivorous marsupial. The whole assemblage appear,
from the joint observations of Professor Owen and Dr. Falconer,
to indicate a low grade of quadruped, probably of the marsupial
type. They were, for the most part, diminutive, the two largest
not much exceeding our common hedgehog and polecat in size.

Next anterior in age are the mammalia of the Lower Oolite
of Stonesfield, of which four species are known, also very small
and probably marsupial, with one exception, the *Stereognathus
ooliticus*, which, according to Professor Owen's conjecture, may
have been a hoofed quadruped and placental, though, as we
have only half of the lower jaw with teeth, and the molars are
unlike any living type, such an opinion is of course hazarded
with due caution.

Still older than the above are some fossil quadrupeds of
small size, found in the Upper Trias of Stuttgart in Germany,
and more lately by Mr. C. Moore in beds of corresponding age
near Frome, which are also of a very low grade, like the living
Myrmecobius of Australia. Beyond this limit our knowledge
of the highest class of vertebrata does not as yet extend into
the past, but the frequent shifting back of the old landmarks,
nearly all of them once supposed in their turn to indicate the
date of the first appearance of warm-blooded quadrupeds on
this planet, should serve as a warning to us not to consider
the goal at present reached by palæontology as one beyond
which they who come after us are never destined to pass.

On the other hand, it may be truly said in favour of progression that after all these discoveries the doctrine is not gain-
said, for the less advanced marsupials precede the more perfect
placental mammalia in the order of their appearance on the
earth.

If the three localities where the most ancient mammalia
have been found—Purbeck, Stonesfield, and Stuttgart—had
belonged all of them to formations of the same age, we might
well have imagined so limited an area to have been peopled
exclusively with pouched quadrupeds, just as Australia now
is, while other parts of the globe were inhabited by placentals,
for Australia now supports one hundred and sixty species of
marsupials, while the rest of the continents and islands are
tenanted by about seventeen hundred species of mammalia,

of which only forty-six are marsupial, namely, the opossums of North and South America. But the great difference of age of the strata in each of these three localities seems to indicate the predominance throughout a vast lapse of time (from the era of the Upper Trias to that of the Purbeck beds) of a low grade of quadrupeds; and this persistency of similar generic and ordinal types in Europe while the species were changing, and while the fish, reptiles, and mollusca were undergoing vast modifications, raises a strong presumption that there was also a vast extension in space of the same marsupial forms during that portion of the Secondary epoch which has been termed " the age of reptiles."

As to the class Reptilia, some of the orders which prevailed when the Secondary rocks were formed are confessedly much higher in their organisation than any of the same class now living. If the less perfect ophidians, or snakes, which now abound on the earth had taken the lead in those ancient days among the land reptiles, and the Deinosaurians had been contemporary with Man, there can be no doubt that the progressionist would have seized upon this fact with unfeigned satisfaction as confirmatory of his views. Now that the order of succession is precisely reversed, and that the age of the *Iguanodon* was long anterior to that of the Eocene *Palæophis* and living boa, while the crocodile is in our own times the highest representative of its class, a retrograde movement in this important division of the vertebrata must be admitted. It may perhaps be accounted for by the power acquired by the placental mammalia, when they became dominant, a power before which the class of vertebrata next below them, as coming most directly in competition with them, may more than any other have given way.

For no less than thirty-four years it had been a received axiom in palæontology that reptiles had never existed before the Permian or Magnesian Limestone period, when at length in 1844 this supposed barrier was thrown down, and Carboniferous reptiles, terrestrial and aquatic, of several genera were brought to light; and discussions are now going on as to whether some remains of an Enaliosaur (perhaps a large *Labyrinthodon*) have not been detected in the coal of Nova Scotia, and whether certain sandstones near Elgin in Scotland, containing the bones of lacertian, crocodilian, and rhynchosaurian reptiles, may not be referable to the " Old Red " or Devonian group. Still, no traces of this class have yet been detected in rocks as ancient as those in which the oldest fish have been found. [Note 38.]

As to fossil representatives of the ichthyic type, the most ancient were not supposed before 1838 to be of a date anterior to the Coal, but they have since been traced back, first to the Devonian, and then to the Silurian rocks. No remains, however, of them or of any vertebrate animal have yet been discovered in the Ordovician strata, rich as these are in invertebrate fossils, nor in the still older Cambrian; so that we seem authorised to conclude, though not without considerable reserve, that the vertebrate type was extremely scarce, if not wholly wanting, in those epochs often spoken of as "primitive," but which, if the Development Theory be true, were probably the last of a long series of antecedent ages in which living beings flourished.

As to the Mollusca, which afford the most unbroken series of geological medals, the highest of that class, the Cephalopoda, abounded in older Silurian times, comprising several hundred species of chambered univalves. Had there been strong prepossessions against the progressive theory, it would probably have been argued that when these cephalopods abounded, and the siphonated gasteropods were absent, a higher order of zoophagous mollusca discharged the functions afterwards performed by an inferior order in the Secondary, Tertiary, and Post-Tertiary seas. But I have never seen this view suggested as adverse to the doctrine of progress, although much stress has been laid on the fact that the Silurian Brachiopoda, creatures of a lower grade, formerly discharged the functions of the existing lamellibranchiate bivalves, which are higher in the scale.

It is said truly that the Ammonite, Orthoceras, and Nautilus of these ancient rocks were of the tetrabranchiate division, and none of them so highly organised as the Belemnite and other dibranchiate cephalopods which afterwards appeared, and some of which now flourish in our seas. Therefore, we may infer that the simplest forms of the Cephalopoda took precedence of the more complex in time. But if we embrace this view, we must not forget that there are living Cephalopoda, such as the Octopods, which are devoid of any hard parts, whether external or internal, and which could leave behind them no fossil memorials of their existence, so that we must make a somewhat arbitrary assumption, namely, that at a remote era, no such Dibranchiata were in being, in order to avail ourselves of this argument in favour of progression. On the other hand, it is true that in the Lower Cambrian not even the shell-bearing tetrabranchiates have yet been discovered.

In regard to plants, although the generalisation above cited of M. Adolphe Brongniart is probably true, there has been a tendency in the advocates of progression to push the inferences deducible from known facts, in support of their favourite dogma, somewhat beyond the limits which the evidence justifies. Dr. Hooker observes, in his recent *Introductory Essay* on the Flora of Australia, that it is impossible to establish a parallel between the successive appearances of vegetable forms in time, and their complexity of structure or specialisation of organs as represented by the successively higher groups in the natural method of classification. He also adds that the earliest recognisable Cryptogams are not only the highest now existing, but have more highly differentiated vegetative organs than any subsequently appearing, and that the dicotyledonous embryo and perfect exogenous wood, with the highest specialised tissue known (the coniferous with glandular tissue), preceded the monocotyledonous embryo and endogenous wood in date of appearance on the globe—facts wholly opposed to the doctrine of progression, and which can only be set aside on the supposition that they are fragmentary evidence of a time farther removed from the origin of vegetation than from the present day.[1] [Note 39.]

It would be an easy task to multiply objections to the theory now under consideration; but from this I refrain, as I regard it not only as a useful, but rather in the present state of science as an indispensable hypothesis, and one which though destined hereafter to undergo many and great modifications will never be overthrown.

It may be thought almost paradoxical that writers who are most in favour of transmutation (Mr. C. Darwin and Dr. J. Hooker, for example) are nevertheless among those who are most cautious, and one would say timid, in their mode of espousing the doctrine of progression; while, on the other hand, the most zealous advocates of progression are oftener than not very vehement opponents of transmutation. We might have anticipated a contrary leaning on the part of both, for to what does the theory of progression point? It supposes a gradual elevation in grade of the vertebrate type in the course of ages from the most simple ichthyic form to that of the placental mammalia and the coming upon the stage last in the order of time of the most anthropomorphous mammalia, followed by the human race—this last thus appearing as an

[1] *Flora of Australia, Introductory Essay*, p. xxxi., London, 1859. Published separately.

integral part of the same continuous series of acts of development, one link in the same chain, the crowning operation as it were of one and the same series of manifestations of creative power. If the dangers apprehended from transmutation arise from the too intimate connection which it tends to establish between the human and merely animal natures, it might have been expected that the progressive development of organisation, instinct, and intelligence might have been unpopular, as likely to pioneer the way for the reception of the less favoured doctrine. But the true explanation of the seeming anomaly is this, that no one can believe in transmutation who is not profoundly convinced that all we know in palæontology is as nothing compared with what we have yet to learn, and they who regard the record as so fragmentary, and our acquaintance with the fragments which are extant as so rudimentary, are apt to be astounded at the confidence placed by the progressionists in data which must be defective in the extreme. But exactly in proportion as the completeness of the record and our knowledge of it are overrated, in that same degree are many progressionists unconscious of the goal towards which they are drifting. Their faith in the fullness of the annals leads them to regard all breaks in the series of organic existence, or in the sequence of the fossiliferous rocks, as proofs of original chasms and leaps in the course of nature—signs of the intermittent action of the creational force, or of catastrophes which devastated the habitable surface. They do not doubt that there is a continuity of plan, but they believe that it exists in the Divine mind alone, and they are therefore without apprehension that any facts will be discovered which would imply a material connection between the outgoing organisms and the incoming ones.

CHAPTER XXI

FOR many years after the promulgation of Lamarck's doctrine of progressive development, geologists were much occupied with the question whether the past changes in the animate and inanimate world were brought about by sudden and paroxysmal action, or gradually and continuously, by causes differing neither in kind not degree from those now in operation.

The anonymous author of *The Vestiges of Creation* published in 1844 a treatise, written in a clear and attractive style, which made the English public familiar with the leading views of Lamarck on transmutation and progression, but brought no new facts or original line of argument to support those views, or to combat the principal objections which the scientific world entertained against them.

No decided step in this direction was made until the publication in 1858 of two papers, one by Mr. Darwin and another by Mr. Wallace, followed in 1859 by Mr. Darwin's celebrated work on *The Origin of Species by Means of Natural Selection; or, the Preservation of favoured Races in the Struggle for Life.* The author of this treatise had for twenty previous years strongly inclined to believe that variation and the ordinary laws of reproduction were among the secondary causes always employed by the Author of nature, in the introduction from time to time of new species into the world, and he had devoted himself patiently to the collecting of facts and making of experiments in zoology and botany, with a view of testing the soundness of the theory of transmutation.   Part of the MS. of his projected work was read to Dr. Hooker as early as 1844 and some of the principal results were communicated to me on several occasions. [Note 40.]   Dr. Hooker and I had repeatedly urged him to publish without delay, but in vain, as he was always unwilling to interrupt the course of his investigations; until at length Mr. Alfred R. Wallace, who had been engaged for years in collecting and studying the animals of the East Indian archipelago, thought out independently for himself one of the most novel and important of Mr. Darwin's theories.   This he embodied in an essay " On the Tendency of Varieties to depart indefinitely from the original Type."   It was written at Ternate in February 1858, and sent to Mr. Darwin with a request that it might be shown to me if thought sufficiently novel and interesting.   Dr. Hooker and I were of opinion that it should be immediately printed, and we succeeded in persuading Mr. Darwin to allow one of the MS. chapters of his *Origin of Species,* entitled " On the Tendency of Species to form Varieties, and on the Perpetuation of Species and Varieties by natural Means of Selection," to appear at the same time.[1]

By reference to these memoirs it will be seen that both writers begin by applying to the animal and vegetable worlds the Malthusian doctrine of population, or its tendency to increase in a geometrical ratio, while food can only be made to augment even locally in an arithmetical one.   There being therefore no room or means of subsistence for a large proportion of the plants and animals which are born into the world, a great number must annually perish.   Hence there is a constant struggle for existence among the individuals which represent each species and the vast majority can never reach the adult state, to say nothing of the multitudes of ova and seeds which are never hatched or allowed to germinate.   Of birds it is estimated that the number of those which die every year equals the aggregate number by which the species to which they respectively belong is on the average permanently represented.

The trial of strength which must decide what individuals are to survive and what to succumb occurs in the season when the means of subsistence are fewest, or enemies most numerous,

[1] See *Proceedings of Linnæan Society,* 1858.

or when the individuals are enfeebled by climate or other causes; and it is then that those varieties which have any, even the slightest, advantage over others come off victorious. They may often owe their safety to what would seem to a casual observer a trifling difference, such as a darker or lighter shade of colour rendering them less visible to a species which preys upon them, or sometimes to attributes more obviously advantageous, such as greater cunning or superior powers of flight or swiftness of foot. These peculiar qualities and faculties, bodily and instinctive, may enable them to outlive their less favoured rivals, and being transmitted by the force of inheritance to their offspring will constitute new races, or what Mr. Darwin calls " incipient species." If one variety, being in other respects just equal to its competitors, happens to be more prolific, some of its offspring will stand a greater chance of being among those which will escape destruction, and their descendants, being in like manner very fertile, will continue to multiply at the expense of all less prolific varieties.

As breeders of domestic animals, when they choose certain varieties in preference to others to breed from, speak technically of their method as that of " selecting," Mr. Darwin calls the combination of natural causes, which may enable certain varieties of wild animals or plants to prevail over others of the same species, " natural selection."

A breeder finds that a new race of cattle with short horns or without horns may be formed in the course of several generations by choosing varieties having the most stunted horns as his stock from which to breed; so nature, by altering in the course of ages, the conditions of life, the geographical features of a country, its climate, the associated plants and animals, and consequently the food and enemies of a species and its mode of life, may be said, by this means to select certain varieties best adapted for the new state of things. Such new races may often supplant the original type from which they have diverged, although that type may have been perpetuated without modification for countless anterior ages in the same region, so long as it was in harmony with the surrounding conditions then prevailing.

Lamarck, when speculating on the origin of the long neck of the giraffe, imagined that quadruped to have stretched himself up in order to reach the boughs of lofty trees, until by continued efforts and longing to reach higher he obtained an elongated neck. Mr. Darwin and Mr. Wallace simply suppose that, in a season of scarcity, a longer-necked variety, having the advantage in this respect over most of the herd, as being able to browse on foliage out of their reach, survived them and transmitted its peculiarity of cervical conformation to its successors.

By the multiplying of slight modifications in the course of thousands of generations and by the handing down of the newly-acquired peculiarities by inheritance, a greater and greater divergence from the original standard is supposed to be effected, until what may be called a new species, or in a greater lapse of time a new genus will be the result.

Every naturalist admits that there is a general tendency in animals and plants to vary; but it is usually taken for granted, though he have no means of proving the assumption to be true, that there are certain limits beyond which each species cannot pass under any circumstances or in any number of generations. Mr. Darwin and Mr. Wallace say that the opposite hypothesis, which assumes that every species is capable of varying indefinitely from its original type, is not a whit more arbitrary, and has this manifest claim to be preferred, that it will account for a multitude of phenomena which the ordinary theory is incapable of explaining.

We have no right, they say, to assume, should we find that a variable species can no longer be made to vary in a certain direction, that it has reached the utmost limit to which it might under more favourable conditions or if more time were allowed be made to diverge from the parent type.

Hybridisation is not considered by Mr. Darwin as a cause of new species, but rather as tending to keep variation within bounds. Varieties which are nearly allied cross readily with each other, and with the parent stock, and such crossing tends to keep the species true to its type, while forms which are less nearly related, although they may intermarry, produce no mule offspring capable of perpetuating their kind.

The competition of races and species, observes Mr. Darwin, is always most severe between those which are most closely allied and which fill nearly the same place in the economy of nature. Hence when the conditions of existence are modified the original stock runs great risk of being superseded by some one of its modified offshoots. The new race or species may not be absolutely superior in the sum of its powers and endowments to the parent stock, and may even be more simple in structure and of a lower grade of intelligence, as well as of organisation, provided on the whole it happens to have some slight advantage over its rivals. Progression, therefore, is not a necessary accompaniment of variation and natural selection, though when a higher organisation happens to be coincident with superior fitness to new conditions, the new species will have greater power and a greater chance of permanently maintaining and extending its ground. One of the principal claims of Mr. Darwin's theory to acceptance is that it enables us to dispense with a law of progression as a necessary accompaniment of variation. It will account equally well for what is called degradation, or a retrograde movement towards a simpler structure, and does not require Lamarck's continual creation of monads; for this was a necessary part of his system, in order to explain how, after the progressive power had been at work for myriads of ages, there were as many beings of the simplest structure in existence as ever.

Mr. Darwin argues, and with no small success, that all true classification in zoology and botany is in fact genealogical, and that community of descent is the hidden bond which naturalists have been unconsciously seeking, while they often imagined that they were looking for some unknown plan of creation.

As the *Origin of Species* [1] is in itself a condensed abstract of a much larger work not yet published [Note 41] I could not easily give an analysis of its contents within narrower limits than those of the original, but it may be useful to enumerate briefly some of the principal classes of phenomena on which the theory of " natural selection " would throw light.

In the first place it would explain, says Mr. Darwin, the unity of type which runs through the whole organic world, and why there is sometimes a fundamental agreement in structure in the same class of beings which is quite independent of their habits of life, for such structure, derived by inheritance from a remote progenitor, has been modified in the course of ages in different ways according to the conditions of existence. It would also explain why all living and extinct beings are united, by complex radiating and circuitous lines of affinity with one another into one grand system; [2] also, there having been a continued extinction of old races and species in progress and a formation of new ones by variation, why in some genera which are largely represented, or to which a great many species belong, many of these are closely but unequally related; also, why there are distinct geographical provinces of species of animals and plants, for after long isolation by physical barriers each fauna

---

[1] *Origin of Species*, p. 121.   [2] *Origin*, p. 498.

and flora by varying continually must become distinct from its ancestral type, and from the new forms assumed by other descendants which have diverged from the same stock.

The theory of indefinite modification would also explain why rudimentary organs are so useful in classification, being the remnants preserved by inheritance of organs which the present species once used—as in the case of the rudiments of eyes in insects and reptiles inhabiting dark caverns, or of the wings of birds and beetles which have lost all power of flight. In such cases the affinities of species are often more readily discerned by reference to these imperfect structures than by others of much more physiological importance to the individuals themselves.

The same hypothesis would explain why there are no mammalia in islands far from continents, except bats, which can reach them by flying; and also why the birds, insects, plants, and other inhabitants of islands, even when specifically unlike, usually agree generically with those of the nearest continent, it being assumed that the original stock of such species came by migration from the nearest land.

Variation and natural selection would also afford a key to a multitude of geological facts otherwise wholly unaccounted for, as for example why there is generally an intimate connection between the living animals and plants of each great division of the globe and the extinct fauna and flora of the Post-Tertiary or Tertiary formations of the same region; as, for example, in North America, where we not only find among the living mollusca peculiar forms foreign to Europe, such as *Gnathodon* and *Fulgur* (a subgenus of *Fusus*), but meet also with extinct species of those same genera in the Tertiary fauna of the same part of the world. In like manner, among the mammalia we find in Australia not only living kangaroos and wombats, but fossil individuals of extinct species of the same genera. So also there are recent and fossil sloths, armadilloes and other Edentata in South America, and living and extinct species of elephant, rhinoceros, tiger, and bear in the great Europeo-Asiatic continent. The theory of the origin of new species by variation will also explain why a species which has once died out never reappears and why the fossil fauna and flora recede farther and farther from the living type in proportion as we trace them back to remoter ages. It would also account for the fact that when we have to intercalate a new set of fossiliferous strata between two groups previously known, the newly discovered fossils serve to fill up gaps between specific or generic types previously familiar to us, supplying often the missing links of the chain, which, if transmutation is accepted, must once have been continuous.

One of the most original speculations in Mr. Darwin's work is derived from the fact that, in the breeding of animals, it is often observed that at whatever age any variation first appears in the parent, it tends to reappear at a corresponding age in the offspring. Hence the young individuals of two races which have sprung from the same parent stock are usually more like each other than the adults. Thus the puppies of the grey-hound and bull-dog are much more nearly alike in their proportions than the grown-up dogs, and in like manner the foals of the carthorse and racehorse than the adult individuals. For the same reason we may understand why the species of the same genus, or genera of the same family, resemble each other more nearly in their embryonic than in their more fully developed state, or how it is that in the eyes of most naturalists the structure of the embryo is even more important in classification than that of the adult, " for the embryo is the animal in its less modified state, and in so far it reveals the structure of its progenitor. In two groups of animals, however much they may at present differ from each other in structure and habits, if they pass through the same or similar embryonic stages, we may feel assured that they have both descended from the same or nearly similar parents, and are therefore in that degree closely related. Thus community in embryonic structure reveals community of descent, however much the structure of the adult may have been modified." [1]

If then there had been a system of progressive development, the successive changes through which the embryo of a species of a high class, a mammifer for example, now passes, may be expected to present us with a picture of the stages through which, in the course of ages, that class of animals has successively passed in advancing from a lower to a higher grade. Hence the embryonic states exhibited one after the other by the human individual bear a certain amount of resemblance to those of the fish, reptile, and bird before assuming those of the highest division of the vertebrata.

Mr. Darwin, after making a laborious analysis of many floras, found that those genera which are represented by a large number of species contain a greater number of variable species,

[1] Darwin, *Origin*, etc., p. 448.

relatively speaking, than the smaller genera or those less numerously represented. This fact he adduces in support of his opinion that varieties are incipient species, for he observes that the existence of the larger genera implies that the manufacturing of species has been active in the period immediately preceding our own, in which case we ought generally to find the same forces still in full activity, more especially as we have every reason to believe the process by which new species are produced is a slow one.[1]

Dr. Hooker tells us that he was long disposed to doubt this result, as he was acquainted with so many variable small genera, but after examining Mr. Darwin's data, he was compelled to acquiesce in his generalisation.[2]

It is one of those conclusions, to verify which requires the investigation of many thousands of species, and to which exceptions may easily be adduced both in the animal and vegetable kingdoms, so that it will be long before we can expect it to be thoroughly tested, and if true, fairly appreciated. Among the most striking exceptions will be some genera still large, but which are beginning to decrease, the conditions favourable to their former predominance having already begun to change. To many, this doctrine of " natural selection," or " the preservation of favoured races in the struggle for life," seems so simple, when once clearly stated, and so consonant with known facts and received principles, that they have difficulty in conceiving how it can constitute a great step in the progress of science. Such is often the case with important discoveries, but in order to assure ourselves that the doctrine was by no means obvious, we have only to refer back to the writings of skilful naturalists who attempted in the earlier part of the nineteenth century to theorise on this subject, before the invention of this new method of explaining how certain forms are supplanted by new ones and in what manner these last are selected out of innumerable varieties and rendered permanent.

Dr. Hooker on the Theory of " Creation by Variation " as applied to the Vegetable Kingdom

Of Dr. Hooker, whom I have often cited in this chapter, Mr. Darwin has spoken in the Introduction to his *Origin of Species,* as one " who had, for fifteen years, aided him in every

---

[1] *Origin of Species*, ch. ii., p. 56.
[2] *Introductory Essay* to *Flora of Australia*, p. vi.

possible way, by his large stores of knowledge, and his excellent judgment." This distinguished botanist published his *Introductory Essay to the Flora of Australia* in December 1859, the year after the memoir on "Natural Selection" was communicated to the Linnæan Society, and a month after the appearance of the *Origin of Species*.

Having, in the course of his extensive travels, studied the botany of arctic, temperate, and tropical regions, and written on the flora of India, which he had examined at all heights above the sea from the plains of Bengal to the limits of perpetual snow in the Himalaya, and having specially devoted his attention to "geographical varieties," or those changes of character which plants exhibit when traced over wide areas and seen under new conditions; being also practically versed in the description and classification of new plants, from various parts of the world, and having been called upon carefully to consider the claims of thousands of varieties to rank as species, no one was better qualified by observation and reflection to give an authoritative opinion on the question, whether the present vegetation of the globe is or is not in accordance with the theory which Mr. Darwin has proposed. We cannot but feel, therefore, deeply interested when we find him making the following declaration:

"The mutual relations of the plants of each great botanical province, and, in fact, of the world generally, is just such as would have resulted if variation had gone on operating throughout indefinite periods, in the same manner as we see it act in a limited number of centuries, so as gradually to give rise in the course of time, to the most widely divergent forms."

In the same essay, this author remarks, "The element of mutability pervades the whole Vegetable Kingdom; no class, nor order, nor genus of more than a few species claims absolute exemption from it, whilst the grand total of unstable forms, generally assumed to be species, probably exceeds that of the stable." Yet he contends that species are neither visionary, nor even arbitrary creations of the naturalist, but realities, though they may not remain true for ever. The majority of them, he remarks, are so far constant, " within the range of our experience," and their forms and characters so faithfully handed down through thousands of generations, that they admit of being treated as if they were permanent and immutable. But the range of " our experience " is so limited, that it will " not account for a single fact in the present geographical distribution, or origin of any one species of plant, nor for the amount of variation it has undergone, nor will it indicate the time when it first appeared, nor the form it had when created." [1]

To what an extent the limits of species are indefinable, is evinced, he says, by the singular fact that, among those botanists who believe them to be immutable, the number of flowering plants is by some assumed to be 80,000, and by others over 150,000. The general limitation of species to certain areas suggests the idea that each of them, with all their varieties, have sprung from a common parent and have spread in various directions from a common centre. The frequency also of the grouping of genera within certain geographical limits is in favour of the same law, although the migration of species may sometimes cause apparent exceptions to the rule and make the same types appear to have originated independently at different spots.[1]

Certain genera of plants, which, like the brambles, roses, and willows in Europe, consist of a continuous series of varieties between the terms of which no intermediate forms can be intercalated, may be supposed to be newer types and on the increase, and therefore undergoing much variation; whereas genera which present no such perplexing gradations may be of older date and may have been losing species and varieties by extinction. In this case, the annihilation of intermediate forms which once existed makes it an easy task to distinguish those which remain.

It had usually been supposed by the advocates of the immutability of species that domesticated races, if allowed to run wild, always revert to their parent type. Mr. Wallace had said in reply that a domesticated species, if it loses the protection of Man, can only stand its ground in a wild state by resuming those habits and recovering those attributes which it may have lost when under domestication. If these faculties are so much enfeebled as to be irrecoverable it will perish; if not and if it can adapt itself to the surrounding conditions, it will revert to the state in which Man first found it: for in one, two, or three thousand years, which may have elapsed since it was originally tamed, there will not have been time for such geographical, climatal, and organic changes as would only be suited to a new race or a new and allied species.

But in regard to plants Dr. Hooker questions the fact of reversion. According to him, species in general do not readily vary, but when they once begin to do so the new varieties, as

[1] Hooker, *Introductory Essay to Flora of Australia.*   [2] *Ibid.*, p. 13.

every horticulturist knows, show a great inclination to go on departing more and more from the old stock. As the best marked varieties of a wild species occur on the confines of the area which it inhabits, so the best marked varieties of a cultivated plant are those last produced by the gardener. Cabbages, for example, wall fruits, and cereal, show no disposition, when neglected, to assume the characters of the wild states of these plants. Hence the difficulty of determining what are the true parent species of most of our cultivated plants. Thus the finer kinds of apples, if grown from seed, degenerate and become crabs, but in so doing they do not revert to the original wild crab-apple, but become crab states of the varieties to which they belong.[1]

It would lead me into too long a digression were I to attempt to give a fuller analysis of this admirable essay; but I may add that none of the observations are more in point, as bearing on the doctrine of what Hooker terms " creation by variation," than the great extent to which the internal characters and properties of plants, or their physiological constitution, are capable of being modified, while they exhibit externally no visible departure from the normal form. Thus, in one region a species may possess peculiar medicinal qualities which it wants in another, or it may be hardier and better able to resist cold. The average range in altitude, says Hooker, of each species of flowering plant in the Himalayan Mountains, whether in the tropical, temperate, or Alpine region, is 4000 feet, which is equivalent to twelve degrees of isothermals of latitude. If an individual of any of these species be taken from the upper limits of its range and carried to England, it is found to be better able to stand our climate than those from the lower or warmer stations. When several of these internal or physiological modifications are accompanied by variation in size, habits of growth, colour of the flowers, and other external characters, and these are found to be constant in successive generations, botanists may well begin to differ in opinion as to whether they ought to regard them as distinct species or not.

Alternation of Generations

Hitherto, no rival hypothesis has been proposed as a substitute for the doctrine of transmutation; for what we term " independent creation," or the direct intervention of the

[1] *Introductory Essay* to *Flora of Australia*, p. ix.

Supreme Cause, must simply be considered as an avowal that we deem the question to lie beyond the domain of science.

The discovery by Steenstrup of alternate generation enlarges our views of the range of metamorphosis through which a species may pass, so that some of its stages (as when a *Sertularia* and a *Medusa* interchange) deviate so far from others as to have been referred by able zoologists to distinct genera, or even families. But in all these cases the organism, after running through a certain cycle of change, returns to the exact point from which it set out, and no new form or species is thereby introduced into the world. The only secondary cause therefore which has as yet been even conjecturally brought forward, to explain how in the ordinary course of nature a new specific form may be generated is, as Lamarck declared, " variation," and this has been rendered a far more probable hypothesis by the way in which "natural selection" is shown to give intensity and permanency to certain varieties.

Independent Creation

When I formerly advocated the doctrine that species were primordial creations and not derivative, I endeavoured to explain the manner of their geographical distribution, and the affinity of living forms to the fossil types nearest akin to them in the Tertiary strata of the same part of the globe, by supposing that the creative power, which originally adapts certain types to aquatic and others to terrestrial conditions, has at successive geological epochs introduced new forms best suited to each area and climate, so as to fill the places of those which may have died out.

In that case, although the new species would differ from the old (for these would not be revived, having been already proved by the fact of their extinction to be incapable of holding their ground), still they would resemble their predecessors generically. For, as Mr. Darwin states in regard to new races, those of a dominant type inherit the advantages which made their parent species flourish in the same country, and they likewise partake in those general advantages which made the genus to which the parent species belonged a large genus in its own country.

We might therefore, by parity of reasoning, have anticipated that the creative power, adapting the new types to the new combination of organic and inorganic conditions of a given region, such as its soil, climate, and inhabitants, would introduce new modifications of the old types—marsupials, for example, in Australia, new sloths and armadilloes in South America, new heaths at the Cape, new roses in the northern and new calceolarias in the southern hemisphere. But to this line of argument Mr. Darwin and Dr. Hooker reply that when animals or plants migrate into new countries, whether assisted by man or without his aid, the most successful colonisers appertain by no means to those types which are most allied to the old indigenous species. On the contrary it more frequently happens that members of genera, orders, or even classes, distinct and foreign to the invaded country, make their way most rapidly and become dominant at the expense of the endemic species. Such is the case with the placental quadrupeds in Australia, and with horses and many foreign plants in the pampas of South America, and numberless instances in the United States and elsewhere which might easily be enumerated. Hence the transmutationists infer that the reason why these foreign types, so peculiarly fitted for these regions, have never before been developed there is simply that they were excluded by natural barriers. But these barriers of sea or desert or mountain could never have been of the least avail had the creative force acted independently of material laws or had it not pleased the Author of Nature that the origin of new species should be governed by some secondary causes analogous to those which we see preside over the appearance of new varieties, which never appear except as the offspring of a parent stock very closely resembling them.

CHAPTER XXII

OBJECTIONS TO THE HYPOTHESIS OF TRANSMUTATION CONSIDERED

Statement of Objections to the Hypothesis of Transmutation founded on the Absence of intermediate Forms—Genera of which the Species are closely allied—Occasional Discovery of the missing Links in a Fossil State—Davidson's Monograph on the Brachiopoda—Why the Gradational Forms, when found, are not accepted as Evidence of Transmutation—Gaps caused by Extinction of Races and Species—Vast Tertiary Periods during which this Extinction has been going on in the Fauna and Flora now existing—Genealogical Bond between Miocene and Recent Plants and Insects—Fossils of Oeningen—Species of Insects in Britain and North America represented by distinct Varieties—Falconer's Monograph on living and fossil Elephants—Fossil Species and Genera of the Horse Tribe in North and South America—Relation of the Pliocene Mammalia of North America, Asia, and Europe—Species of Mammalia, though less persistent than the Mollusca, change slowly—Arguments for and against Transmutation derived from the Absence of Mammalia in Islands—Imperfection of the Geological Record—Intercalation of newly discovered Formation of intermediate Age in the chronological Series—Reference of the St. Cassian Beds to the Triassic Periods—Discovery of new organic Types—Feathered Archæpteryx of the Oolite.

Theory of Transmutation—Absence of Intermediate Links

THE most obvious and popular of the objections urged against the theory of transmutation may be thus expressed: If the extinct species of plants and animals of the later geological periods were the progenitors of the living species, and gave origin to them by variation and natural selection, where are all the intermediate forms, fossil and living, through which the lost types must have passed during their conversion into the living ones? And why do we not find almost everywhere passages between the nearest allied species and genera, instead of such strong lines of demarcation and often wide intervening gaps?

We may consider this objection under two heads:—

First. To what extent are the gradational links really wanting in the living creation or in the fossil world, and how far may we expect to discover such as are missing by future research?

Secondly. Are the gaps more numerous than we ought to anticipate, allowing for the original defective state of the geological records, their subsequent dilapidation and our slight acquaintance with such parts of them as are extant, and allowing also for the rate of extinction of races and species now going on, and which has been going on since the commencement of the Tertiary period?

First. As to the alleged absence of intermediate varieties connecting one species with another, every zoologist and botanist who has engaged in the task of classification has been occasionally thrown into this dilemma—if I make more than one species in this group, I must, to be consistent, make a great many. Even in a limited region like the British Isles this embarrassment is continually felt.

Scarcely any two botanists, for example, can agree as to the number of roses, still less as to how many species of bramble we possess. Of the latter genus, *Rubus*, there is one set of forms respecting which it is still a question whether it ought to be regarded as constituting three species or thirty-seven. Mr. Bentham adopts the first alternative and Mr. Babington the second, in their well-known treatises on British plants.

We learn from Dr. Hooker that at the antipodes, both in New Zealand and Australia, this same genus *Rubus* is represented by several species rich in individuals and remarkable for their variability. When we consider how, as we extend our knowledge of the same plant over a wider area, new geographical varieties commonly present themselves, and then endeavour to imagine the number of forms of the genus *Rubus* which may now exist, or probably have existed, in Europe and in regions intervening between Europe and Australia, comprehending all which may have flourished in Tertiary and Post-Tertiary periods, we shall perceive how little stress should be laid on arguments founded on the assumed absence of missing links in the flora as it now exists.

If in the battle of life the competition is keenest between closely allied varieties and species, as Mr. Darwin contends, many forms can never be of long duration, nor have a wide range, and these must often pass away without leaving behind them any fossil memorials. In this manner we may account for many breaks in the series which no future researches will ever fill up.

Davidson on Fossil Brachiopoda

It is from fossil conchology more than from any other department of the organic world that we may hope to derive traces of a transition from certain types to others, and fossil memorials of all the intermediate shades of form. We may especially hope to gain this information from the study of some of the lower groups, such as the Brachiopoda, which are persistent in type, so that the thread of our inquiry is less likely to be interrupted by breaks in the sequence of the fossiliferous rocks. The splendid monograph just concluded by Mr. Davidson on the British Brachiopoda, illustrates, in the first place, the tendency of certain generic forms in this division of the mollusca to be persistent throughout the whole range of geological time yet known to us; for the four genera, *Rhynchonella, Crania, Discina,* and *Lingula,* have been traced through the Silurian, Devonian, Carboniferous, Permian, Jurassic, Cretaceous, Tertiary, and Recent periods, and still retain in the existing seas the identical shape and character which they exhibited in the earliest formations. On the other hand, other Brachiopoda have gone through in shorter periods a vast series of transformations, so that distinct specific and even generic names have been given to the same varying form, according to the different aspects and characters it has put on in successive sets of strata.

In proportion as materials of comparison have accumulated, the necessity of uniting species previously regarded as distinct under one denomination has become more and more apparent. Mr. Davidson, accordingly, after studying not less than 260 reputed species from the British Carboniferous rocks, has been obliged to reduce that number to 100, to which he has added 20 species either entirely new or new to the British strata; but he declares his conviction that, when our knowledge of these 120 Brachiopoda is more complete, a further reduction of species will take place.

Speaking of one of these forms, which he calls *Spirifer trigonalis,* he says that it is so dissimilar to another extreme of the series, *S. crassa,* that in the first part of his memoir (published some ten years ago) he described them as distinct, and the idea of confounding them together must, he admits, appear absurd to those who have never seen the intermediate links, such as are presented by *S. bisulcata,* and at least four others with their varieties, most of them shells formerly recognised as distinct by the most eminent palæontologists, but respecting which these same authorities now agree with Mr. Davidson in uniting them into one species.[1]

The same species has sometimes continued to exist under slightly modified forms throughout the whole of the Ordovician and Silurian as well as the entire Devonian and Carboniferous periods, as in the case of the shell generally known as *Leptæna*

[1] Monograph on British Brachiopoda, Palæontographical Society, p. 222.

rhomboidalis, Wahlenberg. No less than fifteen commonly received species are demonstrated by Mr. Davidson by the aid of a long series of transitional forms, to appertain to this one type; and it is acknowledged by some of the best writers that they were induced on purely theoretical grounds to give distinct names to some of the varieties now suppressed, merely because they found them in rocks so widely remote in time that they deemed it contrary to analogy to suppose that the same species could have endured so long: a fallacious mode of reasoning, analogous to that which leads some zoologists and botanists to distinguish by specific names slight varieties of living plants and animals met with in very remote countries, as in Europe and Australia, for example; it being assumed that each species has had a single birthplace or area of creation, and that they could not by migration have gone from the northern to the southern hemisphere across the intervening tropics.

Examples are also given by Mr. Davidson of species which pass from the Devonian into the Carboniferous, and from that again into the Permian rocks. The vast longevity of such specific forms has not been generally recognised in consequence of the change of names which they have undergone when derived from such distant formations, as when *Atrypa unguicularis* assumes, when derived from a Carboniferous rock, the name of *Spirifer Urei*, besides several other synonyms, and then, when it reaches the Permian period, takes the name of *Spirifer Clannyana*, King; all of which forms the author of the monograph, now under consideration, asserts to be one and the same.

No geologist will deny that the distance of time which separates some of the eras above alluded to, or the dates of the earliest and latest appearances of some of the fossils above mentioned, must be reckoned by millions of years. According to Mr. Darwin's views, it is only by having at our command the records of such enormous periods that we can expect to be able to point out the gradations which unite very distinct specific forms. But the advocate of transmutation must not be disappointed if, when he has succeeded in obtaining some of the proofs which he was challenged to produce, they make no impression on the mind of his opponent. All that will be conceded is that specific variation in the Brachiopoda, at least, has a wider range than was formerly suspected. So long as several allied species were brought nearer and nearer to each other, considerable uneasiness might have been felt as to the reality of species in general, but when fifteen or more are once fairly merged in one group, constituting in the aggregate a single species, one and indivisible, and capable of being readily distinguished from every other group at present known, all misgivings are at an end. Implicit trust in the immutability of species is then restored, and the more insensible the shades from one extreme to the other, in a word, the more complete the evidence of transition, the more nugatory does the argument derived from it appear. It then simply resolves itself into one of those exceptional instances of what is called a protean form.

Thirty years ago a great London dealer in shells, himself an able naturalist, told me that there was nothing he had so much reason to dread, as tending to depreciate his stock in trade, as the appearance of a good monograph on some large genus of mollusca; for, in proportion as the work was executed in a philosophical spirit, it was sure to injure him, every reputed species pronounced to be a mere variety becoming from that time unsaleable. Fortunately, so much progress has since been made in England in estimating the true ends and aims of science, that specimens indicating a passage between forms usually separated by wide gaps, whether in the Recent or fossil fauna, are eagerly sought for, and often more prized than the mere normal or typical forms.

It is clear that the more ancient the existing mollusca, or the farther back into the past we can trace the remains of shells still living, the more easy it becomes to reconcile with the doctrine of transmutation the distinctness in character of the majority of living species. For, what we want is time, first, for the gradual formation, and then for the extinction of races and allied species, occasioning gaps between the survivors.

In the year 1830 I announced, on the authority of M. Deshayes, that about one-fifth of the mollusca of the Falunian or Upper Miocene strata of Europe, belonged to living species. Although the soundness of that conclusion was afterwards called in question by two or three eminent conchologists (and by the late M. Alcide d'Orbigny among others), it has since been confirmed by the majority of living naturalists and is well borne out by the copious evidence on the subject laid before the public in the magnificent work edited by Dr. Hoernes, and published under the auspices of the Austrian Government, *On the Fossil Shells of the Vienna Basin.*

The collection of Tertiary shells from which those descriptions and beautiful figures were taken is almost unexampled for the fine state of preservation of the specimens, and the care with which all the varieties have been compared. It is now admitted that about one-third of these Miocene forms, univalves and bivalves included, agree specifically with living mollusca, so that much more than the enormous interval which divides the Miocene from the Recent period must be taken into our account when we speculate on the origin by transmutation of the shells now living, and the disappearance by extinction of intermediate varieties and species.

Miocene Plants and Insects related to Recent Species

Geologists were acquainted with about three hundred species of marine shells from the Falunian strata on the banks of the Loire, before they knew anything of the contemporary insects and plants. At length, as if to warn us against inferring from negative evidence the poverty of any ancient set of strata in organic remains proper to the land, a rich flora and entomological fauna was suddenly revealed to us characteristic of Central Europe during the Upper Miocene period. This result followed the determination of the true position of the Oeningen beds in Switzerland, and of certain formations of " Brown Coal " in Germany.

Professor Heer, who has described nearly five hundred species of fossil plants from Oeningen, besides many more from other Miocene localities in Switzerland,[1] estimates the phanerogamous species which must have flourished in Central Europe at that time at 3000, and the insects as having been more numerous in the same proportion as they now exceed the plants in all latitudes. This European Miocene flora was remarkable for the preponderance of arborescent and shrubby evergreens, and comprised many generic types no longer associated together in any existing flora or geographical province. Some genera, for example, which are at present restricted to America, co-existed in Switzerland with forms now peculiar to Asia, and with others at present confined to Australia.

Professor Heer has not ventured to identify any of this vast assemblage of Miocene plants and insects with living species, so far at least as to assign to them the same specific names, but he presents us with a list of what he terms homologous forms, which are so like the living ones that he supposes the one to have been derived genealogically from the others. He hesitates indeed as to the manner of the transformation or the

[1] Heer, *Flora tertiaria Helvetiæ*, 1859; and Gaudin's French translation, with additions, 1861.

precise nature of the relationship, "whether the changes were brought about by some influence exerted continually for ages, or whether at some given moment the old types were struck with a new image."

Among the homologous plants alluded to are forty species, of which both the leaves and fruits are preserved, and thirty others, known at present by their leaves only. In the first list we find many American types, such as the tulip tree (*Liriodendron*), the deciduous cypress (*Taxodium*), the red maple and others, together with Japanese forms, such as a cinnamon, which is very abundant. And what is worthy of notice, some of these fossils so closely allied to living plants occur not only in the Upper, but even some few of them as far back in time as the Lower Miocene formations of Switzerland and Germany, which are probably as distant from the Upper Miocene or Oeningen beds as are the latter from our own era.

Some of the fossil plants to which Professor Heer has given new names have been regarded as Recent species by other eminent naturalists. Thus, one of the trees allied to the elm Unger had called *Planera Richardi,* a species which now flourishes in the Caucasus and Crete. Professor Heer had attempted to distinguish it from the living tree by the greater size of its fruit, but this character he confessed did not hold good, when he had an opportunity (1861) of comparing all the varieties of the living *Planera Richardi* which Dr. Hooker laid before him in the rich herbarium of Kew.

As to the " homologous insects " of the Upper Miocene period in Switzerland, we find among them, mingled with genera now wholly foreign to Europe, some very familiar forms, such as the common glowworm, *Lampyris noctiluca,* Linn., the dung-beetle, *Geotrupes stercorarius,* Linn., the ladybird, *Coccinella septem-punctata,* Linn., the ear-wig, *Forficula auricularia,* Linn., some of our common dragon-flies, as *Libellula depressa,* Linn., the honey-bee, *Apis mellifera,* Linn., the cuckoo spittle insect, *Aphrophora spumaria,* Linn., and a long catalogue of others, to all of which Professor Heer had given new names, but which some entomologists may regard as mere varieties until some stronger reasons are adduced for coming to a contrary opinion.

Several of the insects above enumerated, like the common ladybird, are well known at present to have a very wide range over nearly the whole of the Old World, for example, without varying, and might therefore be expected to have been persistent throughout many successive changes of the earth's surface and climate. Yet we may fairly anticipate that even the most constant types will have undergone some modifications in passing from the Miocene to the Recent epoch, since in the former period the geography and climate of Europe, the height of the Alps, and the general fauna and flora were so different from what they now are. But the deviation may not exceed that which would generally be expressed by what is called a well-marked variety.

Before I pass on to another topic, it may be well to answer a question which may have occurred to the reader; how it happens that we remained so long ignorant of the vegetation and insects of the Upper Miocene period in Europe? The answer may be instructive to those who are in the habit of underrating the former richness of the organic world wherever they happen to have no evidence of its condition. A large part of the Upper Miocene insects and plants alluded to have been met with at Oeningen, near the Lake of Constance, in two or three spots embedded in thinly laminated marls, the entire thickness of which scarcely exceeds 3 or 4 feet, and in two quarries of very limited dimensions. The rare combination of causes which seems to have led to the faithful preservation of so many treasures of a perishable nature in so small an area, appear to have been the following: first, a river flowing into a lake; secondly, storms of wind, by which leaves and sometimes the boughs of trees were torn off and floated by the stream into the lake; thirdly, mephitic gases rising from the lake, by which insects flying over its surface were occasionally killed: and fourthly, a constant supply of carbonate of lime in solution from mineral springs, the calcareous matter when precipitated to the bottom mingling with fine mud and thus forming the fossiliferous marls.

Species of Insects in Britain and North America, represented by distinct Varieties

If we compare the living British insects with those of the American continent, we frequently find that even those species which are considered to be identical, are nevertheless varieties of the European types. I have noticed this fact when speaking of the common English butterfly, *Vanessa atalanta*, or " red admiral," which I saw flying about the woods of Alabama in mid-winter. I was unable to detect any difference myself, but all the American specimens which I took to the British

Museum were observed by Mr. Doubleday to exhibit a slight peculiarity in the colouring of a minute part of the anterior wing,[1] a character first detected by Mr. T. F. Stephens, who has also discovered that similar slight, but equally constant variations, distinguish other Lepidoptera now inhabiting the opposite sides of the Atlantic, insects which, nevertheless, he and Mr. Westwood and the late Mr. Kirby, have always agreed to regard as mere varieties of the same species.

Mr. T. V. Wollaston, in treating of the variation of insects in maritime situations and small islands, has shown how the colour, growth of the wings, and many other characters, undergo modification under the influence of local conditions, continued for long periods of time;[2] and Mr. Brown has lately called our attention to the fact that the insects of the Shetland Isles present slight deviations from the corresponding types occurring in Great Britain, but far less marked than those which distinguish the American from the European varieties.[3] In the case of Shetland, Mr. Brown remarks, a land communication may well be supposed to have prevailed with Scotland at a more modern era than that between Europe and America. In fact, we have seen that Shetland can hardly fail to have been united with Scotland after the commencement of the glacial period (*see* map, Fig. 41); whereas a communication between the north of Europe by Iceland and Greenland (which, as before stated, once enjoyed a genial climate) must have been anterior to the glacial epoch. A much larger isolation, and the impossibility of varieties formed in the two separated areas crossing with each other, would account, according to Mr. Darwin's theory, for the much wider divergence observed in the specific types of the two regions.

The reader will remember that at the commencement of the Glacial period there was scarcely any appreciable difference between the molluscous fauna and that now living. When therefore the events of the Glacial period, as described in the earlier part of this volume, are duly pondered on, and when we reflect that in the Upper Miocene period the living species of mollusca constitute only one-third of the whole fauna, we see clearly by how high a figure we must mulitiply the time in order to express the distance between the Miocene period and our own days.

[1] Lyell's *Second Visit to the United States*, vol. ii., p. 293.
[2] Wollaston, *On the Variation of Species*, etc., London, 1856.
[3] *Transactions of Northern Entomological Society*, 1862.

Species of Mammalia Recent and Fossil—Proboscidians

But it may perhaps be said that the mammalia afford more conspicuous examples than do the mollusca, insects, or plants of the wide gaps which separate species and genera, and that if in this higher class such a multitude of transitional forms had ever existed as would be required to unite the Tertiary and Recent species into one series or net-work of allied or transitional forms, they could not so entirely have escaped observation whether in the fossil or living fauna. A zoologist who entertains such an opinion would do well to devote himself to the study of some one genus of mammalia, such as the elephant, rhinoceros, hippopotamus, bear, horse, ox, or deer; and after collecting all the materials he can get together respecting the extinct and Recent species, decide for himself whether the present state of science justifies his assuming that the chain could never have been continuous, the number of the missing links being so great.

Among the extinct species formerly contemporary with man, no fossil quadruped has so often been alluded to in this work as the mammoth, *Elephas primigenius*. From a monograph on the proboscidians by Dr. Falconer, it appears that this species represents one extreme of a type of which the Pliocene *Mastodon borsoni* represents the other. Between these extremes there are already enumerated by Dr. Falconer no less than twenty-six species, some of them ranging as far back in time as the Miocene period, others still living, like the Indian and African forms. Two of these species, however, he has always considered as doubtful, *Stegodon ganesa*, probably a mere variety of one of the others, and *Elephas priscus* of Goldfuss, founded partly on specimens of the African elephant, assumed by mistake to be fossil, and partly on some aberrant forms of *E. antiquus*.

The first effect of the intercalation of so many intermediate forms between the two most divergent types, has been to break down almost entirely the generic distinction between *Mastodon* and *Elephas*. Dr. Falconer, indeed, observes that *Stegodon* (one of several subgenera which he has founded) constitutes an intermediate group, from which the other species diverge through their dental characters, on the one side into the mastodons, and on the other into the Elephants.[1] The next result

[1] *Quart. Jour. Geol. Soc.*, vol. xiii., 1857, p. 314.

is to diminish the distance between the several members of each of these groups.

Dr. Falconer has discovered that no less than four species of elephant were formerly confounded together under the title of *Elephas primigenius,* whence its supposed ubiquity in Pleistocene times, or its wide range over half the habitable globe. But even when this form has been thus restricted in its specific characters, it has still its geographical varieties; for the mammoth's teeth brought from America may in most instances, according to Dr. Falconer, be distinguished from those proper to Europe. On this American variety Dr. Leidy has conferred the name of *E. americanus.* Another race of the same mammoth (as determined by Dr. Falconer) existed, as we have seen, before the Glacial period, or at the time when the buried forest of Cromer and the Norfolk cliffs was deposited; and the Swiss geologists have lately found remains of the mammoth in their country, both in pre-glacial and post-glacial formations.

Since the publication of Dr. Falconer's monograph, two other species of elephant, *E. mirificus,* Leidy, and *E. imperator,* have been obtained from the Pliocene formations of the Niobrara Valley in Nebraska, one of which, however, may possibly be found hereafter to be the same as *E. columbi,* Falc. A remarkable dwarf species also (*Elephas melitensis*) has been discovered, belonging, like the existing *E. africanus,* to the group *Loxodon.* This species has been established by Dr. Falconer on remains found by Captain Spratt, R.N., in a cave in Malta.[1]

How much the difficulty of discriminating between the fossil representatives of this genus may hereafter augment, when all the species with their respective geographical varieties are known, may be inferred from the following fact:—Professor H. Schlegel, in a recently published memoir, endeavours to show that the living elephant of Sumatra agrees with that of Ceylon, but is a distinct species from that of Continental India, being distinguishable by the number of its dorsal vertebræ and ribs, the form of its teeth, and other characteristics.[2] Dr. Falconer, on the other hand, considers these two living species as mere geographical varieties, the characters referred to not being constant, as he has ascertained, on comparing different individuals of *E. indicus* in different parts of Bengal in which the ribs vary from nineteen to twenty, and different varieties of *E. africanus* in which they vary from twenty to twenty-one.

[1] *Proceedings of the Geological Society,* London, 1862.
[2] Schlegel, *Natural History Review,* No. 5, 1862, p. 72.

An inquiry into the various species of the genus *Rhinoceros*, recent and fossil, has led Dr. Falconer to analogous results, as might be inferred from what was said in Chapter X., and as a forthcoming memoir by the same writer will soon more fully demonstrate.

Among the fossils brought in 1858 by Mr. Hayden from the Niobrara Valley, Dr. Leidy describes a rhinoceros so like the Asiatic species, *R. indicus*, that he at first referred it to the same, and, what is most singular, he remarks generally of the Pliocene fauna of that part of North America that it is far more related in character to the Pleistocene and Recent fauna of Europe than to that now inhabiting the American continent.

It seems indeed more and more evident that when we speculate in future on the pedigree of any extinct quadruped which abounds in the drift or caverns of Europe, we shall have to look to North and South America as a principal source of information. Thirty years ago, if we had been searching for fossil types which might fill up a gap between two species or genera of the horse tribe (or great family of the Solipedes), we might have thought it sufficient to have got together as ample materials as we could obtain from the continents of Europe, Africa, and Asia. We might have presumed that as no living representative of the equine family, whether horse, ass, zebra, or quagga, had been furnished by North or South America when those regions were first explored by Europeans, a search in the transatlantic world for fossil species might be dispensed with. But how different is the prospect now opening before us! Mr. Darwin first detected the remains of a fossil horse during his visit to South America, since which two other species have been met with on the same continent, while in North America, in the valley of the Nebraska alone, Mr. Hayden, besides a species not distinguishable from the domestic horse, has obtained, according to Dr. Leidy, representatives of five other fossil genera of Solipedes. These he names, *Hipparion, Protohippus, Merychippus, Hypohippus,* and *Parahippus.* On the whole, no less than twelve equine species, belonging to seven genera (including the Miocene *Anchitherium* of Nebraska), being already detected in the Tertiary and Post-Tertiary formations of the United States.[1]

Professors Unger[2] and Heer[3] have advocated, on botanical

---

[1] *Proceedings of Academy of Natural Science,* Philadelphia, for 1858, p. 89
[2] *Die versunkene Insel Atlantis.*
[3] *Flora tertiaria Helvetiæ.*

grounds, the former existence of an Atlantic continent during some part of the Tertiary period, as affording the only plausible explanation that can be imagined, of the analogy between the Miocene flora of Central Europe and the existing flora of Eastern America. Professor Oliver, on the other hand, after showing how many of the American types found fossil in Europe are common to Japan, inclines to the theory, first advanced by Dr. Asa Gray, that the migration of species, to which the community of types in the eastern states of North America and the Miocene flora of Europe is due, took place when there was an overland communication from America to eastern Asia between the fiftieth and sixtieth parallels of latitude, or south of Behring Straits, following the direction of the Aleutian islands.[1] By this course they may have made their way, at any epoch, Miocene, Pliocene, or Pleistocene, antecedently to the glacial epoch, to Mongolia, on the east coast of northern Asia.

We have already seen that a large proportion of the living quadrupeds of Mongolia (34 out of 48) are specifically identical with those at present inhabiting the continent of Western Europe and the British Isles.

A monograph on the hippopotamus, bear, ox, stag, or any other genus of mammalia common in the European drift or caverns, might equally well illustrate the defective state of the materials at present at our command. We are rarely in possession of one perfect skeleton of any extinct species, still less of skeletons of both sexes, and of different ages. We usually know nothing of the geographical varieties of the Pleistocene and Pliocene species, least of all, those successive changes of form which they must have undergone in the preglacial epoch between the Upper Miocene and Pleistocene eras. Such being the poverty of our palæontological data, we cannot wonder that osteologists are at variance as to whether certain remains found in caverns are of the same species as those now living; whether, for example, the *Talpa fossilis* is really the common mole, the *Meles morreni* the common badger, *Lutra antiqua* the otter of Europe, *Sciurus priscus* the squirrel, *Arctomys primigenia* the marmot, *Myoxus fossilis* the dormouse, Schmerling's *Felis engihoulensis* the European lynx, or whether *Ursus spelæus* and *Ursus priscus* are not extinct races of the living brown bear (*Ursus arctos*).

If at some future period all the above-mentioned species should be united with their allied congeners, it cannot fail to

[1] Oliver, Lecture at the Royal Institution, March 7, 1862.

enlarge our conception of the modifications which a species is capable of undergoing in the course of time, although the same form may appear absolutely immutable within the narrow range of our experience.

Longevity of Species in the Mammalia

In the *Principles of Geology*, in 1833,[1] I stated that the longevity of species in the class mollusca exceeded that in the mammalia. It has been since found that this generalisation can be carried much farther, and that in fact the law which governs the changes in organic being is such that the lower their place in a graduated scale, or the simpler their structure, the more persistent are they in form and organisation. I soon became aware of the force of this rule in the class mollusca, when I first attempted to calculate the numerical proportion of Recent species in the Newer Pliocene formations as compared to the Older Pliocene, and of them again as contrasted with the Miocene; for it appeared invariably that a greater number of the lamellibranchs could be identified with living species than of the gasteropods, and of these last a greater number in the lower division, that of entire-mouthed univalves, than in that of the siphonated. In whatever manner the changes have been brought about, whether by variation and natural selection, or by any other causes, the rate of change has been greater where the grade of organisation is higher.

It is only, therefore, where there is a full representation of all the principal orders of mollusca, or when we compare those of corresponding grade, that we can fully rely on the percentage test, or on the proportion of Recent to extinct species as indicating the relation of two groups to the existing fauna.

The foraminifera which exemplify the lowest stage of animal existence exhibit, as we learn from the researches of Dr. Carpenter and of Messrs. Jones and Parker, extreme variability in their specific forms, and yet these same forms are persistent throughout vast periods of time, exceeding, in that respect, even the brachiopods before mentioned.

Dr. Hooker observes, in regard to plants of complex floral structure, that they manifest their physical superiority in a greater extent of variation and in thus better securing a succession of race, an attribute which in some senses he regards

[1] 1st ed., vol. iii., pp. 48 and 140.

as of a higher order than that indicated by mere complexity or specialisation of organ.[1]

As one of the consequences of this law, he says that species, genera, and orders are, on the whole, best limited in plants of higher grade, the dicotyledons better than the monocotyledons, and the *Dichlamydeæ* better than the *Achlamydeæ*.

Mr. Darwin remarks, " We can, perhaps, understand the apparently quicker rate of change in terrestrial, and in more highly organised productions, compared with marine and lower productions, by the more complex relations of the higher beings to their organic and inorganic conditions of life." [2]

If we suppose the mammalia to be more sensitive than are the inferior classes of the vertebrata, to every fluctuation in the surrounding conditions, whether of the animate or inanimate world, it would follow that they would oftener be. called upon to adapt themselves by variation to new conditions, or if unable to do so, to give place to other types. This would give rise to more frequent extinction of varieties, species, and genera, whereby the surviving types would be better limited, and the average duration of the same unaltered specific types would be lessened.

Absence of Mammalia in Islands considered in reference to Transmutation

But if mammalia vary upon the whole at a more rapid rate than animals lower in the scale of being, it must not be supposed that they can alter their habits and structures readily, or that they are convertible in short periods into new species. The extreme slowness with which such changes of habits and organisation take place, when new conditions arise, appears to be well exemplified by the absence even of small warm-blooded quadrupeds in islands far from continents, however well such islands may be fitted by their dimensions to support them.

Mr. Darwin has pointed to this absence of mammalia as favouring his views, observing that bats, which are the only exceptions to the rule, might have made their way to distant islands by flight, for they are often met with on the wing far out at sea. Unquestionably, the total exclusion of quadrupeds in general, which could only reach such isolated habita-

---

[1] *Introductory Essay to the Flora of Australia*, p. vii.
[2] *Origin of Species*, 3rd ed., p. 340.

tions by swimming, seems to imply that nature does not dispense with the ordinary laws of reproduction when she peoples the earth with new forms; for if causes purely immaterial were alone at work, we might naturally look for squirrels, rabbits, polecats, and other small vegetable feeders and beasts of prey, as often as for bats, in the spots alluded to.

On the other hand, I have found it difficult to reconcile the antiquity of certain islands, such as those of the Madeiran Archipelago, and those of still larger size in the Canaries, with the total absence of small indigenous quadrupeds, for, judging by ancient deposits of littoral shells, now raised high above the level of the sea, several of these volcanic islands (Porto Santo and the Grand Canary among others) must have existed ever since the Upper Miocene period. But, waiving all such claims to antiquity, it is at least certain that since the close of the Newer Pliocene period, Madeira, and Porto Santo have constituted two separate islands, each in sight of the other, and each inhabited by an assemblage of land shells (*Helix, Pupa, Clausilia,* etc.), for the most part different or proper to each island. About thirty-two fossil species have been obtained in Madeira, and forty-two in Porto Santo, only five of the whole being common to both islands. In each the living land-shells are equally distinct, and correspond, for the most part, with the species found fossil in each island respectively.

Among the fossil species, one or two appear to be entirely extinct, and a larger number have disappeared from the fauna of the Madeiran Archipelago, though still extant in Africa and Europe. Many which were amongst the most common in the Pliocene period, have now become the scarcest, and others formerly scarce, are now most numerously represented. The variety-making force has been at work with such energy—perhaps we ought to say, has had so much time for its development—that almost every isolated rock within gun-shot of the shores has its peculiar living forms, or those very marked races to which Mr. Lowe, in his excellent description of the fauna, has given the name of "sub-species."

Since the fossil shells were embedded in sand near the coast, these volcanic islands have undergone considerable alterations in size and shape by the wasting action of the waves of the Atlantic beating incessantly against the cliffs, so that the evidence of a vast lapse of time is derivable from inorganic as well as from organic phenomena.

During this period no mammalia, not even of small species,

excepting bats, have made their appearance, whether in Madeira and Porto-Santo or in the larger and more numerous islands of the Canarian group. It might have been expected, from some expressions met with here and there in the *Origin of Species*, though not perhaps from a fair interpretation of the whole tenor of the author's reasoning, that this dearth of the highest class of vertebrata is inconsistent with the powers of mammalia to accommodate their habits and structures to new conditions. Why did not some of the bats, for example, after they had greatly multiplied, and were hard pressed by a scarcity of insects on the wing, betake themselves to the ground in search of prey, and, gradually losing their wings, become transformed into non-volant Insectivora? Mr. Darwin tells me that he has learnt that there is a bat in India which has been known occasionally to devour frogs. One might also be tempted to ask, how it has happened that the seals which swarmed on the shores of Madeira and the Canaries, before the European colonists arrived there, were never induced, when food was scarce in the sea, to venture inland from the shores, and begin in Teneriffe, and the Grand Canary especially, and other large islands, to acquire terrestrial habits, venturing first a few yards inland, and then farther and farther until they began to occupy some of the " places left vacant in the economy of nature." During these excursions, we might suppose some varieties, which had the skin of the webbed intervals of their toes less developed, to succeed best in walking on the land, and in the course of several generations they might exchange their present gait or manner of shuffling along and jumping by aid of the tail and their fin-like extremities, for feet better adapted for running.

It is said that one of the bats in the island of Palma (one of the Canaries) is of a peculiar species, and that some of the *Cheiroptera* of the Pacific islands are even of peculiar genera. If so, we seem, on organic as well as on geological grounds, to be precluded from arguing that there has not been time for great divergence of character. We seem also entitled to ask why the bats and rodents of Australia, which are spread so widely among the marsupials over that continent, have never, under the influence of the principle of progression, been developed into higher placental types, since we have now ascertained that that continent was by no means unfitted to sustain such mammalia, for these when once introduced by Man have run wild and become naturalised in many parts. The following answers may perhaps be offered to the above criticisms of some of Mr. Darwin's theoretical views.

First, as to the bats and seals: they are what zoologists call aberrant and highly specialised types, and therefore precisely those which might be expected to display a fixity and want of pliancy in their organisation, or the smallest possible aptitude for deviating in new directions towards new structures, and the acquisition of such altered habits as a change from aquatic to terrestrial or from volant to non-volant modes of living would imply.

Secondly, the same powers of flight which enabled the first bats to reach Madeira or the Canaries, would bring others from time to time from the African continent, which, mixing with the first emigrants and crossing with them, would check the formation of new races, or keep them true to the old types, as is found to be actually the case with the birds of Madeira and the Bermudas.

This would happen the more surely, if, as Mr. Darwin has endeavoured to prove, the offspring of races slightly varying are usually more vigorous than the progeny of parents of the same race, and would be more prolific, therefore, than the insular stock which had been for a long time breeding in and in.

The same cause would tend in a still more decided manner to prevent the seals from diverging into new races or " incipient species," because they range freely over the wide ocean, and, may therefore have continual intercourse with all other individuals of their species.

Thirdly, as to peculiar species, and even genera of bats in islands, we are perhaps too little acquainted at present with all the species and genera of the neighbouring continents to be able to affirm, with any degree of confidence, that the forms supposed to be peculiar do not exist elsewhere: those of the Canaries in Africa, for example. But what is still more important, we must bear in mind how many species and genera of Pleistocene mammalia have everywhere become extinct by causes independent of Man. It is always possible, therefore, that some types of Cheiroptera, originally derived from the main land, have survived in islands, although they have gradually died out on the continents from whence they came; so that it would be rash to infer that there has been time for the creation, whether by variation or other agency, of new species or genera in the islands in question.

As to the Rodents and Cheiroptera of Australia, we are as

yet too ignorant of the Pleistocene and Pliocene fauna of that part of the world, to be able to decide whether the introduction of such forms dates from a remote geological time. We know, however, that, before the Recent period, that continent was peopled with large kangaroos, and other herbivorous and carnivorous marsupials, of species long since extinct, their remains having been discovered in ossiferous caverns. The preoccupancy of the country by such indigenous tribes may have checked the development of the placental Rodents and Cheiroptera, even were we to concede the possibility of such forms being convertible by variation and progressive development into higher grades of mammalia.

Imperfection of the Geological Record [Note 42]

When treating in the eighth chapter of the dearth of human bones in alluvium containing flint implements in abundance, I pointed out that it is not part of the plan of Nature to write everywhere, and at all times, her autobiographical memoirs. On the contrary, her annals are local and exceptional from the first, and portions of them are afterwards ground into mud, sand, and pebbles, to furnish materials for new strata. Even of those ancient monuments now forming the crust of the earth, which have not been destroyed by rivers and the waves of the sea, or which have escaped being melted by volcanic heat, three-fourths lie submerged beneath the ocean, and are inaccessible to Man; while of those which form the dry land, a great part are hidden for ever from our observation by mountain masses, thousands of feet thick, piled over them.

Mr. Darwin has truly said that the fossiliferous rocks known to geologists consist, for the most part, of such as were formed when the bottom of the sea was subsiding. This downward movement protects the new deposits from denudation, and allows them to accumulate to a great thickness; whereas sedimentary matter, thrown down where the sea-bottom is rising, must almost invariably be swept away by the waves as fast as the land emerges.

When we reflect, therefore, on the fractional state of the annals which are handed down to us, and how little even these have as yet been studied, we may wonder that so many geologists should attribute every break in the series of strata and every gap in the past history of the organic world to catastrophes and convulsions of the earth's crust or to leaps made by the creational force from species to species, or from class to class. For it is clear that, even had the series of monuments been perfect and continuous at first (an hypothesis quite opposed to the analogy of the working of causes now in action), it could not fail to present itself to our eyes in a broken and disconnected state.

Those geologists who have watched the progress of discovery during the last half century can best appreciate the extent to which we may still hope by future exertion to fill up some of the wider chasms which now interrupt the regular sequence of fossiliferous rocks. The determination, for example, of late years of the true place of the Hallstadt and St. Cassian beds on the N. and S. flanks of the Austrian Alps, has revealed to us, for the first time, the marine fauna of a period (that of the Upper Trias) of which, until lately, but little was known. In this case, the palæontologist is called upon suddenly to inter-calcate about 800 species of Mollusca and Radiata, between the fauna of the Lower Lias and that of the Middle Trias. The period in question was previously believed, even by many a philosophical geologist, to have been comparatively barren of organic types. In England, France, and northern Germany, the only known strata of Upper Triassic date had consisted almost entirely of fresh or brackish-water beds, in which the bones of terrestrial and amphibious reptiles were the most characteristic fossils. The new fauna was, as might have been expected, in part peculiar, not a few of the species of Mollusca being referable to new genera; while some species were common to the older, and some to the newer rocks. On the whole, the new forms have helped greatly to lessen the discordance, not only between the Lias and Trias, but also generally between Palæozoic and Mesozoic formations. Thus the genus *Orthoceras* has been for the first time recognised in a Mesozoic deposit, and with it we find associated, for the first time, large Ammonites with foliated lobes, a form never seen before below the Lias; also the Ceratites, a family of Cephalopods never before met with in the Upper Trias, and never before in the same stratum with such lobed Ammonites.

We can now no longer doubt that should we hereafter have an opportunity of studying an equally rich marine fauna of the age of the Lower Trias (or Bunter Sandstein), the marked hiatus which still separates the Triassic and Permian eras would almost disappear.

Archæopteryx macrurus, Owen.—I could readily add a copious list of minor deposits, belonging to the Primary, Secondary and Tertiary series, which we have been called upon in like manner to intercalate in the course of the last quarter of a century into the chronological series previously known; but it would lead me into too long a digression. I shall therefore content myself with pointing out that it is not simply new formations which are brought to light from year to year, reminding us of the elementary state of our knowledge of palæontology, but new types also of structure are discovered in rocks whose fossil contents were supposed to be peculiarly well known.

.The last and most striking of these novelties is " the feathered fossil " from the lithographic stone of Solenhofen.

Until the year 1858, no well-determined skeleton of a bird had been detected in any rocks older than the Tertiary. In that year, Mr. Lucas Barrett found in the Cambridge Greensand of the Cretaceous series, the femur, tibia, and some other bones of a swimming bird, supposed by him to be of the gull tribe. His opinion as to the ornithic character of the remains was afterwards confirmed by Professor Owen.

The *Archæopteryx macrurus,* Owen, recently acquired by the British Museum, affords a second example of the discovery of the osseous remains of a bird in strata older than the Eocene. It was found in the great quarries of lithographic limestone at Solenhofen in Bavaria, the rock being a member of the Upper Oolite.

It was at first conjectured in Germany, before any experienced osteologist had had an opportunity of inspecting the original specimen, that this fossil might be a feathered Pterodactyl (flying reptiles having been often met with in the same stratum), or that it might at least supply some connecting links between a reptile and a bird. But Professor Owen, in a memoir lately read to the Royal Society (November 20, 1862), has shown that it is unequivocally a bird, and that such of its characters as are abnormal are by no means strikingly reptilian. The skeleton was lying on its back when embedded in calcareous sediment, so that the ventral part is exposed to view. It is about 1 foot 8 inches long, and 1 foot across, from the apex of the right to that of the left wing. The furculum, or merry-thought, which is entire, marks the fore part of the trunk; the ischium, scapula, and most of the wing and leg bones are preserved, and there are impressions of the quill feathers and of down on the body. The vanes and shafts of the feathers can be seen by the naked eye.

Fourteen long quill feathers diverge on each side of the metacarpal and phalangial bones, and decrease in length from 6 inches to 1 inch. The wings have a general resemblance to those of gallinaceous birds. The tarso-metatarsal, or drumstick, exhibits at its distal end a trifid articular surface supporting three toes, as in birds. The furculum, pelvis, and bones of the tail are in their natural position. The tail consists of twenty vertebræ, each of which supports a pair of plumes. The length of the tail with its feathers is 11½ inches, and its breadth 3½. It is obtusely truncated at the end. In all living birds the tail-feathers are arranged in fan-shaped order and attached to a coccygean bone, consisting of several vertebræ united together, whereas in the embryo state these same vertebræ are distinct. The greatest number is seen in the ostrich, which has eighteen caudal vertebræ in the fœtal state, which are reduced to nine in the adult bird, many of them having been anchylosed together. Professor Owen therefore considers the tail of the *Archæopteryx* as exemplifying the persistency of what is now an embryonic character. The tail, he remarks, is essentially a variable organ; there are long-tailed bats and short-tailed bats, long-tailed rodents and short-tailed rodents, long-tailed pterodactyls and short-tailed pterodactyls.

The *Archæopteryx* differs from all known birds, not only in the structure of its tail, but in having two, if not three, digits in the hand; but there is no trace of the fifth digit of the winged reptile.

The conditions under which the skeleton occurs are such, says Professor Owen, as to remind us of the carcass of a gull which has been a prey to some Carnivore, which had removed all the soft parts, and perhaps the head, nothing being left but the bony legs and the indigestible quill-feathers. But since Professor Owen's paper was read, Mr. John Evans, whom I have often had occasion to mention in the earlier chapters of this work, seems to have found what may indicate a part of the missing cranium. He has called our attention to a smooth protuberance on the otherwise even surface of the slab of limestone which seems to be the cast of the brain or interior of the skull. Some part even of the cranial bone itself appears to be still buried in the matrix. Mr. Evans has pointed out the resemblance of this cast to one taken by himself from the cranium of a crow, and still more to that of a jay, observing that in the fossil the median line which separates the two hemispheres of the brain is visible.

To conclude, we may learn from this valuable relic how rashly the existence of Birds at the epoch of the Secondary rocks has been questioned, simply on negative evidence, and secondly, how many new forms may be expected to be brought to light in strata with which we are already best acquainted, to say nothing of the new formations which geologists are continually discovering.

CHAPTER XXIII

ORIGIN AND DEVELOPMENT OF LANGUAGES AND SPECIES COMPARED [Note 43]

Aryan Hypothesis and Controversy—The Races of Mankind change more slowly than their Languages—Theory of the gradual Origin of Languages—Difficulty of defining what is meant by a Language as distinct from a Dialect—Great Number of extinct and living Tongues —No European Language a Thousand Years old—Gaps between Languages, how caused—Imperfection of the Record—Changes always in Progress—Struggle for Existence between Rival Terms and Dialects—Causes of Selection—Each Language formed slowly in a single Geographical Area—May die out gradually or suddenly— Once lost can never be revived—Mode of Origin of Languages and Species a Mystery—Speculations as to the Number of original Languages or Species unprofitable.

THE supposed existence, at a remote and unknown period, of a language conventionally called the Aryan, has of late years been a favourite subject of speculation among German philologists, and Professor Max Müller has given us lately the most improved version of this theory, and has set forth the various facts and arguments by which it may be defended, with his usual perspicuity and eloquence. He observes that if we know nothing of the existence of Latin—if all historical documents previous to the fifteenth century had been lost—if tradition even was silent as to the former existence of a Roman empire, a mere comparison of the Italian, Spanish, Portuguese, French, Wallachian, and Rhætian dialects would enable us to say that at some time there must have been a language from which these six modern dialects derive their origin in common. Without this supposition it would be impossible to account for their structure and composition, as, for example, for the forms of the auxiliary verb " to be," all evidently varieties of one common type, while it is equally clear that no one of the six affords the original form from which the others could have been borrowed. So also in none of the six languages do we find the elements of which these verbal and other forms could have been composed; they must have been handed down as relics from a former period, they must have existed in some antecedent language, which we know to have been the Latin.

But, in like manner, he goes on to show, that Latin itself,

as well as Greek, Sanscrit, Zend (or Bactrian), Lithuanian, old Sclavonic, Gothic, and Armenian are also eight varieties of one common and more ancient type, and no one of them could have been the original from which the others were borrowed. They have all such an amount of mutual resemblance as to point to a more ancient language, the Aryan, which was to them what Latin was to the six Romance languages. The people who spoke this unknown parent speech, of which so many other ancient tongues were off-shoots, must have migrated at a remote era to widely separated regions of the old world, such as Northern Asia, Europe, and India south of the Himalaya.[1]

The soundness of some parts of this Aryan hypothesis has lately been called in question by Mr. Crawfurd, on the ground that the Hindoos, Persians, Turks, Scandinavians, and other people referred to as having derived not only words but grammatical forms from an Aryan source, belong each of them to a distinct race, and all these races have, it is said, preserved their peculiar characters unaltered from the earliest dawn of history and tradition. If, therefore, no appreciable change has occurred in three or four thousand years, we should be obliged to assume a far more remote date for the first branching off of such races from a common stock than the supposed period of the Aryan migrations, and the dispersion of that language over many and distant countries.

But Mr. Crawfurd has, I think, himself helped us to remove this stumbling-block, by admitting that a nation speaking a language allied to the Sanscrit (the oldest of the eight tongues alluded to), once probably inhabited that region situated to the north-west of India, which within the period of authentic history has poured out its conquering hordes over a great extent of Western Asia and Eastern Europe. The same people, he says, may have acted the same part in the long, dark night which preceded the dawn of tradition.[2] These conquerors may have been few in number when compared to the populations which they subdued. In such cases the new settlers, although reckoned by tens of thousands, might merge in a few centuries into the millions of subjects which they ruled. It is an acknowledged fact that the colour and features of the Negro or European are entirely lost in the fourth generation, provided that no fresh infusion of one or other of the two races takes place. The distinctive physical features, therefore, of the Aryan con-

---

[1] Max Müller, *Comparative Mythology*. Oxford Essays, 1856.
[2] Crawfurd, *Transactions of the Ethnological Society*, vol. i. 1861.

querors might soon wear out and be lost in those of the nations they overran; yet many of the words, and, what is more in point, some of the grammatical forms of their language, might be retained by the masses which they had governed for centuries, these masses continuing to preserve the same features of race which had distinguished them long before the Aryan invasions.

There can be no question that if we could trace back any set of cognate languages now existing to some common point of departure, they would converge and meet sooner in some era of thé past than would the existing races of mankind; in other words, races change much more slowly than languages. But, according to the doctrine of transmutation, to form a new species would take an incomparably longer period than to form a new race. No language seems ever to last for a thousand years, whereas many a species seems to have endured for hundreds of thousands. A philologist, therefore, who is contending that all living languages are derivative and not primordial, has a great advantage over a naturalist who is endeavouring to inculcate a similar theory in regard to species.

It may not be uninstructive, in order fairly to appreciate the vast difficulty of the task of those who advocate transmutation in natural history, to consider how hard it would be even for a philologist to succeed, if he should try to convince an assemblage of intelligent but illiterate persons that the language spoken by them, and all those talked by contemporary nations, were modern inventions, moreover that these same forms of speech were still constantly undergoing change, and none of them destined to last for ever.

We will suppose him to begin by stating his conviction, that the living languages have been gradually derived from others now extinct, and spoken by nations which had immediately preceded them in the order of time, and that those again had used forms of speech derived from still older ones. They might naturally exclaim, " How strange it is that you should find records of a multitude of dead languages, that a part of the human economy which in our own time is so remarkable for its stability, should have been so inconstant in bygone ages! We all speak as our parents and grandparents spoke before us, and so, we are told, do the Germans and French. What evidence is there of such incessant variation in remoter times? and, if it be true, why not imagine that when one form of speech was lost, another was suddenly and supernaturally created by a gift of tongues or confusion of languages, as at the building of the Tower of Babel? Where are the memorials of all the inter-mediate dialects, which must have existed, if this doctrine of perpetual fluctuation be true? And how comes it that the tongues now spoken do not pass by insensible gradations the one into the other, and into the dead languages of dates im-mediately antecedent?

" Lastly, if this theory of indefinite modifiability be sound, what meaning can be attached to the term language, and what definition can be given of it so as to distinguish a language from a dialect? "

In reply to this last question, the philologist might confess that the learned are not agreed as to what constitutes a language as distinct from a dialect. Some believe that there are 4000 living languages, others that there are 6000, so that the mode of defining them is clearly a mere matter of opinion. Some con-tend, for example, that the Danish, Norwegian, and Swedish form one Scandinavian tongue, others that they constitute three different languages, others that the Danish and Norwegian are one—mere dialects of the same language, but that Swedish is distinct.

The philologist, however, might fairly argue that this very ambiguity was greatly in favour of his doctrine, since if lan-guages had all been constantly undergoing transmutation, there ought often to be a want of real lines of demarcation between them. He might, however, propose that he and his pupils should come to an understanding that two languages should be regarded as distinct whenever the speakers of them are unable to converse together, or freely to exchange ideas, whether by word or writing. Scientifically speaking, such a test might be vague and unsatisfactory, like the test of species by their capability of producing fertile hybrids; but if the pupil is persuaded that there are such things in nature as distinct languages, whatever may have been their origin, the definition above suggested might be of practical use, and enable the teacher to proceed with his argument.

He might begin by undertaking to prove that none of the languages of modern Europe were a thousand years old. No English scholar, he might say, who has not specially given himself up to the study of Anglo-Saxon, can interpret the documents in which the chronicles and laws of England were written in the days of King Alfred, so that we may be sure that none of the English of the nineteenth century could converse with the subjects of that monarch if these last could now be restored to life. The difficulties encountered would not arise merely from the intrusion of French terms, in consequence of the Norman conquest, because that large portion of our language (including the articles, pronouns, etc.), which is Saxon has also undergone great transformations by abbreviation, new modes of pronunciation, spelling, and various corruptions, so as to be unlike both ancient and modern German. They who now speak German, if brought into contact with their Teutonic ancestors of the ninth century, would be quite unable to converse with them, and, in like manner, the subjects of Charlemagne could not have exchanged ideas with the Goths of Alaric's army, or with the soldiers of Arminius in the days of Augustus Cæsar. So rapid indeed has been the change in Germany, that the epic poem called the Nibelungen Lied, once so popular, and only seven centuries old, cannot now be enjoyed, except by the erudite.

If we then turn to France, we meet again with similar evidence of ceaseless change. There is a treaty of peace still extant a thousand years old, between Charles the Bald and King Louis of Germany (dated A.D. 841), in which the German king takes an oath in what was the French tongue of that day, while the French king swears in the German of the same era; and neither of these oaths would now convey a distinct meaning to any but the learned in these two countries. So also in Italy, the modern Italian cannot be traced back much beyond the time of Dante, or some six centuries before our time. Even in Rome, where there had been no permanent intrusion of foreigners, such as the Lombard settlers of German origin in the plains of the Po, the common people of the year 1000 spoke quite a distinct language from that of their Roman ancestors or their Italian descendants, as is shown by the celebrated chronicle of the monk Benedict, of the convent of St. Andrea on Mount Soracte, written in such barbarous Latin, and with such strange grammatical forms, that it requires a profoundly skilled linguist to decipher it.[1]

Having thus established the preliminary fact, that none of the tongues now spoken were in existence ten centuries ago, and that the ancient languages have passed through many a transitional dialect before they settled into the forms now in use, the philologist might bring forward proofs of the great numbers both of lost and living forms of speech.

Strabo tells us that in his time, in the Caucasus alone (a chain of mountains not longer than the Alps, and much narrower),

---

[1] See G. Pertz, Monumenta Germanica, vol. iii.

there were spoken at least seventy languages. At the present period the number, it is said, would be still greater if all the distinct dialects of those mountains were reckoned. Several of these Caucasian tongues admit of no comparison with any known living or lost Asiatic or European language. Others which are not peculiar are obsolete forms of known languages, such as the Georgian, Mongolian, Persian, Arabic, and Tartarian. It seems that as often as conquering hordes swept over that part of Asia, always coming from the north and east, they drove before them the inhabitants of the plains, who took refuge in some of the retired valleys and high mountain fastnesses, where they maintained their independence, as do the Circassians in our time in spite of the power of Russia.

In the Himalayan Mountains, from Assam to its extreme north-western limit, and generally in the more hilly parts of British India, the diversity of languages is surprisingly great, impeding the advance of civilisation and the labours of the missionary. In South America and Mexico, Alexander Humboldt reckoned the distinct tongues by hundreds, and those of Africa are said to be equally numerous. Even in China, some eighteen provincial dialects prevail, almost all deviating so much from others that the speakers are not mutually intelligible, and besides these there are other distinct forms of speech in the mountains of the same empire.

The philologist might next proceed to point out that the geographical relations of living and dead languages favour the hypothesis of the living ones having been derived from the extinct, in spite of our inability, in most instances, to adduce documentary evidence of the fact or to discover monuments of all the intermediate and transitional dialects which must have existed. Thus he would observe that the modern Romance languages are spoken exactly where the ancient Romans once lived or ruled, and the Greek of our days where the older classical Greek was formerly spoken. Exceptions to this rule might be detected, but they would be explicable by reference to colonisation and conquest.

As to the many and wide gaps sometimes encountered between the dead and living languages, we must remember that it is not part of the plan of any people to preserve memorials of their forms of speech expressly for the edification of posterity. Their MSS. and inscriptions serve some present purpose, are occasional and imperfect from the first, and are rendered more fragmentary in the course of time, some being intentionally destroyed, others lost by the decay of the perishable materials on which they are written; so that to question the theory of all known languages being derivative on the ground that we can rarely trace a passage from the ancient to the modern through all the dialects which must have flourished one after the other in the intermediate ages, implies a want of reflection on the laws which govern the recording as well as the obliterating processes.

But another important question still remains to be considered, namely, whether the trifling changes which can alone be witnessed by a single generation, can possibly represent the working of that machinery which, in the course of many centuries, has given rise to such mighty revolutions in the forms of speech throughout the world. Everyone may have noticed in his own lifetime the stealing in of some slight alterations of accent, pronunciation or spelling, or the introduction of some words borrowed from a foreign language to express ideas of which no native term precisely conveyed the import. He may also remember hearing for the first time some cant terms or slang phrases, which have since forced their way into common use, in spite of the efforts of the purist. But he may still contend that, " within the range of his experience," his language has continued unchanged, and he may believe in its immutability in spite of minor variations. The real question, however, at issue is, whether there are any limits to this variability. He will find on farther investigation, that new technical terms are coined almost daily in various arts, sciences, professions, and trades, that new names must be found for new inventions, that many of these acquire a metaphorical sense, and then make their way into general circulation, as " stereotyped," for instance, which would have been as meaningless to the men of the seventeenth century as would the new terms and images derived from steamboat and railway travelling to the men of the eighteenth.

If the numerous words, idioms, and phrases, many of them of ephemeral duration, which are thus invented by the young and old in various classes of society, in the nursery, the school, the camp, the fleet, the courts of law and the study of the man of science or literature, could all be collected together and put on record, their number in one or two centuries might compare with the entire permanent vocabulary of the language. It becomes, therefore, a curious subject of inquiry, what are the laws which govern not only the invention, but also the " selection " of some of these words or idioms, giving them currency in preference to others?—for as the powers of the human memory are limited, a check must be found to the endless increase and multiplication of terms, and old words must be dropped nearly as fast as new ones are put into circulation. Sometimes the new word or phrase, or a modification of the old ones, will entirely supplant the more ancient expressions, or, instead of the latter being discarded, both may flourish together, the older one having a more restricted use.

Although the speakers may be unconscious that any great fluctuation is going on in their language—although when we observe the manner in which new words and phrases are thrown out, as if at random or in sport, while others get into vogue, we may think the process of change to be the result of mere chance —there are nevertheless fixed laws in action, by which, in the general struggle for existence, some terms and dialects gain the victory over others. The slightest advantage attached to some new mode of pronouncing or spelling, from considerations of brevity or euphony, may turn the scale, or more powerful causes of selection may decide which of two or more rivals shall triumph and which succumb. Among these are fashion, or the influence of an aristocracy, whether of birth or education, popular writers, orators, preachers—a centralised government organising its schools expressly to promote uniformity of diction, and to get the better of provincialisms and local dialects. Between these dialects, which may be regarded as so many " incipient languages," the competition is always keenest when they are most nearly allied, and the extinction of any one of them destroys some of the links by which a dominant tongue may have been previously connected with some other widely distinct one. It is by the perpetual loss of such intermediate forms of speech that the great dissimilarity of the languages which survive is brought about. Thus, if Dutch should become a dead language, English and German would be separated by a wider gap.

Some languages which are spoken by millions, and spread over a wide area, will endure much longer than others which have never had a wide range, especially if the tendency to incessant change in one of these dominant tongues is arrested for a time by a standard literature. But even this source of stability is insecure, for popular writers themselves are great innovators, sometimes coining new words, and still oftener new expressions and idioms, to embody their own original conceptions and sentiments, or some peculiar modes of thought and feeling characteristic of their age. Even when a language is regarded with superstitious veneration as the vehicle of divine truths and

religious precepts, and which has prevailed for many generations, it will be incapable of permanently maintaining its ground.
Hebrew had ceased to be a living language before the Christian era.
Sanscrit, the sacred language of the Hindoos, shared the same
fate, in spite of the veneration in which the Vedas are still held,
and in spite of many a Sanscrit poem once popular and national.

The Christians of Constantinople and the Morea still hear
the New Testament and their liturgy read in ancient Greek,
while they speak a dialect in which Paul might have preached
in vain at Athens. So in the Catholic Church, the Italians pray
in one tongue and talk another. Luther's translation of the
Bible acted as a powerful cause of " selection," giving at once
to one of many competing dialects (that of Saxony) a prominent
and dominant position in Germany; but the style of Luther
has, like that of our English Bible, already become somewhat
antiquated.

If the doctrine of gradual transmutation be applicable to
languages, all those spoken in historical times must each of them
have had a closely allied prototype; and accordingly, whenever
we can thoroughly investigate their history, we find in them some
internal evidence of successive additions by the invention of new
words or the modification of old ones.   Proofs also of borrowing
are discernible, letters being retained in the spelling of some
words which have no longer any meaning as they are now pronounced—no connection with any corresponding sounds. Such
redundant or silent letters, once useful in the parent speech,
have been aptly compared by Mr. Darwin to rudimentary organs
in living beings, which, as he interprets them, have at some
former period been more fully developed, having had their proper
functions to perform in the organisation of a remote progenitor.

If all known languages are derivative and not primordial
creations, they must each of them have been slowly elaborated in
a single geographical area.   No one of them can have had two
birthplaces.   If one were carried by a colony to a distant region,
it would immediately begin to vary unless frequent intercourse
was kept up with the mother country.   The descendants of the
same stock, if perfectly isolated, would in five or six centuries,
perhaps sooner, be quite unable to converse with those who
remained at home, or with those who may have migrated to some
distant region, where they were shut out from all communication
with others speaking the same tongue.

A Norwegian colony which settled in Iceland in the ninth
century, maintained its independence for about 400 years,

during which time the old Gothic which they at first spoke became corrupted and considerably modified. In the meantime the natives of Norway, who had enjoyed much commercial intercourse with the rest of Europe, acquired quite a new speech, and looked on the Icelandic as having been stationary, and as representing the pure Gothic original of which their own was an offshoot.

A German colony in Pennsylvania was cut off from frequent communication with Europe for about a quarter of a century, during the wars of the French Revolution between 1792 and 1815. So marked had been the effect even of this brief and imperfect isolation, that when Prince Bernhard of Saxe-Weimar travelled among them a few years after the peace, he found the peasants speaking as they had done in Germany in the preceding century,[1] and retaining a dialect which at home had already become obsolete.

Even after the renewal of the German emigration from Europe, when I travelled in 1841 among the same people in the retired valleys of the Alleghanies, I found the newspapers full of terms half English and half German, and many an Anglo-Saxon word which had assumed a Teutonic dress, as "fencen," to fence, instead of umzäunen, "flauer" for flour, instead of mehl, and so on. What with the retention of terms no longer in use in the mother country, and the borrowing of new ones from neighbouring states, there might have arisen in Pennsylvania in five or six generations, but for the influx of newcomers from Germany, a mongrel speech equally unintelligible to the Anglo-Saxon and to the inhabitants of the European fatherland.

If languages resemble species in having had each their "specific centre" or single area of creation, in which they have been slowly formed, so each of them is alike liable to slow or to sudden extinction. They may die out very gradually in consequence of transmutation, or abruptly by the extermination of the last surviving representatives of the unaltered type. We know in what century the last Dodo perished, and we know that in the seventeenth century the language of the Red Indians of Massachusetts, into which Father Eliot had translated the Bible, and in which Christianity was preached for several generations, ceased to exist, the last individuals by whom it was spoken having at that period died without issue.[2] But if just before that event

---

[1] *Travels of Prince Bernhard of Saxe-Weimar, in North America, in 1825, and 1826*, p. 123.

[2] Lyell, *Travels in North America*, vol. i. p. 260. 1845.

the white man had retreated from the continent, or had been swept off by an epidemic, those Indians might soon have repeopled the wilderness, and their copious vocabulary and peculiar forms of expression might have lasted without important modification to this day. The extinction, however, of languages in general is not abrupt, any more than that of species. It will also be evident from what has been said, that a language which has once died out can never be revived, since the same assemblage of conditions can never be restored even among the descendants of the same stock, much less simultaneously among all the surrounding nations with whom they may be in contact.

We may compare the persistency of languages, or the tendency of each generation to adopt without change the vocabulary of its predecessor, to the force of inheritance in the organic world, which causes the offspring to resemble its parents. The inventive power which coins new words or modifies old ones, and adapts them to new wants and conditions as often as these arise, answers to the variety-making power in the animate creation.

Progressive improvement in language is a necessary consequence of the progress of the human mind from one generation to another. As civilisation advances, a greater number of terms are required to express abstract ideas, and words previously used in a vague sense, so long as the state of society was rude and barbarous, gradually acquire more precise and definite meanings, in consequence of which several terms must be employed to express ideas and things, which a single word had before signified, though somewhat loosely and imperfectly.

The farther this subdivision of function is carried, the more complete and perfect the language becomes, just as species of higher grade have special organs, such as eyes, lungs, and stomach, for seeing, breathing, and digesting, which in simpler organisms are all performed by one and the same part of the body.[1]

When we had satisfied ourselves that all the existing languages, instead of being primordial creations, or the direct gifts of a supernatural Power, have been slowly elaborated, partly by the modification of pre-existing dialects, partly by borrowing terms at successive periods from numerous foreign sources, and partly by new inventions made some of them deliberately, and some casually and as it were fortuitously—when we have discovered the principal causes of selection, which have guided the adoption or rejection of rival names for the same things and ideas,

[1] *See* Herbert Spencer's *Psychology* and *Scientific Essays.*

rival modes of pronouncing the same words and provincial dialects competing one with another—we are still very far from comprehending all the laws which have governed the formation of each language.

It was a profound saying of William Humboldt, that " Man is Man only by means of speech, but in order to invent speech he must be already Man." Other animals may be able to utter sounds more articulate and as varied as the click of the Bushman, but voice alone can never enable brute intelligence to acquire language.

When we consider the complexity of every form of speech spoken by a highly civilised nation, and discover that the grammatical rules and the inflections which denote number, time, and equality are usually the product of a rude state of society— that the savage and the sage, the peasant and man of letters, the child and the philosopher, have worked together, in the course of many generations, to build up a fabric which has been truly described as a wonderful instrument of thought, a machine, the several parts of which are so well adjusted to each other as to resemble the product of one period and of a single mind—we cannot but look upon the result as a profound mystery, and one of which the separate builders have been almost as unconscious as are the bees in a hive of the architectural skill and mathematical knowledge which is displayed in the construction of the honeycomb.

In our attempts to account for the origin of species, we find ourselves still sooner brought face to face with the working of a law of development of so high an order as to stand nearly in the same relation as the Deity himself to man's finite understanding, a law capable of adding new and powerful causes, such as the moral and intellectual faculties of the human race, to a system of nature which had gone on for millions of years without the intervention of any analogous cause. If we confound " Variation " or " Natural Selection " with such creational laws, we deify secondary causes or immeasurably exaggerate their influence.

Yet we ought by no means to undervalue the importance of the step which will have been made, should it hereafter become the generally received opinion of men of science (as I fully expect it will), that the past changes of the organic world have been brought about by the subordinate agency of such causes as " Variation " and " Natural Selection." All our advances in the knowledge of Nature have consisted of such steps as these, and we must not be discouraged because greater mysteries remain behind wholly inscrutable to us.

If the philologist is asked whether in the beginning of things there was one or five, or a greater number of languages, he may answer that, before he can reply to such a question, it must be decided whether the origin of Man was single, or whether there were many primordial races. But he may also observe, that if mankind began their career in a rude state of society, their whole vocabulary would be limited to a few words, and that if they then separated into several isolated communities, each of these would soon acquire an entirely distinct language, some roots being lost and others corrupted and transformed beyond the possibility of subsequent identification, so that it might be hopeless to expect to trace back the living and dead languages to one starting point, even if that point were of much more modern date than we have now good reason to suppose. In like manner it may be said of species, that if those first formed were of very simple structure, and they began to vary and to lose some organs by disuse and acquire new ones by development, they might soon differ as much as so many distinctly created primordial types. It would therefore be a waste of time to speculate on the number of original monads or germs from which all plants and animals were subsequently evolved, more especially as the oldest fossiliferous strata known to us may be the last of a long series of antecedent formations, which once contained organic remains. It was not till geologists ceased to discuss the condition of the original nucleus of the planet, whether it was solid or fluid, and whether it owed its fluidity to aqueous or igneous causes, that they began to achieve their great triumphs; and the vast progress which has recently been made in showing how the living species may be connected with the extinct by a common bond of descent, has been due to a more careful study of the actual state of the living world, and to those monuments of the past in which the relics of the animate creation of former ages are best preserved and least mutilated by the hand of time.

CHAPTER XXIV

BEARING OF THE DOCTRINE OF TRANSMUTATION ON THE ORIGIN
OF MAN, AND HIS PLACE IN THE CREATION

Whether Man can be regarded as an Exception to the Rule if the Doctrine
of Transmutation be embraced for the rest of the Animal Kingdom—
Zoological Relations of Man to other Mammalia—Systems of Classification—Term Quadrumanous, why deceptive—Whether the Structure of the Human Brain entitles Man to form a distinct Sub-class
of the Mammalia—Intelligence of the lower Animals compared to
the Intellect and Reason of Man—Grounds on which Man has been
referred to a distinct Kingdom of Nature—Immaterial Principle
common to Man and Animals—Non-discovery of intermediate Links
among Fossil Anthropomorphous Species—Hallam on the compound
Nature of Man, and his Place in the Creation—Great Inequality of
mental Endowment in different Human Races and Individuals
developed by Variation and ordinary Generation—How far a corresponding Divergence in physical Structure may result from the Working of the same Causes—Concluding Remarks.

SOME of the opponents of transmutation, who are well versed
in Natural History, admit that though that doctrine is unten-
able, it is not without its practical advantages as a " useful
working hypothesis," often suggesting good experiments and
observations and aiding us to retain in the memory a multitude
of facts respecting the geographical distribution of genera and
species, both of animals and plants, the succession in time of
organic remains, and many other phenomena which, but for
such a theory, would be wholly without a common bond of
relationship.

It is in fact conceded by many eminent zoologists and
botanists, as before explained, that whatever may be the nature
of the species-making power or law, its effects are of such a
character as to imitate the results which variation, guided by
natural selection, would produce, if only we could assume with
certainty that there are no limits to the variability of species.
But as the anti-transmutationists are persuaded that such
limits do exist, they regard the hypothesis as simply a pro-
visional one, and expect that it will one day be superseded by
another cognate theory, which will not require us to assume the
former continuousness of the links which have connected the
past and present states of the organic world, or the outgoing
with the incoming species.

In like manner, many of those who hesitate to give in their full adhesion to the doctrine of progression, the other twin branch of the development theory, and who even object to it, as frequently tending to retard the reception of new facts supposed to militate against opinions solely founded on negative evidence, are nevertheless agreed that on the whole it is of great service in guiding our speculations. Indeed it cannot be denied that a theory which establishes a connection between the absence of all relics of vertebrata in the oldest fossiliferous rocks, and the presence of man's remains in the newest, which affords a more than plausible explanation of the successive appearance in strata of intermediate age of the fish, reptile, bird, and mammal, has no ordinary claims to our favour as comprehending the largest number of positive and negative facts gathered from all parts of the globe, and extending over countless ages, that science has perhaps ever attempted to embrace in one grand generalisation.

But will not transmutation, if adopted, require us to include the human race in the same continuous series of developments, so that we must hold that Man himself has been derived by an unbroken line of descent from some one of the inferior animals? We certainly cannot escape from such a conclusion without abandoning many of the weightiest arguments which have been urged in support of variation and natural selection considered as the subordinate causes by which new types have been gradually introduced into the earth. Many of the gaps which separate the most nearly allied genera and orders of mammalia are, in a physical point of view, as wide as those which divide Man from the mammalia most nearly akin to him, and the extent of his isolation, whether we regard his whole nature or simply his corporeal attributes, must be considered before we can discuss the bearing of transmutation upon his origin and place in the creation.

Systems of Classification

In order to qualify ourselves to judge of the degree of affinity in physical organisation between Man and the lower animals, we cannot do better than study those systems of classification which have been proposed by the most eminent teachers of natural history. Of these an elaborate and faithful summary has recently been drawn up by the late Isidore Geoffroy St. Hilaire, which the reader will do well to consult.[1]

[1] *Histoire Naturelle Générale des Règnes organiques*, Paris, vol. ii., 1856.

He begins by passing in review numerous schemes of classification, each of them having some merit, and most of them having been invented with a view of assigning to Man a separate place in the system of Nature, as, for example, by dividing animals into rational and irrational, or the whole organic world into three kingdoms, the human, the animal, and the vegetable—an arrangement defended on the ground that Man is raised as much by his intelligence above the animals as are these by their sensibility above plants. Admitting that these schemes are not unphilosophical, as duly recognising the double nature of Man (his moral and intellectual, as well as his physical attributes), Isidore G. St. Hilaire observes that little knowledge has been imparted by them. We have gained, he says, much more from those masters of the science who have not attempted any compromise between two distinct orders of ideas, the physical and psychological, and who have confined their attention strictly to Man's physical relation to the lower animals.

Linnæus led the way in this field of inquiry by comparing Man and the apes, in the same manner as he compared these last with the carnivores, ruminants, rodents, or any other division of warm-blooded quadrupeds. After several modifications of his original scheme, he ended by placing Man as one of the many genera in his order Primates, which embraced not only the apes and lemurs, but the bats also, as he found these last to be nearly allied to some of the lowest forms of the monkeys. But all modern naturalists, who retain the order Primates, agree to exclude from it the bats or *Cheiroptera;* and most of them class Man as one of several families of the order Primates. In this, as in most systems of classification, the families of modern zoologists and botanists correspond with the genera of Linnæus.

Blumenbach, in 1779, proposed to deviate from this course, and to separate Man from the apes as an order apart, under the name of Bimana, or two-handed. In making this innovation he seems at first to have felt that it could not be justified without calling in psychological considerations to his aid, to strengthen those which were purely anatomical; for, in the earliest edition of his *Manual of Natural History*, he defined Man to be " animal rationale, loquens, erectum, bimanum," whereas in later editions he restricted himself entirely to the two last characters, namely, the erect position and the two hands, or " animal erectum, bimanum."

The terms " bimanous " and " quadrumanous " had been

already employed by Buffon in 1766, but not applied in a strict zoological classification till so used by Blumenbach. Twelve years later, Cuvier adopted the same order Bimana for the human family, while the apes, monkeys, and lemurs constituted a separate order called Quadrumana.

Respecting this last innovation, Isidore G. St. Hilaire asks, " How could such a division stand, repudiated as it was by the anthropologists in the name of the moral and intellectual supremacy of Man; and by the zoologists, on the ground of its incompatibility with natural affinities and with the true principles of classification? Separated as a group of ordinal value, placed at the same distance from the ape as the latter from the carnivore, Man is at once too near and too distant from the higher mammalia;—too near if we take into account those elevated faculties, which, raising Man above all other organised beings, accord to him not only the first, but a separate place in the creation—too far if we merely consider the organic affinities which unite him with the quadrumana; with the apes especially, which, in a purely physical point of view, approach Man more nearly than they do the lemurs.

" What, then, is this order of Bimana of Blumenbach and Cuvier? An impracticable compromise between two opposite and irreconcilable systems—between two orders of ideas which are clearly expressed in the language of natural history by these two words: the human *kingdom* and the human *family*. It is one of those would-be *via media* propositions which, once seen through, satisfy no one, precisely because they are intended to please everybody; half-truths, perhaps, but also half-falsehoods; for what, in science, is a half-truth but an error? "

Isidore G. St. Hiliare then proceeds to show how, in spite of the great authority of Blumenbach and Cuvier, a large proportion of modern zoologists of note have rejected the order Bimana, and have regarded Man simply as a family of one and the same order, Primates.

Term " Quadrumanous," why deceptive

Even the term " Quadrumanous " has lately been shown by Professor Huxley, in a lecture delivered by him in the spring of 1860-61, which I had the good fortune to hear, to have proved a fertile source of popular delusion, conveying ideas which the great anatomists Blumenbach and Cuvier never entertained themselves, namely, that in the so-called Quadrumana the extremities of the hind-limbs bear a real resemblance to the human hands, instead of corresponding anatomically with the human feet.

As this subject bears very directly on the question, how far Man is entitled, in a purely zoological classification, to rank as an order apart, I shall proceed to cite, in an abridged form, the words of the lecturer above alluded to.[1]

" To gain," he observes, " a precise conception of the resemblances and differences of the hand and foot, and of the distinctive characters of each, we must look below the skin, and compare the bony framework and its motor apparatus in each.

" The foot of Man is distinguished from his hand by—

" 1. The arrangement of the tarsal bones.

" 2. By having a short flexor and a short extensor muscle of the digits.

" 3. By possessing the muscle termed *peronæus longus*.

And if we desire to ascertain whether the terminal division of a limb in other animals is to be called a foot or a hand, it is by the presence or absence of these characters that we must be guided, and not by the mere proportions, and greater or lesser mobility of the great toe, which may vary indefinitely without any fundamental alteration in the structure of the foot. Keeping these considerations in mind, let us now turn to the limbs of the Gorilla. The terminal division of the fore-limb presents no difficulty—bone for bone, and muscle for muscle, are found to be arranged precisely as in Man, or with such minute differences as are found as varieties in Man. The Gorilla's hand is clumsier, heavier, and has a thumb somewhat shorter in proportion than that of Man; but no one has ever doubted its being a true hand.

" At first sight, the termination of the hind-limb of the Gorilla looks very hand-like, and as it is still more so in the lower apes, it is not wonderful that the appellation ' Quadrumana,' or four-handed creatures, adopted from the older anatomists by Blumenbach, and unfortunately rendered current by Cuvier, should have gained such wide acceptance as a name for the ape order. But the most cursory anatomical investigation at once proves that the resemblance of the so-called ' hindhand ' to a true hand is only skin deep, and that, in all essential respects, the hind-limb of the Gorilla is as truly terminated by

[1] Professor Huxley's third lecture " On the Motor Organs of Man compared with those of other Animals," delivered in the Royal School of Mines, in Jermyn Street (March 1861), has been embodied with the rest of the course in his work entitled *Evidence as to Man's Place in Nature.*

a foot as that of Man. The tarsal bones, in all important circumstances of number, disposition, and form, resemble those of Man. The metatarsals and digits, on the other hand, are proportionally longer and more slender, while the great toe is not only proportionally shorter and weaker, but its metatarsal bone is united by a far more movable joint with the tarsus. At the same time, the foot is set more obliquely upon the leg than in Man.

" As to the muscles, there is a short flexor, a short extensor, and a peronæus longus, while the tendons of the long flexors of the great toe and of the other toes are united together and into an accessory fleshy bundle.

" The hind-limb of the Gorilla, therefore, ends in a true foot with a very movable great toe. It is a prehensile foot, if you will, but is in no sense a hand: it is a foot which differs from that of Man in no fundamental character, but in mere proportions—degree of mobility—and secondary arrangement of its parts.

" It must not be supposed, however, that because I speak of these differences as not fundamental, that I wish to underrate their value. They are important enough in their way, the structure of the foot being in strict correlation with that of the rest of the organism; but after all, regarded anatomically, the resemblances between the foot of Man and the foot of the Gorilla are far more striking and important than the differences." [1]

After dwelling on some points of anatomical detail, highly important, but for which I have not space here, the Professor continues:—" Throughout all these modifications, it must be recollected that the foot loses no one of its essential characters. Every monkey and lemur exhibits the characteristic arrangement of tarsal bones, possesses a short flexor and short extensor muscle, and a peronæus longus. Varied as the proportions and appearance of the organ may be, the terminal division of the hind-limb remains in plan and principle of construction a foot, and never in the least degree approaches a hand." [2] For these reasons, Professor Huxley rejects the term " Quadrumana," as leading to serious misconception, and regards Man as one of the families of the Primates. This method of classification he shows to be equally borne out by an appeal to another character on which so much reliance has always been placed in classification, as affording in the mammalia the most trustworthy indications of affinity, namely, the dentition.

[1] Professor Huxley, *ibid.*   [2] *Ibid.*

" The number of teeth in the Gorilla and all the Old World monkeys, except the lemurs, is thirty-two, the same as in Man, and the general pattern of their crowns the same. But besides other distinctions, the canines in all but Man project in the upper or lower jaws almost like tusks. But all the American apes have four more teeth in their permanent set, or thirty-six in all, so that they differ in this respect more from the Old World apes than do these last from Man."

If therefore, by reference to this character, we place Man in a separate order, we must make several orders for the apes, monkeys, and lemurs, and so, in regard to the structure of the hands and feet before alluded to, " the Gorilla differs far more from some of the quadrumana than he differs from Man." Indeed, Professor Huxley contends that there is more difference between the hand and foot of the Gorilla and those of the Orang, one of the anthropomorphous apes, than between those of the Gorilla and Man, for " the thumb of the Orang differs by its shortness and by the absence of any special long flexor muscle from that of the Gorilla more than it differs from that of Man." The carpus also of the Orang, like that of most lower apes, contains nine bones, while in the Gorilla, as in Man and the Chimpanzee, there are only eight." Other characters are also given to show that the Orang's foot separates it more widely from the Gorilla than that of the Gorilla separates that ape from Man. In some of the lower apes, the divergence from the human type of hand and foot, as well as from those of the Gorilla, is still greater, as, for example, in the spider-monkey and marmoset.[1]

If the muscles, viscera, or any other part of the animal fabric, including the brain, be compared, the results are declared to be similar.

Whether the Structure of the Human Brain entitles Man to form a distinct Sub-class of the Mammalia

In consequence of these and many other zoological considerations, the order Bimana had already been declared, in 1856, by Isidore G. St. Hilaire in his history of the science above quoted " to have become obsolete," even though sanctioned by the great names of Blumenbach and Cuvier. But in opposition to the new views Professor Owen announced, the year after the publication of G. St. Hilaire's work, that he had been

[1] Huxley, *ibid.*, p. 29.

led by purely anatomical considerations to separate Man from the other Primates and from the mammalia generally as a distinct *sub-class*, thus departing farther from the classification of Blumenbach and Cuvier than they had ventured to do from that of Linnæus.

The proposed innovation was based chiefly on three cerebral characters belonging, it was alleged, exclusively to Man and thus described in the following passages of a memoir communicated to the Linnæan Society in 1857, in which all the mammalia were divided, according to the structure of the brain, into four sub-classes, represented by the kangaroo, the beaver, the ape, and Man respectively:—

" In Man, the brain presents an ascensive step in development, higher and more strongly marked than that by which the preceding sub-class was distinguished from the one below it. Not only do the cerebral hemispheres overlap the olfactory lobes and cerebellum, but they extend in advance of the one and farther back than the other. Their posterior development is so marked that anatomists have assigned to that part the character of a third lobe; it is peculiar to the genus *Homo*, and equally peculiar is the ' posterior horn of the lateral ventricle ' and the ' hippocampus minor ' which characterises the hind-lobe of each hemisphere. The superficial grey matter of the cerebrum, through the number and depth of its convolutions, attains its maximum of extent in Man.

" Peculiar mental powers are associated with this highest form of brain, and their consequences wonderfully illustrate the value of the cerebral character; according to my estimate of which I am led to regard the genus *Homo* as not merely a representative of a distinct order, but of a distinct sub-class of the mammalia, for which I propose the name of ' *Archencephala*.' " [1]

The above definition is accompanied in the same memoir by the following note:—" Not being able to appreciate, or conceive, of the distinction between the psychical phenomena of a chimpanzee and of a Boschisman, or of an Aztec with arrested brain-growth, as being of a nature so essential as to preclude a comparison between them, or as being other than a difference of degree, I cannot shut my eyes to the significance of that all-pervading similitude of structure—every tooth, every bone, strictly homologous—which makes the determination of the difference between *Homo* and *Pithecus* the anatomist's

[1] Owen, *Proceedings of the Linnæan Society*, London, vol. viii., p. 20.

difficulty; and therefore, with every respect for the author of the *Records of Creation*,[1] I follow Linnæus and Cuvier in regarding mankind as a legitimate subject of zoological comparison and classification."

To illustrate the difference between the human and Simian brain, Professor Owen gave figures of the negro's brain as represented by Tiedemann, an original one of a South American monkey, *Midas rufimanus*, and one of the chimpanzee (Fig. 54), from a memoir published in 1849 by MM. Schroeder van der Kolk and M. Vrolik.[2]

The selection of the last-mentioned figure was most unfortunate, for three years before, M. Gratiolet, the highest authority in cerebral anatomy of our age, had, in his splendid work on *The Convolutions of the Brain in Man and the Primates* (Paris, 1854), pointed out that, though this engraving faithfully expressed the cerebral foldings as seen on the surface, it gave a very false idea of the relative position of the several parts of the brain, which, as very commonly happens in such preparations, had shrunk and greatly sunk down by their own weight.[3]

Anticipating the serious mistakes which would arise from this inaccurate representation of the brain of the ape, published under the auspices of men so deserving of trust as the two above-named Dutch anatomists, M. Gratiolet thought it expedient, by way of warning to his readers, to repeat their incorrect figures (Figs. 54 and 55), and to place by the side of them two correct views (Figs. 56 and 57) of the brain of the same ape. By reference to these illustrations, as well as to Fig. 58, the reader will see not only the contrast of the relative position of the cerebrum and cerebellum, as delineated in the natural as well as in the distorted state, but also the remarkable general correspondence between the chimpanzee brain and that of the human subject in everything save in size. The human brain (Fig. 58) here given, by Gratiolet, is that of an African bushwoman, called the Hottentot Venus, who was exhibited formerly in London, and who died in Paris.

Respecting this striking analogy of cerebral structure in Man and the apes, Gratiolet says, in the work above cited:

[1] The late Archbishop of Canterbury, Dr. Sumner.
[2] *Comptes rendus de l'Académie Royale des Sciences*, Amsterdam, vol. xiii.
[3] Gratiolet's words are: " Les plis cérébraux du chimpanzé y sont fort bien étudiés, malheureusement le cerveau qui leur a servi de modèle était profondément affaissé, aussi la forme générale du cerveau est-elle rendue, dans leurs planches, d'une manière tout-à-fait fausse." *Ibid.*, p. 18.

Upper surface of brain of Chimpanzee, distorted (from Schroeder van der Kolk and Vrolik).

A Left cerebral hemisphere.
B Right ditto.
C Cerebellum displaced.

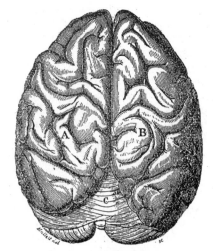

Fig. 54.

Side view of same (from Schroeder van der Kolk and Vrolik), showing at *e* the extension of the displaced cerebellum beyond the cerebrum at *d*.

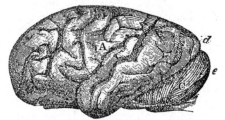

Fig. 55.

Correct side view of Chimpanzee's brain (from Gratiolet), showing the backward extension of the cerebrum at *d*, beyond the cerebellum at *e*.

f f Fissure of Sylvius.

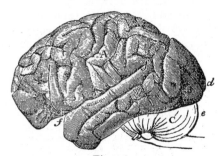

Fig. 56.

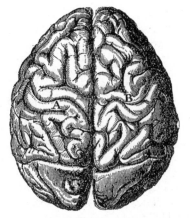

Fig. 57.

Correct view of upper surface of Chimpanzee's brain (from Gratiolet),
in which the cerebrum covers and conceals the cerebellum.

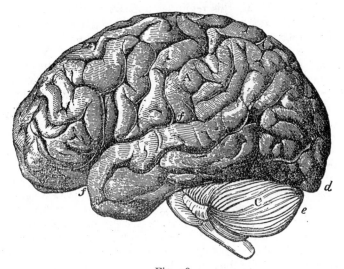

Fig. 58.

Side view of human brain (from Gratiolet), namely, that of the bushwoman called the Hottentot Venus.

A Left cerebral hemisphere.   c Cerebellum.
                    f f Fissure of Sylvius.

Scale of the five Figures, from 54 to 58, half the diameter of the
natural size.

"The convoluted brain of Man and the smooth brain of the marmoset resemble each other by the quadruple character of a rudimentary olfactory lobe, a posterior lobe *completely covering the cerebellum*, a well-defined fissure of Sylvius (*ff*, Fig. 56), and lastly a posterior horn in the lateral ventricle. These characters are not met with together except in Man and the apes."[1]

In reference to the other figure of a monkey given by Professor Owen, namely, that of the *Midas*, one of the marmosets, he states, in 1857 as he had done in 1837, that the posterior part of the cerebral hemispheres " extends, as in most of the quadrumana, *over the greater part of the cerebellum*."[2]   In 1859, in his Rede Lecture, delivered to the University of Cambridge, the same illustrations of the ape's brain were given, namely, that of the *Midas* and the distorted one of the Dutch anatomists already cited (Fig. 54).[3]

Two years later, Professor Huxley, in a memoir *On the Zoological Relations of Man with the Lower Animals*, took occasion to refer to Gratiolet's warning, and to cite his criticism on the Dutch plates;[4] but this reminder appears to have been overlooked by Professor Owen, who six months later came out with a new paper on *The Cerebral Character of Man and the Ape*, in which he repeated the incorrect representation of Schroeder van der Kolk and Vrolik, associating it with Tiedemann's figure of a negro's brain, expressly to show the relative and different extent to which the cerebellum is overlapped by the cerebrum in the two cases respectively.[5]   In the ape's brain as thus depicted, the portion of the cerebellum left uncovered is greater than in the lemurs, the lowest type of Primates, and almost as large as in the rodentia, or some of the lowest grades of the mammalia.

When the Dutch naturalists above mentioned found their figures so often appealed to as authority, by one the weight of whose opinion on such matters they well knew how to appreciate, they resolved to do their best towards preventing the public from being misled. Accordingly, they addressed to the Royal Academy of Amsterdam a memoir *On the brain of an Orang-outang* which had just died in the Zoological Gardens

[1] Gratiolet, *ibid.*   Avant-propos, p. 2, 1854.
[2] *Proceedings of the Linnæan Society*, 1857, p. 18, note, and *Philosophical Transactions*, 1837, p. 93.
[3] *See* Appendix M.
[4] Huxley, *Natural History Review*, January 7, 1861, p. 76.
[5] *Annals and Magazine of Natural History*, vol. vii., 1861, p. 456, and Pl. XX.

of that city.[1] The dissection of this ape, in 1861, fully bore out the general conclusions at which they had previously arrived in 1849, as to the existence both in the human and the simian brain of the three characters, which Professor Owen had represented as exclusively appertaining to Man, namely, the occipital or posterior lobe, the hippocampus minor, and the posterior cornu. These last two features consist of certain cavities and furrows in the posterior lobes, which are caused by the foldings of the brain, and are only visible when it is dissected. MM. Schroeder van der Kolk and Vrolik took this opportunity of candidly confessing that M. Gratiolet's comments on the defects of their two figures (Figs. 54 and 55) were perfectly just, and they expressed regret that Professor Owen should have overstated the differences existing between the brain of Man and the Quadrumana, " led astray, as they supposed, by his zeal to combat the Darwinian theory respecting the transformation of species," a doctrine against which they themselves protested strongly, saying that it belongs to a class of speculations which are sure to be revived from time to time, and are always " peculiarly seductive to young and sanguine minds." [2]

As the two memoirs before alluded to by us, the one by Mr. Darwin on *Natural Selection,* and the other by Mr. Wallace *On the Tendency of Varieties to depart indefinitely from the original Type,* did not appear till 1858, a year after Professor Owen's classification of the mammalia, and as Darwin's *Origin of Species* was not published till another year had elapsed, we cannot accept the explanation above offered to us of the causes which led the founder of the sub-class Archencephala to seek for new points of distinction between the human and simian brains; but the Dutch anatomists may have fallen into this anachronism by having just read, in the paper by Professor Owen in the *Annals,* some prefatory allusions to " the Vestiges of Creation," " Natural Selection, and the question whether man be or be not a descendant of the ape."

The number of original and important memoirs to which this discussion on the cerebral relations of Man to the Primates has already given rise in less than five years, must render the controversy for ever memorable in the history of Comparative Anatomy.[3]

[1] This paper is reprinted in the original French in the *Natural History Review,* vol. ii., 1862, p. 111.   [2] *Ibid.,* p. 114.

[3] Rolleston, *Natural History Review,* April 1861. Huxley, on " Brain of *Ateles,"* *Proc. Zool. Soc.,* 1861. " Flower, Posterior Lobe in Quadrumana," etc., *Philosophical Transactions,* 1862. *Id.* " Javan *Loris," Proc. Zool. Soc.,* 1862. *Id.* on " Anatomy of *Pithecia," ibid.,* 1862.

In England alone, no less than fifteen genera of the Primates (the subjects having been almost all furnished by that admirable institution the Zoological Gardens of London) have been anatomically examined, and they include nearly all the leading types of structure of the Old and New World apes and monkeys, from the most anthropoid form to that farthest removed from Man; in other words, from the Chimpanzee to the Lemur. These are—

Troglodytes (Chimpanzee).	*Cebus* (Capuchin Monkey).
Pithecus (Orang).	*Pithecia* (Saki).
Hylobates (Gibbon).	*Nyctipithecus* (Douricouli).
Semnopithecus.	*Hapale* (Marmoset).
Cercopithecus.	
Macacus.	*Otolicnus.*
Cynocephalus (Baboon).	*Stenops.*
Ateles (Spider Monkey).	*Lemur.*

In July 1861 Mr. Marshall, in a paper on the brain of a young Chimpanzee, which he had dissected immediately after its death, gave a series of photographic drawings, showing that when the parts are all in a fresh state, the posterior lobe of the cerebrum, instead of simply covering the cerebellum, is prolonged backwards beyond it even to a greater extent than in Gratiolet's figure, 56, and, what is more in point, in a greater degree relatively speaking (at least in the young state of the animal) than in Man. In fact, " the projection is to the extent of about one-ninth of the total length of the cerebrum, whereas the average excess of overlapping is only one-eleventh in the human brain." [1]

The same author gives an instructive account of the manner in which displacement and distortion take place when such brains are preserved in spirits as in the ordinary preparations of the anatomist.

Mr. Flower, in a recent paper on the posterior lobe of the cerebrum in the Quadrumana,[2] remarks, that although Tiedemann had declared himself unable in 1821 to detect the hippocampus minor or the posterior cornu of the lateral ventricle in the brain of a *Macacus* dissected by him, Cuvier, nevertheless, mentions the latter as characteristic of Man and the apes, and

[1] Marshall, *Natural History Review*, July 1861. *See also* on this subject Professor Rolleston on the slight degree of backward extension of the cerebrum in some races of Man. *Medical Times*, October 1862, p. 419.
[2] *Philosophical Transactions*, 1862, p. 185.

M. Serres in his well-known work on the brain in 1826, has shown in at least four species of apes the presence of both the hippocampus minor and the posterior cornu.

Tiedemann had expressly stated that "the third or hinder lobe in the ape covered the cerebellum as in Man,"[1] and as to his negative evidence in respect to the internal structure of that lobe, it can have no weight whatever against the positive proofs obtained to the contrary by a host of able observers. Even before Tiedemann's work was published, Kuhl had dissected, in 1820, the brain of the spider-monkey (*Ateles beelzebuth*), and had given a figure of a long posterior cornu to the lateral ventricle, which he had described as such.[2]

The general results arrived at by the English anatomists already cited, and by Professor Rolleston in various papers on the same subject, have thus been briefly stated by Professor Huxley:—

"Every lemur which has yet been examined has its cerebellum partially visible from above, and its posterior lobe, with the contained posterior cornu and hippocampus minor, more or less rudimentary. Every marmoset, American monkey, Old World monkey, baboon, or man-like ape, on the contrary, has its cerebellum entirely hidden, and possesses a large posterior cornu, with a well-developed hippocampus minor.

"In many of these creatures, such as the Saimiri (*Chrysothrix*), the cerebral lobes overlap and extend much farther behind the cerebellum in proportion than they do in Man."[3]

It is by no means pretended that these conclusions of British observers as to the affinity in cerebral structure of Man and the Primates are new, but on the contrary that they confirm the inductions previously made by the principal continental teachers of the last and present generations, such as Tiedemann, Cuvier, Serres, Leuret, Wagner, Schroeder van der Kolk, Vrolik, Gratiolet, and others.

At a late meeting of the British Association (1862), Professor Owen read a paper "On the brain and limb characters of the Gorilla as contrasted with those of Man"[4] in which, he observes, that in the gorilla the cerebrum "extends over the cerebellum, not beyond it." This statement, although slightly at variance with one published the year before (1861) by Professor Huxley,

---

[1] Tiedemann, *Icones cerebri Simiarum*, etc., p. 48.
[2] *Beiträge zur Zoologie*, etc., Frankfurt am Main, 1820.
[3] Huxley, *Evidence as to Man's place in Nature*, p. 97.
[4] *Medical Times and Gazette*, October 1862, p. 373.

who maintains that it does project beyond, is interesting as correcting the description of the same brain given by Professor Owen in that year, in a lecture to the Royal Institution, in which a considerable part of the cerebellum of the gorilla was represented as uncovered.[1] In the same memoir, it is remarked that in the Maimon Baboon the cerebrum not only covers but " extends backwards even beyond the cerebellum." [2] This baboon, therefore, possesses a posterior lobe, according to every description yet given of such a lobe, including a new definition of the same lately proposed by Professor Owen. For the posterior lobe was formerly considered to be that part of the cerebrum which covers the cerebellum, whereas Professor Owen defines it as that part which covers the posterior third of the cerebellum, and extends beyond it.

We may, therefore, consider the attempt to distinguish the brain of Man from that of the ape on the ground of newly-discovered cerebral characters, presenting differences in kind, as virtually abandoned by its originator, and if the sub-class Archencephala is to be retained, it must depend on differences in degree, as, for example, the vast increase of the brain in Man, as compared with that of the highest ape, " in absolute size, and the still greater superiority in relative size to the bulk and weight of the body." [3]

If we ask why this character, though well known to Cuvier and other great anatomists before our time, was not considered by them to entitle Man, physically considered, to claim a more distinct place in the group called Primates than that of a separate order, or, according to others, a separate genus or family only, we shall find the answer thus concisely stated by Professor Huxley in his new work, before cited:—

" So far as I am aware, no human cranium belonging to an adult man has yet been observed with a less cubical capacity than 62 cubic inches, the smallest cranium observed in any race of men, by Morton, measuring 63 cubic inches; while on the other hand, the most capacious gorilla skull yet measured has a content of not more than $34\frac{1}{2}$ cubic inches. Let us assume, for simplicity's sake, that the lowest man's skull has twice the capacity of the highest gorilla's. No doubt this is a very striking difference, but it loses much of its apparent systematic

---

[1] *Athenæum*, Report of Royal Institution Lecture, March 23, 1861, and reference to it by Professor Owen as to Gorilla, *ibid.*, March 30, p. 434.

[2] For Report of Professor Owen's Cambridge British Association paper, see *Medical Times*, October 11, 1862, p. 373.

[3] Owen, *ibid.*, p. 373.

value, when viewed by the light of certain other equally indubitable facts respecting cranial capacities.

" The first of these is, that the difference in the volume of the cranial cavity of different races of mankind is far greater, absolutely, than that between the lowest man and the highest ape, while, relatively, it is about the same; for the largest human skull measured by Morton contained 114 cubic inches, that is to say, had very nearly double the capacity of the smallest, while its absolute preponderance of over 50 cubic inches is far greater than that by which the lowest adult male human cranium surpasses the largest of the gorillas ($62 - 34\frac{1}{2}$ $= 27\frac{1}{2}$). Secondly, the adult crania of gorillas which have as yet been measured, differ among themselves by nearly one-third, the maximum capacity being 34.5 cubic inches, the minimum 24 cubic inches; and, thirdly, after making all due allowance for difference of size, the cranial capacities of some of the lower apes fall nearly as much relatively below those of the higher apes, as the latter fall below Man." [1]

Are we then to conclude that differences in mental power have no intimate connection with the comparative volume of the brain? We cannot draw such an inference, because the highest and most civilised races of Man exceed in the average of their cranial capacity the lowest races, the European brain, for example, being larger than that of the negro, and somewhat more convoluted and less symmetrical, and those apes, on the other hand, which approach nearest to Man in the form and volume of their brain being more intelligent than the Lemurs, or still lower divisions of the mammalia, such as the Rodents and Marsupials, which have smaller brains. But the extraordinary intelligence of the elephant and dog, so far exceeding that of the larger part of the Quadrumana, although their brains are of a type much more remote from the human, may serve to convince us how far we are as yet from understanding the real nature of the dependence of intellectual superiority on cerebral structure.

Professor Rolleston, in reference to this subject, remarks, that " even if it were to be proved that the differences between Man's brain and that of the ape are differences entirely of quantity, there is no reason, in the nature of things, why so many and such weighty differences in degree should not amount to a difference in kind.

" Differences of degree and differences of kind are, it is true,

[1] Huxley, *Evidence as to Man's place in Nature*, London, 1863, p. 78.

mutually exclusive terms in the language of the schools; but whether they are so also in the laboratory of Nature, we may very well doubt." [1]

The same physiologist suggests, that as there is considerable plasticity in the human frame, not only in youth and during growth, but even in the adult, we ought not always to take for granted, as some advocates of the development theory seem to do, that each advance in psychical power depends on an improvement in bodily structure, for why may not the soul, or the higher intellectual and moral faculties, play the first instead of the second part in a progressive scheme?

Intelligence of the lower Animals compared to that of Man

Ever since the days of Leibnitz, metaphysicians who have attempted to draw a line of demarcation between the intelligence of the lower animals and that of Man, or between instinct and reason, have experienced difficulties analogous to those which the modern anatomist encounters when he tries to distinguish the brain of an ape from that of Man by some characters more marked than those of mere size and weight, which vary so much in individuals of the same species, whether simian or human.

Professor Agassiz, after declaring that as yet we scarcely possess the most elementary information requisite for a scientific comparison of the instincts and faculties of animals with those of Man, confesses that he cannot say in what the mental faculties of a child differ from those of a young chimpanzee. He also observes, that " the range of the passions of animals is as extensive as that of the human mind, and I am at a loss to perceive a difference of kind between them, however much they may differ in degree and in the manner in which they are expressed. The gradations of the moral faculties among the higher animals and Man are, moreover, so imperceptible, that to deny to the first a certain sense of responsibility and consciousness would certainly be an exaggeration of the difference between animals and Man. There exists, besides, as much individuality within their respective capabilities among animals as among Man, as every sportsman, or every keeper of menageries, or every farmer and shepherd can testify, who has had a large experience

[1] Report of a Lecture delivered at the Royal Institution by Professor George Rolleston " On the Brain of Man and Animals," *Medical Gazette*, March 15, 1862, p. 262.

with wild, or tamed, or domesticated animals. This argues strongly in favour of the existence in every animal of an immaterial principle, similar to that which, by its excellence and superior endowments, places Man so much above animals. Yet the principle exists unquestionably, and whether it be called soul, reason, or instinct, it presents, in the whole range of organised beings, a series of phenomena closely linked together, and upon it are based not only the higher manifestations of the mind, but the very permanence of the specific differences which characterise every organ. Most of the arguments of philosophy in favour of the immortality of Man apply equally to the permanency of this principle in other living beings." [1]

Professor Huxley, when commenting on a passage in Professor Owen's memoir, above cited, argues that there is a unity in psychical as in physical plan among animated beings, and adds, that although he cannot go so far as to say that " the determination of the difference between *Homo* and *Pithecus* is the anatomist's difficulty," yet no impartial judge can doubt that the roots, as it were, of those great faculties which confer on Man his immeasurable superiority above all other animate things are traceable far down into the animate world. The dog, the cat, and the parrot, return love for our love and hatred for our hatred. They are capable of shame and of sorrow, and though they may have no logic nor conscious ratiocination, no one who has watched their ways can doubt that they possess that power of rational cerebration which evolves reasonable acts from the premises furnished by the senses—a process which takes fully as large a share as conscious reason in human activity. [2]

Grounds for referring Man to a distinct Kingdom of Nature

Few if any of the authors above cited, while they admit so fully the analogy which exists between the faculties of Man and the inferior animals, are disposed to underrate the enormous gap which separates Man from the brutes, and if they scarcely allow him to be referable to a distinct order, and much less to a separate sub-class, on purely physical grounds, it does not follow that they would object to the reasoning of M. Quatrefages, who says, in his work on the *Unity of the Human Species*, that

[1] Contributions to the *Natural History of the United States of North America*, vol. i., part i., pp. 60, 64.
[2] *Natural History Review*, No. 1, January 1861, p. 68.

Man must form a kingdom by himself if once we permit his moral and intellectual endowments to have their due weight in classification.

As to his organisation, he observes, " We find in the mammalia nearly absolute identity of anatomical structure, bone for bone, muscle for muscle, nerve for nerve—similar organs performing like functions. It is not by a vertical position on his feet, the *os sublime* of Ovid, which he shares with the penguin, nor by his mental faculties, which, though more developed, are fundamentally the same as those of animals, nor by his powers of perception, will, memory, and a certain amount of reason, nor by articulate speech, which he shares with birds and some mammalia, and by which they express ideas comprehended not only by individuals of their own species but often by Man, nor is it by the faculties of the heart, such as love and hatred, which are also shared by quadrupeds and birds, but it is by something completely foreign to the mere animal, and belonging exclusively to Man, that we must establish a separate kingdom for him (p. 21). These distinguishing characters," he goes on to say, " are the abstract notion of good and evil, right and wrong, virtue and vice, or the moral faculty, and a belief in a world beyond ours, and in certain mysterious beings, or a Being of a higher nature than ours, whom we ought to fear or revere; in other words, the religious faculty."—P. 23.

By these two attributes the moral and the religious, not common to man and the brutes, M. Quatrefages proposes to distinguish the human from the animal kingdom.

But he omits to notice one essential character, which Dr. Sumner, the late Archbishop of Canterbury, brought out in strong relief fifty years ago in his *Records of Creation*. " There are writers," he observes, " who have taken an extraordinary pleasure in levelling the broad distinction which separates Man from the Brute Creation. Misled to a false conclusion by the infinite variety of Nature's productions, they have described a chain of existence connecting the vegetable with the animal world, and the different orders of animals one with another, so as to rise by an almost imperceptible gradation from the tribe of Simiæ to the lowest of the human race, and from these upwards to the most refined. But if a comparison were to be drawn, it should be taken, not from the upright form, which is by no means confined to mankind, nor even from the vague term reason, which cannot always be accurately separated from instinct, but from that power of progressive and improvable reason, which is Man's peculiar and exclusive endowment.

" It has been sometimes alleged, and may be founded on fact, that there is less difference between the highest brute animal and the lowest savage than between the savage and the most improved Man. But, in order to warrant the pretended analogy, it ought to be also true that this lowest savage is no more capable of improvement than the Chimpanzee or Orang-outang.

" Animals," he adds, " are born what they are intended to remain. Nature has bestowed upon them a certain rank, and limited the extent of their capacity by an impassable decree. Man she has empowered and obliged to become the artificer of his own rank in the scale of beings by the peculiar gift of improvable reason." [1]

We have seen that Professor Agassiz, in his *Essay on Classification*, above cited, speaks of the existence in every animal of " an immaterial principle similar to that which, by its excellence and superior endowments, places man so much above animals; " and he remarks, " that most of the arguments of philosophy in favour of the immortality of Man, apply equally to the permanency of this principle in other living beings."

Although the author has no intention by this remark to impugn the truth of the great doctrine alluded to, it may be well to observe, that if some of the arguments in favour of a future state are applicable in common to Man and the lower animals, they are by no means those which are the weightiest and most relied on. It is no doubt true that, in both, the identity of the individual outlasts many changes of form and structure which take place during the passage from the infant to the adult state, and from that to old age, and the loss again and again of every particle of matter which had entered previously into the composition of the body during its growth, and the substitution of new elements in their place, while the individual remains always the same, carries the analogy a step farther. But beyond this we cannot push the comparison. We cannot imagine this world to be a place of trial and moral discipline for any of the inferior animals, nor can any of them derive comfort and happiness from faith in a hereafter. To Man alone is given this belief, so consonant to his reason, and so congenial to the religious sentiments implanted by nature in his soul, a doctrine which tends to raise him morally and

[1] *Records of Creation*, vol. ii., chap. ii., 2nd ed., 1816.

intellectually in the scale of being, and the fruits of which are, therefore, most opposite in character to those which grow out of error and delusion.

The opponents of the theory of transmutation sometimes argue that, if there had been a passage by variation from the lower Primates to Man, the geologist ought ere this to have detected some fossil remains of the intermediate links of the chain. But what we have said respecting the absence of gradational forms between the Recent and Pliocene mammalia may serve to show the weakness in the present state of science of any argument based on such negative evidence, especially in the case of Man, since we have not yet reached those pages of the great book of nature, in which alone we have any right to expect to find records of the missing links alluded to. The countries of the anthropomorphous apes are the tropical regions of Africa, and the islands of Borneo and Sumatra, lands which may be said to be quite unknown in reference to their Pliocene and Pleistocene mammalia. Man is an old-world type, and it is not in Brazil, the only equatorial region where ossiferous caverns have yet been explored, that the discovery, in a fossil state, of extinct forms allied to the human, could be looked for. Lund, a Danish naturalist, found in Brazil, not only extinct sloths and armadilloes, but extinct genera of fossil monkeys, but all of the American type, and, therefore, widely departing in their dentition and some other characters from the Primates of the old world.

At some future day, when many hundred species of extinct quadrumana may have been brought to light, the naturalist may speculate with advantage on this subject; at present we must be content to wait patiently, and not to allow our judgment respecting transmutation to be influenced by the want of evidence, which it would be contrary to analogy to look for in Pleistocene deposits in any districts, which as yet we have carefully examined. For, as we meet with extinct kangaroos and wombats in Australia, extinct llamas and sloths in South America, so in equatorial Africa, and in certain islands of the East Indian Archipelago, may we hope to meet hereafter with lost types of the anthropoid Primates, allied to the gorilla, chimpanzee, and orang-outang. [Note 44.]

Europe, during the Pliocene period, seems not to have enjoyed a climate fitting it to be the habitation of the quadrumanous mammalia; but we no sooner carry back our researches into Miocene times, where plants and insects, like those of Œningen, and shells, like those of the Faluns of the Loire, would imply a warmer temperature both of sea and land, than we begin to discover fossil apes and monkeys north of the Alps and Pyrenees. Among the few species already detected, two at least belong to the anthropomorphous class. One of these, the *Dryopithecus* of Lartet, a gibbon or long-armed ape, about equal to man in stature, was obtained in the year 1856 in the Upper Miocene strata at Sansan, near the foot of the Pyrenees in the South of France, and one bone of the same ape is reported to have been since procured from a deposit of corresponding age at Eppelsheim, near Darmstadt, in a latitude answering to that of the southern counties of England.[1] But according to the doctrine of progression it is not in these Miocene strata, but in those of Pliocene and Pleistocene date, in more equatorial regions, that there will be the greatest chance of discovering hereafter some species more highly organised than the gorilla and chimpanzee.

The only reputed fossil monkey of Eocene date, namely, that found in 1840 at Kyson, in Suffolk, and so determined by Professor Owen, has recently been pronounced by the same anatomist, after re-examination, and when he had ampler materials at his command, to be a pachyderm.

M. Rütimeyer,[2] however, an able osteologist, referred to in the earlier chapters of this work, has just announced the discovery in Eocene strata, in the Swiss Jura, of a monkey allied to the lemurs, but as he has only obtained as yet a small fragment of a jaw with three molar teeth, we must wait for fuller information before we confidently rely on the claims of his *Cœnopithecus lemuroides* to take rank as one of the Primates.

Hallam on Man's place in the Creation

Hallam, in his *Literature of Europe,* after indulging in some profound reflections on " the thoughts of Pascal," and the theological dogmas of his school respecting the fallen nature of Man, thus speaks of Man's place in the creation:—" It might be wandering from the proper subject of these volumes if we were to pause, even shortly, to inquire whether, while the creation of a world so full of evil must ever remain the most inscrutable of mysteries, we might not be led some way in

---

[1] Owen, *Geologist,* November 1862.
[2] Rutimeyer, *Eocene Säugethiere,* Zürich, 1862.

tracing the connection of moral and physical evil in mankind, with his place in that creation, and especially, whether the law of continuity, which it has not pleased his Maker to break with respect to his bodily structure, and which binds that, in the unity of one great type, to the lower forms of animal life by the common conditions of nourishment, reproduction, and self-defence, has not rendered necessary both the physical appetites and the propensities which terminate in self; whether again the superior endowments of his intellectual nature, his susceptibility of moral emotion, and of those disinterested affections which, if not exclusively, he far more intensely possesses than an inferior being—above all, the gifts of conscience and a capacity to know God, might not be expected, even beforehand, by their conflict with the animal passions, to produce some partial inconsistencies, some anomalies at least, which he could not himself explain in so compound a being. Every link in the long chain of creation does not pass by easy transition into the next. There are necessary chasms, and, as it were, leaps from one creature to another, which, though not exceptions to the law of continuity, are accommodations of it to a new series of being. If Man was made in the image of God, he was also made in the image of an ape. The framework of the body of him who has weighed the stars and made the lightning his slave, approaches to that of a speechless brute, who wanders in the forests of Sumatra. Thus standing on the frontier land between animal and angelic natures, what wonder that he should partake of both! " [1]

The law of continuity here spoken of, as not being violated by occasional exceptions, or by leaps from one creature to another, is not the law of variation and natural selection above explained (Chap. XXI.), but that unity of plan supposed to exist in the Divine Mind, whether realised or not materially and in the visible creation, of which the " links do not pass by an easy transition " the one into the other, at least as beheld by us.

Dr. Asa Gray, an eminent American botanist, to whom we are indebted for a philosophical essay of great merit on the *Origin of Species by Variation and Natural Selection*, has well observed, when speaking of the axiom of Leibnitz, " Natura non agit saltatim," that nature secures her ends and makes her distinctions, on the whole, manifest and real, but without any important breaks or long leaps. " We need not wonder that gradations between species and varieties should occur, or that

[1] Hallam, *Introduction to the Literature of Europe*, etc., vol. iv., p. 162.

genera and other groups should not be absolutely limited, though they are represented to be so in our systems. The classifications of the naturalist define abruptly where nature more or less blends. Our systems are nothing if not definite."

The same writer reminds us that " plants and animals are so different, that the difficulty of the ordinary observer would be to find points of comparison, whereas, with the naturalist, it is all the other way. All the broad differences vanish one by one as we approach the lower confines of the animal and vegetable kingdoms, and no absolute distinction whatever is now known between them." [1]

The author of an elaborate review of Darwin's *Origin of Species*, himself an accomplished geologist, declares that if we embrace the doctrine of the " continuous variation of all organic forms from the lowest to the highest, including Man as the last link in the chain of being, there must have been a transition from the instinct of the brute to the noble mind of Man; and in that case, " where," he asks, " are the missing links, and at what point of his progressive improvement did Man acquire the spiritual part of his being, and become endowed with the awful attribute of immortality? " [2]

Before we raise objections of this kind to a scientific hypothesis, it would be well to pause and inquire whether there are no analogous enigmas in the constitution of the world around us, some of which present even greater difficulties than that here stated. When we contemplate, for example, the many hundred millions of human beings who now people the earth, we behold thousands who are doomed to helpless imbecility, and we may trace an insensible gradation between them and the half-witted, and from these again to individuals of perfect understanding, so that tens of thousands must have existed in the course of ages, who in their moral and intellectual condition, have exhibited a passage from the irrational to the rational, or from the irresponsible to the responsible. Moreover we may infer from the returns of the Registrar General of births and deaths in Great Britain, and from Quetelet's statistics of Belgium, that one-fourth of the human race die in early infancy, nearly one-tenth before they are a month old; so that we may safely affirm that millions perish on the earth in every century,

---

[1] Gray, *Natural Selection not inconsistent with Natural Theology*, Trübner & Co., London, 1861, p. 55.
[2] Physical Theories of the Phenomena of Life, *Fraser's Magazine*, July 1860, p. 88.

in the first few hours of their existence. To assign to such individuals their appropriate psychological place in the creation is one of the unprofitable themes on which theologians and metaphysicians have expended much ingenious speculation.

The philosopher, without ignoring these difficulties, does not allow them to disturb his conviction that "whatever is, is right," nor do they check his hopes and aspirations in regard to the high destiny of his species; but he also feels that it is not for one who is so often confounded by the painful realities of the present, to test the probability of theories respecting the past, by their agreement or want of agreement with some ideal of a perfect universe which those who are opposed to his opinions may have pictured to themselves.

We may also demur to the assumption that the hypothesis of variation and natural selection obliges us to assume that there was an absolutely insensible passage from the highest intelligence of the inferior animals to the improvable reason of Man. The birth of an individual of transcendent genius, of parents who have never displayed any intellectual capacity above the average standard of their age or race, is a phenomenon not to be lost sight of, when we are conjecturing whether the successive steps in advance by which a progressive scheme has been developed may not admit of occasional strides, constituting breaks in an otherwise continuous series of psychical changes.

The inventors of useful arts, the poets and prophets of the early stages of a nation's growth, the promulgators of new systems of religion, ethics, and philosophy, or of new codes of laws, have often been looked upon as messengers from Heaven, and after their death have had divine honours paid to them, while fabulous tales have been told of the prodigies which accompanied their birth. Nor can we wonder that such notions have prevailed when we consider what important revolutions in the moral and intellectual world such leading spirits have brought about; and when we reflect that mental as well as physical attributes are transmissible by inheritance, so that we may possibly discern in such leaps the origin of the superiority of certain races of mankind. In our own time the occasional appearance of such extraordinary mental powers may be attributed to atavism; but there must have been a beginning to the series of such rare and anomalous events. If, in conformity with the theory of progression, we believe mankind to have risen slowly from a rude and humble starting point, such leaps may have successively introduced not only higher and higher forms and grades of intellect, but at a much remoter period may have cleared at one bound the space which separated the highest stage of the unprogressive intelligence of the inferior animals from the first and lowest form of improvable reason manifested by Man.

To say that such leaps constitute no interruption to the ordinary course of nature is more than we are warranted in affirming. In the case of the occasional birth of an individual of superior genius there is certainly no break in the regular genealogical succession; and when all the mists of mythological fiction are dispelled by historical criticism, when it is acknowledged that the earth did not tremble at the nativity of the gifted infant and that the face of heaven was not full of fiery shapes, still a mighty mystery remains unexplained, and it is the *order* of the phenomena, and not their *cause*, which we are able to refer to the usual course of nature.

Dr. Asa Gray, in the excellent essay already cited, has pointed out that there is no tendency in the doctrine of Variation and Natural Selection to weaken the foundations of Natural Theology, for, consistently with the derivative hypothesis of species, we may hold any of the popular views respecting the manner in which the changes of the natural world are brought about. We may imagine " that events and operations in general go on in virtue simply of forces communicated at the first, and without any subsequent interference, or we may hold that now and then, and only now and then, there is a direct interposition of the Deity; or, lastly, we may suppose that all the changes are carried on by the immediate orderly and constant, however infinitely diversified, action of the intelligent, efficient Cause." They who maintain that the origin of an individual, as well as the origin of a species or a genus, can be explained only by the direct action of the creative cause, may retain their favourite theory compatibly with the doctrine of transmutation.

Professor Agassiz, having observed that, " while human thought is consecutive, divine thought is simultaneous," Dr. Asa Gray has replied that, " if divine thought is simultaneous, we have no right to affirm the same of divine action."

The whole course of nature may be the material embodiment of a preconcerted arrangement; and if the succession of events be explained by transmutation, the perpetual adaptation of the organic world to new conditions leaves the argument in favour of design, and therefore of a designer, as valid as ever; " for to do any work by an instrument must require, and therefore presuppose, the exertion rather of more than of less power, than to do it directly." [1]

As to the charge of materialism brought against all forms of the development theory, Dr. Gray has done well to remind us that " of the two great minds of the seventeenth century, Newton and Leibnitz, both profoundly religious as well as philosophical, one produced the theory of gravitation, the other objected to that theory, that it was subversive of natural religion." [2]

It may be said that, so far from having a materialistic tendency, the supposed introduction into the earth at successive geological periods of life—sensation—instinct—the intelligence of the higher mammalia bordering on reason—and lastly the improvable reason of Man himself, presents us with a picture of the ever-increasing dominion of mind over matter.

---

[1] Asa Gray, *Natural Selection not inconsistent with Natural Theology*, Trübner & Co., London, 1861, p. 55.
[2] *Ibid.* p. 31.

NOTES

The classification of the strata above the Chalk, as at present employed by the majority of British geologists, is merely a slight modification of that proposed by Lyell in 1833. The subdivisions generally recognised are as follows (Lake and Rastall, *Textbook of Geology*, London, 1910, p. 438):

Pleistocene	
Pliocene	} Neogene
Miocene	
Oligocene	} Palæogene
Eocene	

This differs chiefly from Lyell's classification in the introduction of the term Oligocene for the upper part of the original Eocene, which was somewhat unwieldy. In the earlier editions of the *Antiquity of Man* and of the *Principles of Geology*, the strata here classed as Pleistocene were designated as Post-pliocene. The term " diluvium," now obsolete in Britain but still lingering on the Continent, is equivalent to Pleistocene. This subdivision is still sometimes separated from the Tertiary, as the *Quaternary* epoch. This, however, is unnecessary and indeed objectionable, as attributing too great importance to relatively insignificant deposits. There is no definite break, either stratigraphical or palæontological, at the top of the Pliocene, and it is most natural to regard the Tertiary epoch as still in progress. Equally unnecessary is the separation of the post-glacial deposits as " Recent," a distinction which still prevails in many quarters, apparently with the sole object of adding another name to an already over-burdened list.

NOTE 2

The table of strata here printed is not that given by Lyell in the later editions of the *Antiquity of Man*. This would have required so much explanation in the light of modern work that it was thought better to abolish it altogether and to substitute an entirely new table, which is to some extent a compromise between the numerous classifications now in vogue. In this form it is only strictly applicable to the British Isles, though the divisions adopted in other countries are generally similar, and in many cases identical.

NOTE 3

A similar succession of forest-beds, five in number, has been observed in the peat of the Fenland, near Ely. Each bed consists for the most part of a single species of tree, and a definite succession of oak, yew, Scotch fir, alder, and willow has been made out. The forest beds are supposed to indicate temporarily drier conditions, due either to changes of climate or to slight uplift of the land, the growth of peat being renewed during periods of damp climate or of depression of the land. (*See* Clement Reid, *Submerged Forests*, Cambridge, 1913.)

NOTE 4

Since the " Stone Age," in the sense in which the term is here employed, obviously occupied an enormous lapse of time and embraced very different stages of culture, it has been found convenient to subdivide it into two primary subdivisions. For these Lord Avebury proposed in 1865 the terms Palæolithic and Neolithic. (*Prehistoric Times*, London, 1865, p. 60.) The first comprises the ages during which man fabricated flint implements solely by chipping, whereas the implements of Neolithic Age are polished by rubbing. But there is another and more fundamental distinction. Palæolithic man was exclusively a hunter, and consequently nomadic in his habits; Neolithic man possessed domesticated animals and cultivated crops. A pastoral and agricultural life implies a settled abode, and these are found, for example, in the lake-villages of Switzerland. The " kitchen-middens " of Denmark also indicate long continuance in one place, in this instance the seashore.

NOTE 5

The famous case of the so-called Temple of Serapis at Pozzuoli, has given rise to a considerable literature. The subject is discussed by Suess at length (*Das Antlitz der Erde*, Vienna, 1888, vol. ii. p. 463, or English translation, *The Face of the Earth*, Oxford, 1904). This author shows that the whole region is highly volcanic, and consequently very liable to disturbance, much relative movement of land and sea having occurred within historic times. Hence the facts here observed cannot be taken as evidence for any general upward or downward movement of wide-spread or universal extent.

NOTE 6

Darwin, *Voyage of the Beagle*, chap. xiv., and a much fuller account in the same author's *Geological Observations on the Volcanic Islands and Parts of South America Visited during the Voyage of H.M.S. Beagle*, chap. ix.

NOTE 7

For a full discussion of the evidence for and against continental elevation and subsidence in general, and as affecting the British Isles and Scandinavia in particular, *see* Sir A. Geikie's Presidential Address to the Geological Society for 1904 (*Proc. Geol. Soc.*, vol. lx., 1904, pp. lxxx.-civ.). Here it is shown that the oldest raised beaches of Scotland are pre-glacial, and the same also holds for the south of Ireland.

NOTE 8

The argument here employed is fallacious, since the mere existence of a distinct beach implies a pause in the movement and a long continuance at one level. It is impossible to form any estimate of the lapse of time necessary for the building up of a beach-terrace. We can only, in some cases, obtain a measure of the time that elapsed between the formation of two successive beaches, as in this instance.

NOTE 9

The " strand lines," or raised beaches of Norway, have given rise to much discussion, of which a summary will be found in the address cited in Note 7.

NOTE 10

A considerable number of skulls and skeletons of the Neanderthal type have now been found in different parts of Southern Europe, extending from Belgium to Gibraltar and Croatia, and it is now known that this type of skull is associated with flint implements of Mousterian Age. (*See* Note 12.)

NOTE 11

The most important discovery of recent years in this connection is that made in Sussex by Mr. C. Dawson and Dr. A. Smith Woodward; this find is described in great detail in the *Quarterly Journal of the Geological Society*, vol. lxix., 1913, pp. 117-151. At a height of about 80 feet above the present level of the River Ouse, at Piltdown, near Uckfield, is a gravel, containing many brown flints of peculiar character, some of which are implements of Chellean or earlier type, associated with some remains of Pleistocene animals and a few of older date, derived from Pliocene deposits. Embedded in this gravel were found fragments of a human skull and lower jaw of very remarkable type, showing in some respects distinctly simian characters, while in other respects it is less ape-like than the Mousterian skulls of Neanderthal and other localties. For this form the name of *Eoanthropus* has been proposed, thus constituting a new genus of the Hominidæ.

NOTE 12

It will be well at this point to give a brief summary of the modern classification of the Palæolithic implement-bearing deposits of Europe. From the labours of many geologists and prehistoric archæologists, especially in France, a definite succession of types of implement has been established, and in some cases it has been found possible to correlate these with actual human remains and with certain well-marked events in the physical history of Pleistocene times, especially with the advance and retreat of ice-sheets. The present state of our knowledge is admirably summarised by Professor Sollas (*Ancient Hunters*, London, 1911), and from that work the following note is condensed.

The stages of Palæolithic culture now recognised are as follows:

Azilian
Magdalenian
Solutrean
Aurignacian
Mousterian
Acheulean
Chellean
Strepyan
Mesvinian

Below the Mesvinian comes the nebulous region of " eoliths," which are not yet definitely proved to be of human workmanship. The Neanderthal skull belongs to the Mousterian stage, but the oldest known definitely human remain, the jaw from the Mauer sands near Heidelberg, may be older than any of these, indeed by some it is assigned to the first interglacial period of Penck and Brückner (*see* Note 32). For figures of the types of implement characterising each period, see *Guide to the Antiquities of the Stone Age in the Department of British and Mediæval Antiquities*, British Museum, 2nd ed., London, 1911, pp. 1-74. This publication gives an admirable summary of recent knowledge on this subject. For an excellent and critical summary of the latest researches on Palæolithic man up till the end of the Aurignacian period, *see* Duckworth, *Prehistoric Man*, Cambridge, 1912. See also note 44.

Note 13

Sir John Evans, K.C.B. (1823-1908), was one of the foremost authorities on prehistoric archæology and a prolific writer on the subject. His best known work is *The Ancient Stone Implements, Weapons, and Ornaments of Great Britain*, 2nd ed., 1897.

Note 14

By the expression " Celtic weapons of the stone period " is presumably meant Neolithic implements, with polished surfaces.

Note 15

It has recently been shown that the growth of peat is a very slow process, and at the present time it is in many places either at a standstill or even in a state of retrogression. In the peat-mosses of Scotland, Lewis has traced nine successive layers, marked by different floras. The lowest of these and another at a higher level are distinctly of an arctic character, the intermediate forest beds, on the other hand, indicate periods of milder climate, when the limit of the growth of trees was at a higher level in Scotland than is now the case. From these facts it is certain that the peat-mosses of Scotland and northern England date back at least as far as the later stages of the glacial period, and indicate at least one mild interglacial episode, when the climate was somewhat warmer than it now is. (*See* Lewis, *Science Progress*, vol. ii., 1907, p. 307.) Hence the statements of the French workmen, here quoted, do not possess much significance.

NOTE 16

Cyrena fluminalis is very abundant in the gravels of an old terrace of the River Cam, at Barnwell, in the suburbs of Cambridge, and also in glacial gravels at Kelsey Hill in Holderness. It is a very remarkable fact that this shell, now an inhabitant of warm regions, should be so abundant in these Pleistocene deposits, in close association with glacial accumulations.

NOTE 17

The implement-bearing deposits of Hoxne, in Suffolk, were investigated with great care by a committee of the British Association, and the results were published in a special and detailed report (" The Relation of Palæolithic Man to the Glacial Epoch," *Report Brit. Ass.*, Liverpool, 1896, pp. 400-15). The deposit consists of a series of lacustrine or fluviatile strata with plant remains, some being arctic in character, resting on Chalky Boulder Clay, and this again on sand. The Palæolithic deposits are all clearly later than the latest boulder-clay of East Anglia, and between their formation and that of the glacial deposits at least two important climatic changes took place, indicating a very considerable lapse of time.

Mention may conveniently be made here of the supposed discovery of the remains of pre-glacial man at Ipswich, which appears to be founded on errors of observation. The boulder-clay above the interment is, according to the best authorities, merely a landslip or flow.

NOTE 18

It has been suggested with a considerable degree of probability, that in Auvergne volcanic eruptions persisted even into historic times. The subject is obscure, depending on the interpretation of difficult passages in two Latin chronicles of the fifth century. The most obvious meaning of both passages would certainly appear to be the occurrence of volcanic eruptions and earthquakes, but attempts have been made to explain them as referring to some artificial conflagration, possibly the burning of a town by an invader. (*See* Bonney, *Volcanoes*, 3rd ed., London, 1913, p. 129.)

NOTE 19

In the early days of glacial geology in Britain, it was commonly accepted that the phenomena could be most satisfactorily explained on the hypothesis of a general submergence of the northern parts of the country to a depth of many hundreds of feet, and this in spite of the original comparison by Agassiz of the glacial deposits of Britain to those of the Alps. In later times, however, a school of geologists arose who attributed the glaciation of Britain to land-ice of the Continental or Greenland type. Of late years this school has been dominant in British geology, with a few notable exceptions, of whom the most important is Professor Bonney. The difficulties presented by both theories are almost equally great, and at the present time, in spite of the vehemence of the supporters of the land-ice theory, it is impossible to hold any dogmatic views on the subject. Against the doctrine of submergence is the absence of glacial deposits in places where they would naturally be expected to occur if the whole of the British Isles north of the Thames and Bristol Channel had been covered by the sea, together with the very general absence of sea-shells in the deposits. The objections to the land-ice hypothesis are largely of a mechanical nature. If we take into account the lateral extent and the thickness that can be assigned to the ice-sheet, we are at once confronted by very considerable difficulties as to the sufficiency of the driving-power behind the ice. Another great difficulty is the shallowness of the North Sea, in which a comparatively thin mass of ice would run aground at almost any point. It has been calculated that the maximum slope of the surface of the ice from Norway to the English coast could not exceed half a degree, and it is therefore difficult to see what force could compel it to move forward at all, much less to climb steep slopes in the way postulated by the extremists of this school.

NOTE 20

The most complete account of the geology of the Norfolk coast is contained in " The Geology of Cromer," by Clement Reid (*Memoir of the Geological Survey*). (*See also* Harmer, " The Pleistocene Period in the Eastern Counties of England," *Geology in the Field, the Jubilee Volume of the Geologists Association*, 1909, chap. iv.). Above the Norwich Crag several more subdivisions are now recognised, and the complete succession of the Pliocene and Pleistocene strata of East Anglia may be summarised as follows:

Pleistocene	Peat and Alluvium
	Gravel Terraces of the present river systems
	Gravels of the old river-systems
	Plateau gravels
	Chalky boulder-clay
	Interglacial sands and gravels and Contorted Drift
	Cromer Till
	Arctic Plant Bed
Pliocene	Cromer Forest Series
	Weybourn Crag
	Chillesford Crag
	Norwich Crag
	Red Crag
	Coralline Crag

NOTE 21

It is now generally agreed that the tree-stumps in the Cromer Forest bed are not in the position of growth. Many of them are upside down or lying on their sides, and they were probably floated into their present position by the waters of a river flowing to the north. This river was a tributary of the Rhine which then flowed for several hundred miles over a plain now forming the bed of the North Sea, collecting all the drainage of eastern England, and debouching into the North Atlantic somewhere to the south of the Faroe Isles. (*See* Harmer, " The Pleistocene Period in the Eastern Counties of England," *Geol. Ass. Jubilee Volume*, London, 1909, pp. 103-123.)

NOTE 22

Of late years an enormous number of characteristic rocks from Norway and Sweden have been recognised in the drifts of Eastern England, as far south as Essex and Middlesex. One of the most easily identifiable types is the well-known Rhombporphyry of the Christiania Fjord, a rock which occurs nowhere else in the world, and is quite unmistakable in appearance. Along with it are many of the distinctive soda-syenites found in the same district, the granites of southern Sweden, and many others. The literature of the subject is very large, but many details may be found in the annual reports of the British Association for the last twenty years.

From a study of these erratics it has been found possible to draw important conclusions as to the direction and sequence of the ice streams which flowed over these regions during the different stages of the glacial period.

NOTE 23

During his first crossing of Greenland from east to west, Nansen attained a height of 9000 feet on a vast expanse of frozen snow, and it is believed that towards the north the surface of this great snow-plateau rises to even greater elevations. The surface of the snow is perfectly clean and free from moraine-material. No rock *in situ* has been seen in the interior of Greenland at a distance greater than 75 miles from the coast.

A great amount of valuable information concerning the glacial conditions of Greenland is to be found in the *Meddelelser om Grön-land*, a Danish publication, but containing many summaries in French or English. For a good account of the phenomena seen in the coastal region of the west coast, *see* Drygalski, *Grönland-Expedition*, a large monograph published by the *Gesellschaft für physischen Erdkunde*, Berlin, 1897.

NOTE 24

The argument is here considerably understated. The southern point of Greenland, Cape Farewell, is in the same latitude as the Shetland Islands and Christiania, and only one degree north of Stockholm; Disko is in about the same latitude as the North Cape. Hence the inhabited portion of Greenland is in the same latitude as Norway and Sweden, both fertile and well-populated countries. Even in Central Norway, in the Gudbrandsdal and Romsdal, thick forests grow up to a height of at least 3000 feet above sea-level, a much greater elevation than trees now attain in the British Isles. This latter fact is probably to be attributed to the protective effect of thick snow lying throughout the winter.

NOTE 25

For a summary of the most recent views as to the classification and succession of the glacial deposits of the British Isles, *see* Lake and Rastall, *Textbook of Geology*, London, 1910, pp. 466-473. Reference may also be made to Jukes-Browne, *The Building of the British Isles*, London, 1912, pp. 430-440.

NOTE 26

Glacier-lakes are fairly common among the fjords of the west coast of Greenland, and illustrate very well what must have been the state of affairs in Glen Roy at the time of formation of the Parallel Roads.

NOTE 27

The high-level shell-bearing deposits of Moel Tryfan, Gloppa, near Oswestry, and Macclesfield, have given rise to much controversy between the supporters of submergence and of land-ice. At Moel Tryfan certain sands and gravels, with erratics, at a height of about 1350 feet, contain abundant marine shells, generally much broken. The northern or seaward face of the hill is much plastered with drift, but none is to be found on the landward side, and it is suggested that the shell-bearing material is the ground-moraine of a great ice-sheet that came in from the Irish Sea, and was forced up on to the Welsh coast, just reaching the watershed, but failing to overtop it. With regard to the explanation by submergence, the great objection is the absence of marine drift on the landward side, which is very difficult to explain if the whole had been submerged sufficiently to allow of normal marine deposits at such a great height. The shell beds of Macclesfield and Gloppa are at a less elevation but of essentially similar character.

The shell-bearing deposits of Moel Tryfan were examined by a committee of the British Association. (See *Rep. Brit. Ass.*, Dover, 1899, pp. 414-423.) At the end of this report is an extensive bibliography.

NOTE 28

During the last forty years the deep-sea dredging expeditions of H.M.S. *Challenger* and others have shown the abundance and variety of animal life at great depths, especially in the Arctic and Antarctic seas. For a recent summary, *see* Murray and Hjort, *The Depths of the Ocean*, London, 1912.

NOTE 29

It is now generally admitted that these shell-beds in Wexford are of Pliocene age, and they therefore have no bearing on the subject under discussion.

NOTE 30

The boulder deposit at Selsey has been described by Mr. Clement Reid (*Quart. Jour. Geol. Soc.*, vol. xlviii., 1892, p. 355). Immediately above the Tertiary beds is a hard greenish clay, full of derived Tertiary fossils and Pleistocene shells with large flints and erratic blocks, some of the latter weighing several tons. They include granite, greenstone, schist, slate, quartzite, and sandstone, and most of them must have been transported for a long distance. Above them are black muds with marine shells, then a shingle beach, and above all the Coombe Rock. (See next note.)

NOTE 31

The Brighton elephant-bed and its equivalent, the Coombe Rock, are fully described by Clement Reid (" On the Origin of Dry Chalk Valleys and the Coombe Rock," *Quart. Jour. Geol. Soc.*, vol. xliii., 1887, p. 364). The Coombe Rock is a mass of unstratified flints and Chalk debris filling the lower parts of the dry valleys (Coombes) of the South Downs and gradually passing into the brick-earth (loam) of the coastal plain. It is clearly a torrential accumulation, and is supposed to have been formed while the Chalk was frozen, thus preventing percolation of water and causing the surface water to run off as strong streams. This must have occurred during some part of the glacial period, which would naturally be a period of heavy precipitation. Of very similar origin is the " Head " of Cornwall, a surface deposit often rich in tinstone and other minerals of economic value. The Coombe Rock has recently been correlated with deposits of Mousterian Age.

NOTE 32

The former extension of the Alpine glaciers and the deposits formed by them have been exhaustively investigated by Penck and Brückner (*Die Alpen im Eiszeitalter*, 3 vols., Leipzig, 1901-9). In this monumental work the authors claim to have established the occurrence of four periods of advance of the ice, to which they give the names of Günz, Mindel, Riss, and Würm glaciations, with corresponding interglacial genial episodes, when the climate was possibly even somewhat warmer than now. Their conclusions and the data on which they are established are summarised by Sollas (*Ancient Hunters*, London, 1911, especially pp. 18-28). For a general account of the glaciers of the Alps and their accompanying phenomena, *see* Bonney, *The Building of the Alps*, London, 1912, pp. 103-151.

NOTE 33

At the time of the maximum advance of the ice, during the Riss period of Penck and Brückner, the terminal moraine of the great glacier of the Rhône extended as far as the city of Lyon, and towards the north-east it became continuous with the similar moraine of the Rhine glacier.

NOTE 34

For the successive phases of advance and retreat of the Alpine glaciers, see the works quoted in Note 32.

NOTE 35

The Loess of Central Europe includes deposits of two different ages. According to Penck the " Older Loess " was formed in the period of warm and dry climate that intervened between the third and fourth glacial episodes, while the " Younger Loess " is post-glacial. Both divisions are for the most part æolian deposits, formed by the redistribution of fine glacial mud originally laid down in water and carried by the wind often to considerable heights. A part, however, of the so-called Loess of northern France, e.g. in the valley of the Somme, is rain-wash, similar in character to the brick-earth of parts of south-eastern England. The Older Loess contains Acheulean implements, while the Younger Loess is of Aurignacian Age.

The greatest development of the Loess is in Central Asia and in China. (See Richthofen, China, Berlin, 1877.) In China the Loess reaches a thickness of several thousand feet, and whole mountain-ranges are sometimes almost completely buried in it. In the deserts of Central Asia the formation of the Loess is still in progress. A very similar deposit, called adobe, is also found in certain parts of the Mississippi valley.

The Loess is a fine calcareous silt or clay of a yellowish colour, quite soft and crumbling between the fingers. However, it resists denudation in a remarkable manner, and in China it often stands up in vertical walls hundreds of feet in height. This property is probably assisted by the presence of numerous fine tubes arranged vertically and lined with calcium carbonate; these are supposed to have been formed in the first place by fibrous rootlets.

NOTE 36

Although highly probable, it cannot yet be regarded as conclusively demonstrated that the Pleistocene glaciations of Europe and of North America were exactly contemporaneous. The ice-sheets in each case radiated from independent centres which were not in the extreme north of either continent, and were not in any way connected with a general polar ice-cap. The European centre was over the Baltic region or the south of Scandinavia, and the American centre in the neighbourhood of Hudson's Bay. The southern margin of the American ice-sheet extended about as far south as latitude 38° N. in the area lying south of the Great Lakes, whereas the North European ice barely passed the limit of 50° N. in Central Europe. This greater southward extension in America was doubtless correlated with the same causes as now produce the low winter temperatures of the eastern states, especially the cold Newfoundland current. The literature of North American glacial geology has now attained colossal dimensions, and it is impossible to give here even a short abstract of the main conclusions. For a general summary reference may be made to Chamberlin and Salisbury, *Geology*, vol. iii.; *Earth History*, London and New York, 1905; or the same authors' *Geology, Shorter Course*, London and New York, 1909.

Note 37

During the last fifty years scarcely any geological subject has given rise to a greater amount of speculation than the cause of the Ice Age, and the solution of the problem is still apparently far off. The theories put forward may for convenience be divided into three groups, namely astronomical, geographical, and meteorological.

As examples of astronomical explanation, we may take the well-known theory of Adhemar and Croll, which is founded on changes in the ellipticity of the earth's orbit. This is expounded and amplified by Sir Robert Ball in his *Cause of an Ice Age*. The weak point of this theory, which is mathematically unassailable, is that it proves too much, and postulates a constant succession of glacial periods throughout earth-history, and for this there is no evidence. The geographical explanations are chiefly founded on supposed changes in the distribution of sea and land, with consequent diversion of cold and warm currents. Another suggestion is that the glaciated areas had undergone elevation into mountain regions, but this is in conflict with evidence for submergence beneath the sea in certain cases. Meteorological hypotheses, such as that of Harmer, founded on a different arrangement of air pressures and wind-directions, seem to offer the most promising field for exploration and future work, but it is clear that much still remains to be explained.

Note 38

The reptile-bearing Elgin Sandstones are of Triassic Age, and they contain a most remarkable assemblage of strange and eccentric forms, especially Anomodont reptiles resembling those found in the Karroo formation of South Africa.

Note 39

The meaning of this statement is not very clear. The Conifers are not dicotyledons: their seeds contain numerous cotyledons, up to twenty in number, and the whole plant, and especially the reproductive system, belongs to a lower stage of development. The argument here employed is therefore fallacious, and in point of fact the different groups actually appeared in the order postulated by the theory of evolution, viz.: (1) Gymnosperms, (2) Monocotyledons, (3) Dicotyledons. *See* Arber, " The Origin of Gymnosperms," *Science Progress*, vol. i., 1906, pp. 222-37.

Note 40

The part of the MS. read to Dr. Hooker in 1844 was undoubtedly the " Essay of 1844," forming the second part of the *Foundations of the Origin of Species*, a volume published by Sir Francis Darwin on the occasion of the Darwin Centenary at Cambridge in 1909. (*See also* Darwin's *Life and Letters*, vol. ii., pp. 16-18.)

Note 41

This projected larger work, which is often referred to in the *Origin of Species*, was never published as such, but Darwin's views on various aspects of evolution were set forth in several later books, such as *The Variation of Animals and Plants under Domestication*, *The Descent of Man*, *Various Contrivances by which Orchids are Fertilised by Insects*, *Movements and Habits of Climbing Plants*, *Insectivorous Plants*, and others.

Note 42

With this section compare the famous chapter with the same title in the *Origin of Species*.

Note 43

No attempt has been made to annotate this chapter, owing to the impossibility of doing so within reasonable compass. Many of the theories here quoted, and the conclusions drawn from them, have not stood the test of time, and recent philological and ethnographical research have clearly shown the danger of attempting to infer the relationships of different peoples from their languages. The modifications undergone by the languages themselves are also subject to influences of such complex character, so largely artificial in their origin, that any attempt to compare them with natural evolution in the organic world must lead to false analogies. The chapter must be regarded as an interesting exposition of one phase of Mid-Victorian scientific thought, but having little real bearing on the subjects discussed in the rest of the book.

Note 44

That the prophecy here given was justified is shown by the discovery in Java in 1891, of the skull and parts of the skeleton of *Pithecanthropus erectus*, a form which, according to the best authorities, must be regarded as in many ways intermediate between man and the apes, though perhaps with more human than ape-like characteristics. For an account of the circumstances of its discovery and a general description of the remains, *see* Sollas, *Ancient Hunters*, London, 1911, pp. 30-39 (with many references). Within the last year or two interest in the ancestry of man has been greatly increased, especially by the Piltdown discovery (see Note 11). This has led to a revision of the whole subject, and the views formerly held have undergone a certain amount of modification. It now seems certain that the different types of culture as represented by the succession of stages given in Note 12 do not correspond to a continuous development of one single race of mankind. There is, undoubtedly, a great break between the Mousterian and Aurignacian. Mousterian or Neanderthal man appears to have become extinct, possibly having been exterminated by a migration of the more highly developed Aurignacian race, which may be regarded as the ancestor of modern man in Europe. It appears, therefore, that the really important line of division comes, not as was formerly thought between Palæolithic and Neolithic, but in the middle of the Palæolithic between Mousterian and Aurignacian. Hence it appears that our classification will in the near future have to undergo revision, since the stages of culture from Aurignacian to Azilian show a much closer affinity to the Neolithic than they do to the earlier Palæolithic. At the present time scarcely sufficient data are available to determine the relationship of *Pithecanthropus* and *Eoanthropus* to the later types of man. For an excellent summary of the most recent views see Thacker " The Significance of the Piltdown Discovery," *Science Progress*, vol. viii. 1913, p. 275.

A CATALOG OF SELECTED DOVER
BOOKS IN ALL FIELDS OF INTEREST

CONCERNING THE SPIRITUAL IN ART, Wassily Kandinsky. Pioneering work by father of abstract art. Thoughts on color theory, nature of art. Analysis of earlier masters. 12 illustrations. 80pp. of text. 5⅜ x 8½.   23411-8

ANIMALS: 1,419 Copyright-Free Illustrations of Mammals, Birds, Fish, Insects, etc., Jim Harter (ed.). Clear wood engravings present, in extremely lifelike poses, over 1,000 species of animals. One of the most extensive pictorial sourcebooks of its kind. Captions. Index. 284pp. 9 x 12.   23766-4

CELTIC ART: The Methods of Construction, George Bain. Simple geometric techniques for making Celtic interlacements, spirals, Kells-type initials, animals, humans, etc. Over 500 illustrations. 160pp. 9 x 12. (Available in U.S. only.)   22923-8

AN ATLAS OF ANATOMY FOR ARTISTS, Fritz Schider. Most thorough reference work on art anatomy in the world. Hundreds of illustrations, including selections from works by Vesalius, Leonardo, Goya, Ingres, Michelangelo, others. 593 illustrations. 192pp. 7⅛ x 10¼.   20241-0

CELTIC HAND STROKE-BY-STROKE (Irish Half-Uncial from "The Book of Kells"): An Arthur Baker Calligraphy Manual, Arthur Baker. Complete guide to creating each letter of the alphabet in distinctive Celtic manner. Covers hand position, strokes, pens, inks, paper, more. Illustrated. 48pp. 8¼ x 11.   24336-2

EASY ORIGAMI, John Montroll. Charming collection of 32 projects (hat, cup, pelican, piano, swan, many more) specially designed for the novice origami hobbyist. Clearly illustrated easy-to-follow instructions insure that even beginning papercrafters will achieve successful results. 48pp. 8¼ x 11.   27298-2

THE COMPLETE BOOK OF BIRDHOUSE CONSTRUCTION FOR WOODWORKERS, Scott D. Campbell. Detailed instructions, illustrations, tables. Also data on bird habitat and instinct patterns. Bibliography. 3 tables. 63 illustrations in 15 figures. 48pp. 5¼ x 8½.   24407-5

BLOOMINGDALE'S ILLUSTRATED 1886 CATALOG: Fashions, Dry Goods and Housewares, Bloomingdale Brothers. Famed merchants' extremely rare catalog depicting about 1,700 products: clothing, housewares, firearms, dry goods, jewelry, more. Invaluable for dating, identifying vintage items. Also, copyright-free graphics for artists, designers. Co-published with Henry Ford Museum & Greenfield Village. 160pp. 8¼ x 11.   25780-0

HISTORIC COSTUME IN PICTURES, Braun & Schneider. Over 1,450 costumed figures in clearly detailed engravings–from dawn of civilization to end of 19th century. Captions. Many folk costumes. 256pp. 8⅜ x 11¾.   23150-X

MY BONDAGE AND MY FREEDOM, Frederick Douglass. Born a slave, Douglass became outspoken force in antislavery movement. The best of Douglass' autobiographies. Graphic description of slave life. 464pp. 5⅜ x 8½. 22457-0

FOLLOWING THE EQUATOR: A Journey Around the World, Mark Twain. Fascinating humorous account of 1897 voyage to Hawaii, Australia, India, New Zealand, etc. Ironic, bemused reports on peoples, customs, climate, flora and fauna, politics, much more. 197 illustrations. 720pp. 5⅜ x 8½. 26113-1

THE PEOPLE CALLED SHAKERS, Edward D. Andrews. Definitive study of Shakers: origins, beliefs, practices, dances, social organization, furniture and crafts, etc. 33 illustrations. 351pp. 5⅜ x 8½. 21081-2

THE MYTHS OF GREECE AND ROME, H. A. Guerber. A classic of mythology, generously illustrated, long prized for its simple, graphic, accurate retelling of the principal myths of Greece and Rome, and for its commentary on their origins and significance. With 64 illustrations by Michelangelo, Raphael, Titian, Rubens, Canova, Bernini and others. 480pp. 5⅜ x 8½. 27584-1

PSYCHOLOGY OF MUSIC, Carl E. Seashore. Classic work discusses music as a medium from psychological viewpoint. Clear treatment of physical acoustics, auditory apparatus, sound perception, development of musical skills, nature of musical feeling, host of other topics. 88 figures. 408pp. 5⅜ x 8½. 21851-1

THE PHILOSOPHY OF HISTORY, Georg W. Hegel. Great classic of Western thought develops concept that history is not chance but rational process, the evolution of freedom. 457pp. 5⅜ x 8½. 20112-0

THE BOOK OF TEA, Kakuzo Okakura. Minor classic of the Orient: entertaining, charming explanation, interpretation of traditional Japanese culture in terms of tea ceremony. 94pp. 5⅜ x 8½. 20070-1

LIFE IN ANCIENT EGYPT, Adolf Erman. Fullest, most thorough, detailed older account with much not in more recent books, domestic life, religion, magic, medicine, commerce, much more. Many illustrations reproduce tomb paintings, carvings, hieroglyphs, etc. 597pp. 5⅜ x 8½. 22632-8

SUNDIALS, Their Theory and Construction, Albert Waugh. Far and away the best, most thorough coverage of ideas, mathematics concerned, types, construction, adjusting anywhere. Simple, nontechnical treatment allows even children to build several of these dials. Over 100 illustrations. 230pp. 5⅜ x 8½. 22947-5

THEORETICAL HYDRODYNAMICS, L. M. Milne-Thomson. Classic exposition of the mathematical theory of fluid motion, applicable to both hydrodynamics and aerodynamics. Over 600 exercises. 768pp. 6⅛ x 9¼. 68970-0

SONGS OF EXPERIENCE: Facsimile Reproduction with 26 Plates in Full Color, William Blake. 26 full-color plates from a rare 1826 edition. Includes "The Tyger," "London," "Holy Thursday," and other poems. Printed text of poems. 48pp. 5¼ x 7. 24636-1

OLD-TIME VIGNETTES IN FULL COLOR, Carol Belanger Grafton (ed.). Over 390 charming, often sentimental illustrations, selected from archives of Victorian graphics—pretty women posing, children playing, food, flowers, kittens and puppies, smiling cherubs, birds and butterflies, much more. All copyright-free. 48pp. 9¼ x 12¼. 27269-9

THE STORY OF THE TITANIC AS TOLD BY ITS SURVIVORS, Jack Winocour (ed.). What it was really like. Panic, despair, shocking inefficiency, and a little heroism. More thrilling than any fictional account. 26 illustrations. 320pp. 5⅜ x 8½.
20610-6

FAIRY AND FOLK TALES OF THE IRISH PEASANTRY, William Butler Yeats (ed.). Treasury of 64 tales from the twilight world of Celtic myth and legend: "The Soul Cages," "The Kildare Pooka," "King O'Toole and his Goose," many more. Introduction and Notes by W. B. Yeats. 352pp. 5⅜ x 8½.
26941-8

BUDDHIST MAHAYANA TEXTS, E. B. Cowell and others (eds.). Superb, accurate translations of basic documents in Mahayana Buddhism, highly important in history of religions. The Buddha-karita of Asvaghosha, Larger Sukhavativyuha, more. 448pp. 5⅜ x 8½.
25552-2

ONE TWO THREE . . . INFINITY: Facts and Speculations of Science, George Gamow. Great physicist's fascinating, readable overview of contemporary science: number theory, relativity, fourth dimension, entropy, genes, atomic structure, much more. 128 illustrations. Index. 352pp. 5⅜ x 8½.
25664-2

EXPERIMENTATION AND MEASUREMENT, W. J. Youden. Introductory manual explains laws of measurement in simple terms and offers tips for achieving accuracy and minimizing errors. Mathematics of measurement, use of instruments, experimenting with machines. 1994 edition. Foreword. Preface. Introduction. Epilogue. Selected Readings. Glossary. Index. Tables and figures. 128pp. 5⅜ x 8½.
40451-X

DALÍ ON MODERN ART: The Cuckolds of Antiquated Modern Art, Salvador Dalí. Influential painter skewers modern art and its practitioners. Outrageous evaluations of Picasso, Cézanne, Turner, more. 15 renderings of paintings discussed. 44 calligraphic decorations by Dalí. 96pp. 5⅜ x 8½. (Available in U.S. only.)
29220-7

ANTIQUE PLAYING CARDS: A Pictorial History, Henry René D'Allemagne. Over 900 elaborate, decorative images from rare playing cards (14th–20th centuries): Bacchus, death, dancing dogs, hunting scenes, royal coats of arms, players cheating, much more. 96pp. 9¼ x 12¼.
29265-7

MAKING FURNITURE MASTERPIECES: 30 Projects with Measured Drawings, Franklin H. Gottshall. Step-by-step instructions, illustrations for constructing handsome, useful pieces, among them a Sheraton desk, Chippendale chair, Spanish desk, Queen Anne table and a William and Mary dressing mirror. 224pp. 8⅛ x 11¼.
29338-6

THE FOSSIL BOOK: A Record of Prehistoric Life, Patricia V. Rich et al. Profusely illustrated definitive guide covers everything from single-celled organisms and dinosaurs to birds and mammals and the interplay between climate and man. Over 1,500 illustrations. 760pp. 7½ x 10⅛.
29371-8

Paperbound unless otherwise indicated. Available at your book dealer, online at **www.doverpublications.com**, or by writing to Dept. GI, Dover Publications, Inc., 31 East 2nd Street, Mineola, NY 11501. For current price information or for free catalogues (please indicate field of interest), write to Dover Publications or log on to **www.doverpublications.com** and see every Dover book in print. Dover publishes more than 500 books each year on science, elementary and advanced mathematics, biology, music, art, literary history, social sciences, and other areas.